SPIN RESONANCE SPECTROSCOPY

SPIN RESONANCE SPECTROSCOPY

Principles and Applications

CHANDRAN KARUNAKARAN

Associate Professor of Chemistry, Biomedical Research Lab,
VHNSN College (Autonomous), Virudhunagar,
Tamilnadu, India

ELSEVIER

Elsevier
Radarweg 29, PO Box 211, 1000 AE Amsterdam, Netherlands
The Boulevard, Langford Lane, Kidlington, Oxford OX5 1GB, United Kingdom
50 Hampshire Street, 5th Floor, Cambridge, MA 02139, United States

Notices
Knowledge and best practice in this field are constantly changing. As new research and experience broaden our understanding, changes in research methods, professional practices, or medical treatment may become necessary.

Practitioners and researchers must always rely on their own experience and knowledge in evaluating and using any information, methods, compounds, or experiments described herein. In using such information or methods they should be mindful of their own safety and the safety of others, including parties for whom they have a professional responsibility.

To the fullest extent of the law, neither the Publisher nor the authors, contributors, or editors, assume any liability for any injury and/or damage to persons or property as a matter of products liability, negligence or otherwise, or from any use or operation of any methods, products, instructions, or ideas contained in the material herein.

Library of Congress Cataloging-in-Publication Data
A catalog record for this book is available from the Library of Congress

British Library Cataloguing-in-Publication Data
A catalogue record for this book is available from the British Library

ISBN: 978-0-12-813608-9

Publisher: Cathleen Sether
Acquisition Editor: John Fedor
Editorial Project Manager: Gabriela D. Capille
Production Project Manager: Nilesh Kumar Shah
Cover Designer: Miles Hitchen

Typeset by TNQ Books and Journals

CONTENTS

LIST OF CONTRIBUTORS

Chandran Karunakaran
VHNSN College (Autonomous), Virudhunagar, India

Murugesan Balamurugan
VHNSN College (Autonomous), Virudhunagar, India

Paulraj Santharaman
VHNSN College (Autonomous), Virudhunagar, India

Raju Rajkumar
VHNSN College (Autonomous), Virudhunagar, India

Murugesan Karthikeyan
VHNSN College (Autonomous), Virudhunagar, India

Dhinesh Kumar Marimuthu
VHNSN College (Autonomous), Virudhunagar, India

PREFACE

This book is primarily intended to provide the fundamentals of the spin resonance spectroscopy including nuclear magnetic resonance (NMR) and electron paramagnetic resonance (EPR), methods, advances, and applications to basic science viz., biology, chemistry, physics, and also medicine. It is hoped that the reader will be able to acquire the knowledge of both the NMR and EPR in a single-source book and the state-of-the-art developments in this field as it stands today. This book covers the syllabus of leading universities and colleges. Thus, this book caters to the needs of students of basic sciences, researchers, and R&D scientists in industries, who wish to understand, enhance, and broaden their knowledge and expertise on the fundamentals, technologies, and applications of spin resonance spectroscopy. The book begins with the introduction of NMR and ESR, theory, instrumentation, and comparison of them. Chapter 1 discusses the parameters obtained from NMR, mechanism of spin—spin splitting, classification of NMR systems, pulse/Fourier transform NMR, relaxation processes and times, Bloch equations, and line width/shape. In Chapter 2, the advances in ^{1}H and ^{13}C NMR viz., double-resonance NMR including nuclear Overhauser effect, single frequency off-resonance decoupling, distortionless enhancement by polarization transfer, insensitive nuclei enhanced by polarization transfer, chemically induced dynamic nuclear polarization, and multidimensional viz., 2D NMR including homonuclear and heteronuclear and 3D NMR are described. In Chapter 3, the applications of NMR to structural elucidation of inorganic compounds especially ^{31}P and ^{19}F NMR are highlighted. Dynamic NMR, solid-state NMR (magic-angle spinning), NMR of paramagnetic molecules, shift and contrast reagents, and magnetic resonance imaging are also discussed.

Chapter 4 presents the basics of EPR, spin Hamiltonian to describe the observed spectra, determination of g-value and factors affecting it, Kramer's degeneracy, zero-field splitting, exchange-coupled EPR, triplet state, free radical EPR, optically detected magnetic resonance, Jahn Teller effect, and spectral simulation software. In Chapter 5, the recent development and advances in EPR viz., multifrequency, high-field EPR, and pulse EPR including electron spin echo envelope modulation, hyperfine sublevel correlation, and multifrequency EPR viz., electron nuclear double resonance, (continous wave and pulsed), TRIPLE, and double electron—electron resonance are discussed. Chapter 6 presents the applications of EPR to various transition metal ions, explore the active site of enzymes, fast flow, rapid freeze-quench, spin trapping, and spin label EPR to detect and monitor the dynamics of unstable free radicals; in vivo EPR and EPR imaging using spin probes are also highlighted. The above chapters thus attempt to provide the essential background knowledge and up-to-date advances in this field. The book should thus serve as introductory text for those who intend to specialize in either the theoretical or practical applications. This textbook is also hoped to be a fruitful launch pad for many careers in spin resonance spectroscopy.

Thanks are due to the contributing authors of this book for their active support and continuous cooperation throughout the course of preparation and all those great and eminent researchers in this field in and around the globe whose valuable contributions made this work possible. It is a real pleasure to acknowledge my mentors Prof. R. Murugesan, Prof. B. Kalyanaraman, Prof. William Antholine, Prof. Brian Benett, and all the teachers for their inspiration and guidance. Many thanks are also due to my collaborative researchers viz., Prof. Brian Volkmann of Medical College of Wisconsin, Prof. John McCraken, Michigan State University, Prof. Anumantha Kanthasamy, Iowa State University, Prof. Narayan Avadhani University of Pennsylvania, Prof. S. Subramanian, IIT-Chennai, Dr. M. Velayutham, University of Pittsburgh, Dr. K.R. Justin Thomas, IIT, Roorkee, Dr. Srigiridhar Kotamraju and Dr. Shasi Kalivendi, IICT, Hyderabad, Dr. B. Kalpana, DIPAS-DRDO, Dr. Sushil Kumar Singh, SSPL-DRDO, Delhi, Prof. Anuradha Dhanasekeran, Anna University, Chennai, Prof. Mainak Das, IIT-Kanpur, Dr. V. Shanmugaiah, Madurai Kamaraj University, Prof. Jonathan Claussen, Iowa State University, Dr. A. Rajendran, Dr. D. Christopher Durairaj, Dr. S. Elango and all colleagues for their input and sharing knowledge, the Principal and Managing Board of our institution for their encouragement and cooperation, Prof. C.S. Gopinath, National Chemical Lab, Pune, and Dr. Sujatha Venkataraman, University of Colorado for giving their suggestions and critical comments to improve the book, and my wife and kids for their perseverance and support during the preparation of the book. I am indebted to my present and past research scholars Dr. S. Rajesh, Dr. U.S.E. Arivudainambi, Dr. S. Rajasingh, Dr. S. Prakash, Dr. P. Gurusamy, Dr. V. Ramesh, Dr. M. Pandiaraj, Dr. T. Madasamy, Dr. M. Balamurugan, Dr. G. Muneeswaran, Dr. P. Santharaman, Mr. R. Rajkumar, Mr. M. Karthikeyan, and Mr. M. Dhinesh Kumar for their contributions enabling me to undertake this task and complete it. Finally, it has been a great experience to work with Elsevier publisher of this book and special thanks to the entire them for their effort in making this book possible.

ABBREVIATIONS

Symbols	Abbreviations
2p	Two pulse
3p	Three pulse
ALS	Amyotrophic lateral sclerosis
B/Bo/H	Magnetic field
bipy	2,2′bipyridine
Boc-MPO/BMPO	5-butoxycarbonyl-5-methyl-1-pyrroline *N*-oxide
CAMELSPIN	Cross-relaxation appropriate for minimolecules emulated by spin-locking
CFT	Crystal field theory
CIDNP	Chemically induced dynamic nuclear polarization
COSY	Correlations spectroscopy
CrOx	Chromium oxalate
CW	Continuous wave
DD	Dipole−dipole interaction
DEER	Double electron−electron resonance
DEPMPO	5-diethoxyphosphoryl-5-methyl-1-pyrroline-*N*-oxide
DEPT	Distortionless enhancement by polarization transfer
DMPO	Dimethyl-1-pyrroline-*N*-oxide
DNMR	Dynamic nuclear magnetic resonance
DPM	Dipivalomethanato
DPPH	Diphenylpicrylhydrazyl
DQ	Double quantum
DQF	Double-quantum filtered
DSA	4,4-dimethyl-4-silapentane-1-ammonium trifluoroacetate
DSS	4,4-dimethyl-4-silapentane-1-sulfonic acid
EDTA	Ethylenediaminetetraacetic acid
EEI	Electron−electron interaction
EFG	Electric field gradient
ELDOR	Electron−electron double resonance
EMPO	5-Ethoxycarbonyl-5-methyl-1-pyrroline *N*-oxide
ENDOR	Electron-nuclear double resonance
EPR	Electron paramagnetic resonance
EPRI	Electron paramagnetic resonance imaging
ESE	Electron spin echoes
ESEEM	Electron spin-echo envelope modulation
ESR	Electron spin resonance
EXSY	Exchange spectroscopy
EZ	Electron Zeeman interaction
FBP	Filtered back projection

Symbols	Abbreviations
FDH II	Formylmethanofuran dehydrogenase
FID	Free induction decay
fMRI	Functional magnetic resonance imaging
FT	Fourier transform
GPC	Glycerophosphorylcholine
GPEA	Glycerylphosphorylethanolamine
HETCOR	Heteronuclear correlation spectroscopy
HF	Hyperfine interaction
hfacac	Hexafluoroacetylacetone anion
HFS	Hyperfine structure
HiPIP	High-potential iron-sulfur protein
HMBC	Heteronuclear multibond connectivity
HMQC	Heteronuclear multiple-quantum correlation
HOMO	Highest occupied molecular orbital
HSQC	Heteronuclear single-quantum coherence
HYSCORE	Hyperfine sublevel correlation
INADEQUATE	Incredible natural abundance double-quantum transfer experiment
INEPT	Insensitive nuclei enhanced by polarization transfer
IR	Inversion recovery
ISC	Intersystem crossing
JT	Jahn–Teller
LCAO-MO	Linear combination of atomic orbital–molecular orbital
LiPc	Lithium phthalocyanine
LUMO	Lowest unoccupied molecular orbital
MAS	Magic angle spinning
MOMD	Microscopic order, macroscopic disorder
MoSophe	Molecular Sophe
MRA	Magnetic resonance angiography
MRE	Magnetic resonance elastography
MRI	Magnetic resonance imaging
MS	Metal-substrate complex
MTH	Maleate tetrahydrate
MTSL	Methanethiosulfonate spin label
MW	Microwave
NIEHS	National Institute of Environmental Health Sciences
NMDR	Nuclear magnetic double resonance
NMR	Nuclear magnetic resonance
NMRI	Nuclear magnetic resonance imaging
NOE	Nuclear Overhauser effect
NOESY	Nuclear Overhauser effect spectroscopy
NQI	Nuclear quadrupole interaction
NQR	Nuclear quadrupole resonance
NZ/NZI	Nuclear Zeeman interaction

Symbols	Abbreviations
o, m, p	*Ortho*, *para*, and *meta*
ODMR	Optically detected magnetic resonance
OEP	Octaethyl porphyrinato
OMRI	Overhauser enhanced magnetic resonance imaging
PBN	Phenyl-*N*-ter-butyl nitrone
PC	Phosphorylcholine
PCBM	Phenyl-C_{61}-butyric acid methyl ester
PD	Proton density
PDT	Photodynamic therapy
PE	Phosphorylethanolamine
PEDRI	Proton—electron double resonance imaging
ppm	Parts per million
PS	Photosensitizer
RF	Radiofrequency
RFQ	Rapid freeze-quench
ROESY	Rotating frame nuclear Overhauser effect spectroscopy
ROS	Reactive oxygen species
RT	Room temperature
SDSL	Site-directed spin labeling
SE	Spin echo
SFORD	Single-frequency off-resonance decoupling
SOC	Spin-orbit coupling
SOD	Superoxide dismutase
SOMO	Singly occupied molecular orbital
SQn	Single-quantum nuclear transition
SR	Saturation recovery
SR-EPR	Saturation recovery electron paramagnetic resonance
SS NMR	Solid-state nuclear magnetic resonance
STE	Stimulated echo
STEAM	Stimulated echo acquisition mode
TE	Echo time
TEMPO	(Tetramethylpiperidin-1-yl)oxyl
TMS	Tetramethylsilane
TOCSY	Total correlation spectroscopy
TR	Repetition time
TR-CW	Time-resolved continuous wave
UDP	Uridine diphosphate
VOTPP	Vanadyl tetraphenylporphyrin
VSEPR	Valence shell electron pair repulsion theory
WT	Wild type
XO	Xanthine oxidase
ZF	Zero-field interaction
ZFS	Zero-field splitting
ZQ	Zero-quantum transition

CHAPTER ONE

Principles of Nuclear Magnetic Resonance and Pulsed Nuclear Magnetic Resonance

Chandran Karunakaran, Raju Rajkumar, Murugesan Balamurugan
VHNSN College (Autonomous), Virudhunagar, India

Contents

Spin Resonance Spectroscopy
ISBN: 978-0-12-813608-9
https://doi.org/10.1016/B978-0-12-813608-9.00001-0

Abstract

Spin resonance spectroscopy deals with the nuclear and electron spin resonance absorptions in the radio and microwave frequencies used for investigating diamagnetic ($I \geq 1/2$) and paramagnetic ($S \geq 1/2$) systems, respectively. This chapter focuses on the theory of nuclear magnetic resonance (NMR), its instrumentation, and comparison with electron paramagnetic resonance spectroscopy. The parameters obtained from NMR viz., chemical shift and spin—spin splitting including their types (first and second orders), characteristics, and mechanisms (dipolar and indirect scalar coupling) are discussed. Furthermore, the classification of NMR systems including chemically and magnetically equivalent nuclei is presented. The simplification techniques for complex NMR spectra are also highlighted. It also includes the principles of pulse and Fourier transform NMR and different methods viz., free induction decay, saturation recovery, spin echo (SE), inversion recovery, and stimulated SE to investigate the relaxation process and times. The time evolution of magnetization is described by Bloch equation. The different lineshapes and factors affecting line width are also discussed.

Keywords: Chemical shift; Coupling constant; FID; Lineshape and line width; NMR spin system; Relaxation times; Spin echo.

1.1 INTRODUCTION

Spin magnetic resonance spectroscopy is broadly classified into two types, viz., (1) nuclear magnetic resonance (NMR) spectroscopy and (2) electron spin resonance spectroscopy (ESR)/electron paramagnetic resonance (EPR)/electron magnetic resonance based on the nuclear spin and electron spin, respectively. NMR was independently discovered in the year 1946 by the American physicists F. Bloch (1905–83) of Stanford University and E.M. Purcell (1912–97) of Harvard University. They shared the 1952 Physics Nobel prize. ESR was first observed by the Russian physicist E. Zavoisky [1a]. It is interesting to note here that the famous Dutch physicist C.J. Gorter [1b] was the first person to look for NMR in bulk matter. Because of his bad luck, he failed to observe resonance as the sample was easily saturated because of its large spin-lattice relaxation time. These two branches viz., NMR and ESR are essentially similar, based on the same principles. However, NMR spectra are observed in the radiofrequency (RF) region and ESR spectra in the microwave region of electromagnetic spectrum. The application of ESR is limited to species containing unpaired electrons such as organic and inorganic free radicals, triplet states, and the ions of the *d-block* and *f-block* elements. The first application of NMR to chemistry was made in 1951 when the NMR spectrum of ethanol was recorded. Since then, NMR technique has almost completely been monopolized by chemists, biochemists, and molecular biologists for structural elucidation, physicists being primarily involved in its theoretical and instrumentation aspects.

NMR spectroscopy takes advantage of the magnetic properties of certain nuclei and records the absorption of energy due to transition between quantized nuclear spin energy levels. In an NMR experiment, the spectrometer is tuned to the frequency of a *particular* magnetic nucleus, viz., ^{1}H, ^{13}C, ^{19}F, etc. It is thus a very powerful technique, the closest analogy being a powerful microscope that allows the chemist to "see" the structure of molecules in solution. Actually, the NMR experiment does not directly show how all the atoms are connected. However, it is up to the chemist to take the information provided by NMR spectra to build a model of the molecule [1–12].

1.2 SPIN MAGNETIC PROPERTIES

NMR is possible owing to the magnetic spin properties of certain nuclei. In addition to charge and mass, several nuclei also possess a property called *nuclear spin*. Because nuclei have a charge, they generate a magnetic field due to its spinning motion and hence magnetic moment.

There are useful empirical rules relating mass number, atomic number (Z), and *nuclear spin quantum number* (I):

Mass number	Z	I
Even	Even	0
Odd	Even or odd	1/2, 3/2, 5/2,...
Even	Odd	1, 2, 3,...

In other words, the rules for determining the net spin of a nucleus are expressed as follows:

1. If the number of neutrons and protons are both even, then the nucleus has no spin (i.e., $I = 0$).
2. If the number of neutrons plus the number of protons is odd, then the nucleus has a half-integer spin (i.e., $I = 1/2, 3/2, 5/2$, etc.).
3. If the number of neutrons and the number of protons are both odd, then the nucleus has an integer spin (i.e., $I = 1, 2, 3$, etc.).

Because NMR depends on the existence of a nuclear spin, nuclei with $I = 0$ have no NMR spectrum (e.g., ^{4}He, ^{12}C, ^{16}O, ^{18}O, ^{32}S). The most important class of nuclei, which are NMR active those with $I = 1/2$ (e.g., ^{1}H, ^{13}C, ^{15}N, ^{19}F) and nuclei with $I > 1/2$ (e.g., ^{11}B, $I = 3/2$; ^{2}H, ^{10}C, ^{14}N, $I = 1$), have *quadrupole moments*, a nonspherical distribution of nuclear charge, which results in broad absorption lines and makes observation of NMR spectra more difficult. The quadrupole moment can even affect the lineshape of neighboring nuclei. For example, the resonances of protons bonded to nitrogen or boron atoms are generally broad in ^{1}H NMR spectra. We shall thus be primarily concerned with nuclei with $I = 1/2$. A list of some common isotopes with $I = 1/2$ is provided in Table 1.1.

Table 1.1 Natural abundances of some common nuclear magnetic resonance–active isotopes with I = 1/2

Isotope	Natural abundance (%)	μ (magnetic moment)[a]	g_N	γ [10^8 rad $T^{-1}s^{-1}$]	υ (MHz at 2.35 T)
^{1}H	100	2.7927	5.585	2.675	100
^{13}C	1.108	0.7022	1.404	0.672	25.14
^{15}N	0.365	0.4036	−0.566	−0.271	10.13
^{19}F	100	2.6273	5.255	2.516	94.07
^{31}P	100	1.1305	2.261	1.083	40.48
^{29}Si	4.71	−0.5555	−1.110	−0.532	19.87

[a]μ in units of nuclear magnetons = 5.05×10^{-27} JT^{-1}.

1.3 THEORY OF NUCLEAR MAGNETIC RESONANCE AND ELECTRON SPIN RESONANCE SPECTROSCOPY

Consider a nucleus of bare proton with nuclear spin $I = 1/2$, so that $m_I = \pm 1/2$, associated with it a magnetic dipole moment μ that is proportional to the angular momentum. Similarly, an isolated electron with $S = 1/2$ and $m_s = \pm 1/2$ also possess magnetic moment.

When the nucleus of magnetic moment μ is placed in a magnetic field B_0, the interaction energy (E) is given by

Nuclear magnetic resonance	Electron spin resonance
$E = -\mu B_0 = -\mu B_0 \cos\theta$	$E = -\mu B_0 \cos\theta$ (1.1)
$\mu = g_N \mu_N I$	$\mu = -g_e \mu_B S$
g_N = nuclear g-factor	g_e = Lande g-factor (g_e is 2.0023)
μ_N = nuclear magneton	μ_B = Bohr magneton
$\mu_N = eh/4\pi m_p$	$\mu_B = eh/4\pi m_e$
e = electronic charge	e = electronic charge
m_p = mass of the proton	m_e = mass of the electron
$E = -g_N \mu_N B_0 I$	$E = g_e \mu_B B_0 S$
$Em_I = -g_N \mu_N B_0 m_I$	$Em_s = g_e \mu_B B_0 m_S$
$E_{1/2} = -g_N \mu_N B_0/2$	$E_{1/2} = g_e \mu_B B_0/2$
$E_{-1/2} = g_N \mu_N B_0/2$	$E_{-1/2} = -g_e \mu_B B_0/2$
$\Delta E = E_{-1/2} - E_{1/2} = g_N \mu_N B_0$	$\Delta E = E_{1/2} - E_{-1/2} = g_e \mu_B B_0 = h\upsilon$

The splitting of the nuclear and electron spin energy levels in an external magnetic field (B_0) is shown above. The diagram is not drawn to the scale. The nuclear splitting should be even smaller than it is shown.

1.3.1 Larmor Frequency

In the presence of applied magnetic field (B_0), the interaction between μ and B_0 produces a force called torque, which causes the nuclear magnetic moment (μ) to precess about z-axis as shown in Fig. 1.1 (similar to gyroscope).

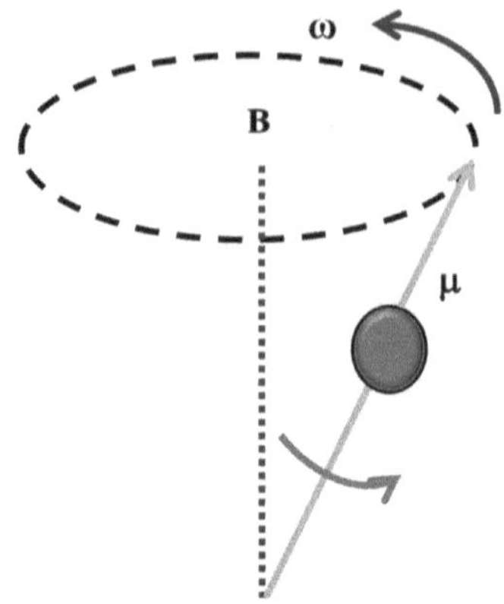

Figure 1.1 Larmor frequency.

This process is called Larmor precession. The angular frequency of precession is known as Larmor frequency (ω) given by

$$\omega_0 = \gamma \mathbf{B_0} \tag{1.2}$$

$$\text{Since, } \omega_0 = 2\pi\upsilon, \quad \upsilon = \gamma \mathbf{B_0}/2\pi \tag{1.3}$$

ω_0 is the precession rate called the Larmor frequency. It is defined as the frequency with which the magnetic dipolar moment of a spin precesses around an external magnetic field (B_0). The constant γ is called the magnetogyric (or) gyromagnetic ratio. It relates the magnetic moment μ and the spin number I for any specific nucleus as follows

$$\gamma = 2\pi\mu/hI \tag{1.4}$$

1.3.2 Resonance Condition

If we irradiate the sample with radio waves (MHz), the proton can absorb the energy and get promoted to the higher energy state. This absorption is called nuclear spin resonance because the frequency of the applied RF radiation and the precession coincide or resonate. Thus, the resonance condition, i.e., ($h\nu = g_N\mu_N B_0$), is achieved by applying an RF oscillating field perpendicular to the external static field. The proton resonance frequencies for different applied field ($\mathbf{B_0}$) strengths (in Gauss) are shown in Table 1.2.

Each nucleus has a characteristic value of γ as shown in Table 1.1. For proton, $\gamma = 2.674 \times 10^4\ G^{-1}\ s^{-1}$. Therefore, for different nuclei with different gyromagnetic ratios, different frequencies must be applied to achieve the resonance condition. The frequencies fall in the RF region of the electromagnetic spectrum. In ESR, the resonant frequencies tend to

Table 1.2 Proton resonance frequency at different magnetic field

Magnetic field (B_0) (Gauss)	Resonance frequency (v) (MHz)
14,100	60
23,500	100
47,000	2000
70,500	300
94,000	400
11,750	500

be much higher, by virtue of the 659-times higher gyromagnetic ratio of an unpaired electron relative to a proton. For example, a typical magnetic field strength used in ESR spectrometers is 0.32 T, with a corresponding resonant frequency of about 9000 MHz (9.8 GHz), whereas in NMR, the resonant frequency is 14 MHz.

1.4 COMPARISON OF NUCLEAR MAGNETIC RESONANCE AND ELECTRON SPIN RESONANCE

Nuclear magnetic resonance (NMR)	Electron spin resonance (ESR)
1. NMR describes the splitting of nuclear spin states in the presence of magnetic field.	ESR is concerned with the magnetically induced splitting of electronic spin states.
2. NMR is an important spectroscopic technique for the study of diamagnetic compounds (all electrons are paired). *NMR-active compounds are diamagnetic.*	ESR is an important spectroscopic technique for the study of paramagnetic species (unpaired electrons) such as free radicals, metal complexes, or triplet excited states of diamagnetic molecules. *ESR active compounds are paramagnetic.*
3. In NMR, the energy required to bring about a transition is low. NMR is observed at lower radio frequencies (MHz) ($\lambda = \sim 2200$ cm). Around 1000 times smaller in NMR.	In ESR, the energy required to bring about a transition is high. ESR usually requires microwave frequency radiation (GHz) ($\lambda = 3.3$ cm).
4. The relaxation times T_1 and T_2 are typically of the order of 0.1–1 s.	The equivalent *electron* relaxation times are a million times shorter, i.e., 0.1–1 μs. (The extremely short relaxation times have important implications on the way in which ESR measurements are carried out).

(Continued)

Nuclear magnetic resonance (NMR)	Electron spin resonance (ESR)
5. NMR is usually measured at room temperature.	The electron paramagnetic resonance (EPR) is measured at low temperature because of the short relaxation times.
6. Signals measured as absorption signal.	Because of large line width, ESR signals are measured as derivative signal.
7. Line width: 0.1 G (or) <1 Hz.	Line width: 1 G (or) $>10^5$ Hz
8. Therefore, resolution in NMR is very high	Resolution in ESR is low
9. However, sensitivity is very low (1 mM)	Sensitivity 1 nM (1 million times greater sensitive than NMR)
10. In NMR, the coupling constant (J, expressed in Hz)	In ESR, the hyperfine coupling (or hyperfine splitting) constant expressed in gauss or MHz.
11. In NMR, different probes are required to record spectra of different magnetic nuclei, viz., ^{1}H, ^{13}C, ^{15}N, ^{19}F.	However, the resonances of different paramagnetic species can be measured in a single run using the same EPR spectrometer.

1.5 NUCLEAR MAGNETIC RESONANCE INSTRUMENTATION

The basis of NMR experiment is to induce a transition between two consecutive nuclear spin energy levels that result when a nucleus is placed in a magnetic field. The resonance condition is given by the following equation, $h\nu = g_N\mu_N B_0$. So, the NMR spectrum can be observed experimentally either

1. by varying the external magnetic field, keeping the frequency fixed (or)
2. by varying the frequency, keeping the magnetic field fixed.

The first approach is often preferred because it is difficult to vary the frequency at a very high level of stability. The basic components of a typical NMR spectrometer (Fig. 1.2) are as follows:

1. *Electromagnet*: It gives a powerful, stable, and homogeneous magnetic field. The field must be constant over the entire area of the sample and over the period of the experiment.
2 *Sweep coil generator*: It supplies a variable current to a secondary magnet. Then, the total applied magnetic field can be varied over a small range.
3. *Sample holder*: The sample container is usually a glass tube (5 mm OD) spun by an air-driven turbine to average the magnetic field over the sample container. The process is often referred to as the spinning of the sample.

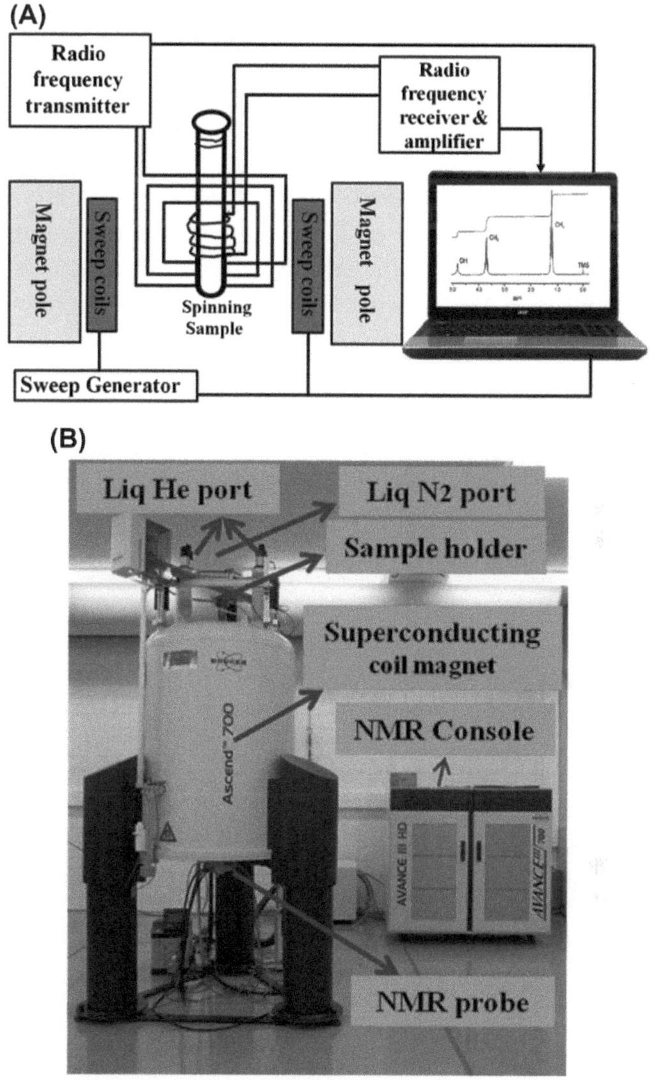

Figure 1.2 (A) Block diagram of nuclear magnetic resonance (NMR) spectrometer and (B) photograph of 700 MHz Brucker spectrometer.

4. *RF transmitter*: It is an oscillator connected to a coil, called the transmitter coil. It generates the RF radiation and transmits the RF energy to the sample. The axis of the coil has to be perpendicular to the magnetic field, i.e., the RF is applied perpendicular to B.

5. *RF receiver*: It is connected to a coil, called the receiver coil, and encircles the sample (Fig. 1.2). Its axis has to be perpendicular to both the magnetic field and the axis of the transmitter coil. The transmitter and receiver coils and the sample holder are constructed into a single unit called *probe*. Often, the necessary small magnetic sweep field is obtained by the use of sweep generator coils built into the NMR probe.
6. *Read out system*: It consists of an RF amplifier, recorder/computer, and other accessories to increase the sensitivity, resolution, and accuracy.

A photograph of NMR spectrometer and schematic representation of these components are shown in Fig. 1.2. At a fixed frequency, the magnetic field at which the resonance occurs for a given nucleus is constant. Therefore, each instrument normally works only for a particular kind of nucleus. Several nuclei can be studied by changing the oscillator, and obviously, the field has also to be changed accordingly. In the case of solids, because the width of lines is expected to be large, this type of NMR setup for investigation cannot be used (see Section 3.5).

1.6 CHEMICAL SHIFT

For bare proton, the field experienced by the ^1H nuclei is the same as the applied field B_0. However, when proton is present in a molecule, the circulating electrons and bonded electrons or other π electrons change (increase or decrease) the magnetic field experienced by the ^1H nuclei; i.e.,

$$B = B_0 \text{ (for bare proton) (no shielding)}$$

$$B = B_0 - B_e \text{ (proton in a molecule) } (B_e \text{ is the shielding field})$$

it may also be represented as

$$B = B_0 (1 - \sigma) \; (\therefore B_e = B_0\sigma) \tag{1.5}$$

σ is called shielding constant.

The Larmor frequency of the precession spins in the presence of applied magnetic field.

$$\omega = \gamma B$$

We know that

$$\omega = 2\pi\upsilon$$

$$\upsilon = \gamma B/2\pi$$

The NMR resonance frequencies obtained from the respective fields of reference (R) and unknown compound (C) are as follows.

$$\upsilon_R = \gamma B_R/2\pi$$
$$\upsilon_C = \gamma B_C/2\pi$$

$$\text{Chemical shift } (\delta, \text{ in ppm}) = (\upsilon_C - \upsilon_R) \times 10^6/\upsilon_R \text{ ppm} \tag{1.6}$$

σ is small typically 10^{-5} for protons and $<10^{-3}$ for other nuclei. δ is defined in terms of a standard reference.

A simple NMR spectrum plots absorption on the y-axis and magnetic field strength on the x-axis (scale is in parts per million (ppm) (δ) and usually goes from 0 to 10 ppm). The different positions of NMR absorptions are described as **chemical shifts** (δ).

A chemical shift is defined as the difference in ppm between the resonance frequency of the observed proton and that of the tetramethylsilane (TMS) hydrogens (Table 1.2). TMS is the most common reference compound used in NMR, and it is set at $\delta = 0$ ppm. The chemical shift in ppm is calculated as the shift downfield from TMS in Hz, divided by the spectrometer operating frequency in MHz. So, if a 60 MHz spectrometer records a proton resonance at a frequency of 426 Hz downfield from TMS, this corresponds to a chemical shift of 426/60 = 7.10 ppm.

$$\text{Chemical shift, ppm } \delta = \frac{(\text{frequency of compound} - \text{frequency of reference}) \times 10^6}{\text{spectrometer frequency (in Hz)}}$$

The **chemical shift** is the position of absorption on the δ scale (in ppm) where the peak occurs.

$$\delta = (\upsilon_C - \upsilon_R) \times 10^6/\upsilon_R \text{ ppm}$$

υ_C—frequency of compound, υ_R—frequency of reference.

The signal shift is very small, ppm, but the great precision with which frequencies can be measured permits the determination of chemical shift to three or more significant figures. The chemical shift can also be expressed as

$$\delta = (\sigma_R - \sigma_C) \times 10^6/(1 - \sigma_R) \tag{1.7}$$

A different chemical shift called τ scale, which is also used by chemists, is as follows

$$\tau = 10.00 - \delta$$

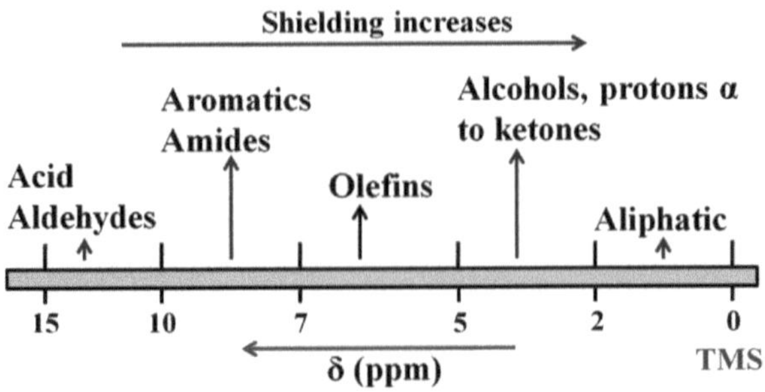

Figure 1.3 Chemical shift values of different functional groups. *TMS*, tetramethylsilane.

For example, the three hydrogens in the CH_3 group are all the same; therefore only one resonance is observed. The OH proton, however, gives a signal at different δ. The different chemical shifts observed for various functional groups are as follows (Fig. 1.3).

The NMR spectra (low resolution, i.e., with no spin–spin splitting) of selected compounds are exhibited as follows to demonstrate the different chemical shift values observed for protons in different molecular environment.

1. Methanol:

   There are two chemically different hydrogens (in $-CH_3$ and $-OH$) in methanol (Fig. 1.4A) to give two resonances at 3.35 and 4.80 ppm, respectively.

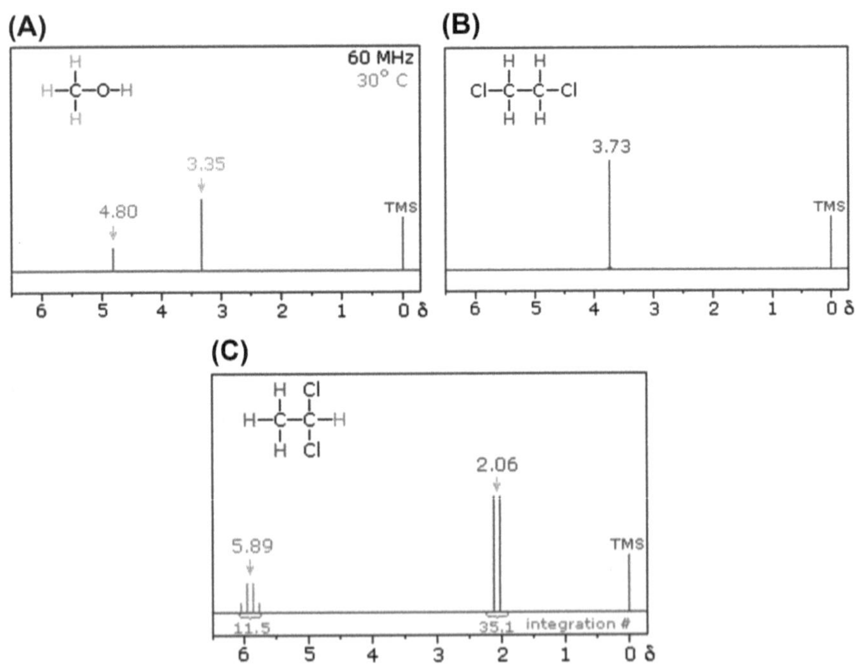

Figure 1.4 Nuclear magnetic resonance spectra of (A) methanol, (B) 1,2-dichloroethane, and (C) 1,1-dichloroethane.

2. 1,2-Dichloroethane:
   Although there are four protons, they are all one type giving one resonance at 3.73 ppm.
3. 1,1-Dichloroethane:
   There are four protons. Three are one type and another one is different type. The high-resolution NMR spectrum is shown in Fig. 1.4C. So, there are two groups of signals.

1.7 NUCLEAR MAGNETIC RESONANCE REFERENCE SAMPLES

The widely used NMR reference samples are (1) TMS, (2) 4,4-dimethyl-4-silapentane-1-sulfonic acid (DSS), and (3) 4,4-dimethyl-4-silapentane-1-ammonium trifluoroacetate (DSA). Their structures and characteristics are described as follows.

1.7.1 Tetramethylsilane

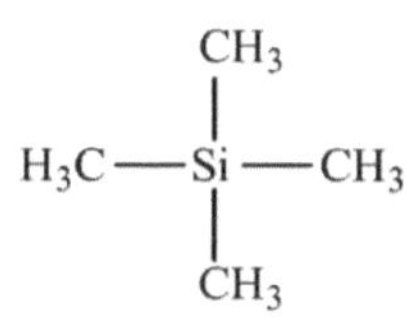

1. It is volatile liquid having boiling point = 26.5°C. It is inert to most of the reagents and is soluble in most of the organic solvent.
2. TMS has 12 protons that are all chemically and magnetically equivalent and four carbons, which are also all equivalent. This means that it gives a single, strong signal in the spectrum, which turns out to be outside the range of most other signals, especially from organic compounds.
3. All the hydrogen atoms in TMS are equivalent and further silicon is more electropositive than carbon. Therefore, it tends to donate electron density to methyl group, thereby increasing their shielding
4. Because its boiling point is low, it can be easily removed from the sample. Hence, pure compound can be recovered.

1.7.2 4,4-Dimethyl-4-silapentane-1-sulfonic Acid

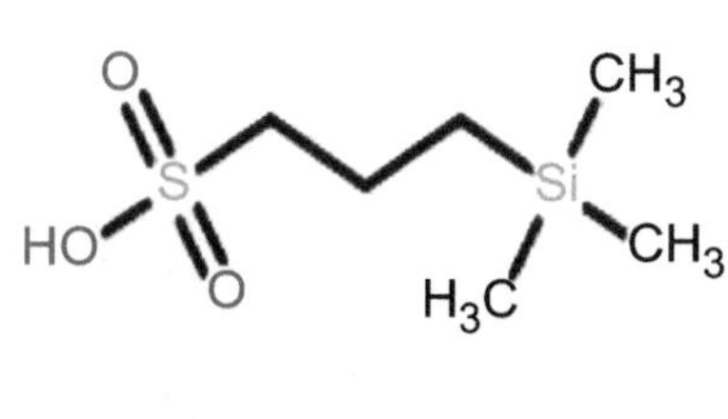

DSS may also be used as a calibration standard, similar to TMS, but with much higher water solubility than TMS.

The low electronegativity of the silicon shields the nine identical methyl protons. The result is a high-intensity proton signal further upfield (at lower chemical shift) than almost all peaks found in naturally occurring organic molecules. The resulting standard peak is easily identified as such and set to chemical shift 0.0.

1.7.3 4,4-Dimethyl-4-silapentane-1-ammonium Trifluoroacetate

For NMR studies in aqueous solution, where TMS is not sufficiently soluble, the reference usually used is DSS $\left(Me_3Si{-}CH_2CH_2{-}SO_3^-Na^+\right)$. However, for aqueous solution of cationic substrates (e.g., amino acids) where there may be interactions between the anionic reference compound and the substrates, an alternative reference standard, DSA ($Me_3Si{-}CH_2CH_2{-}NH_3^+CF_3CO_2^-$), has been used.

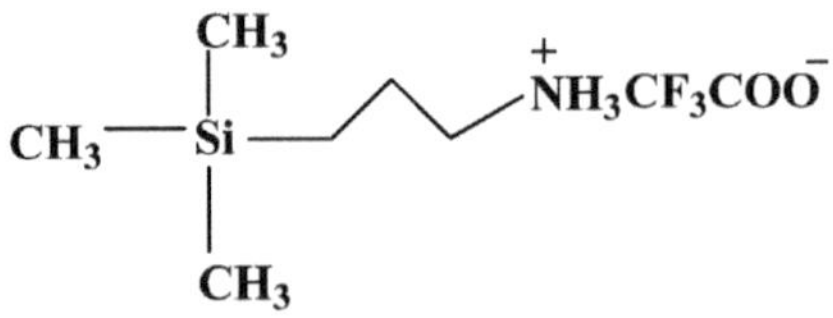

1.7.4 Nuclear Magnetic Resonance Solvents

NMR spectra are almost invariably obtained in solution. The solvents of choice:

1. should have adequate dissolving power.
2. should not associate strongly with solute molecules as this is likely to produce appreciable effects on chemical shifts. This requirement must sometimes be sacrificed to achieve adequate solubility.
3. should be essentially free of interfering signals. Thus for ^{1}H NMR, the best solvents are proton-free.
4. should preferably contain deuterium, ^{2}H. Deuterium is an isotope of hydrogen, which is relatively easy to obtain and incorporate into common solvents in place of hydrogen with insignificant changes to the properties of the solvent. Almost all NMR instruments use deuterium as a convenient "locking" signal to stabilize the magnetic field of the NMR magnet.

Because NMR spectra are recorded for compounds dissolved in a solvent, the signals will be observed for the solvent also and this must be accounted for solving spectral problems. To avoid spectra dominated by the solvent signal, most ^{1}H NMR spectra are recorded in a deuterated solvent. In chloroform solvent, this corresponds to $CDCl_3$, so a singlet signal with small intensity is observed at 7.26 ppm for $CHCl_3$. For methanol solvent, this corresponds to CHD_2OD, so a 1:2:3:2:1 pentet signal (i.e., $2nI + 1 = 2 \times 2 \times 1 + 1$ i.e., $I = 1$ for D) is observed at 3.31 ppm.

1.8 SPIN–SPIN COUPLING

In high-resolution NMR, i.e., high frequency NMR, each peak is often split into multiple (or) cluster of peaks due to magnetic interactions between nonequivalent protons on adjacent carbons; the process is called spin–spin splitting or spin–spin coupling. The coupling constant J (usually in frequency units, Hz) is a measure of magnetic interaction between a pair of protons independent of applied field strength. Nuclei having the same chemical environment or chemical shift are called chemically equivalent. For example, CH_3–CH_3, no spin–spin splitting between two CH_3 group protons. Some nuclei having different chemical shift or different environment are called nonequivalent (for example, $-CH_3$ and $-CH_2$ protons). Such nuclei, which are close to one another, exerts an influence on each other's effective magnetic field to cause spin–spin splitting. If the distance between nonequivalent nuclei is less than or equal to three bond lengths, this effect is observable.

1.8.1 Characteristics of Spin–Spin Coupling

1. Spin–spin coupling (J) is independent of the strength of the applied magnetic field B_0. So increasing the magnetic field strength will increase the chemical shift difference between two peaks in hertz (not ppm), but the coupling constant J will not change.
2. Equivalent nuclei do not interact with each other.
3. Multiplicity is determined by the number of neighboring groups of equivalent nuclei. For protons $I = 1/2$, if n is the number of equivalent nuclei in the neighboring group, the expected multiplicity is $n + 1$ (when the chemical shift difference $\Delta\nu \gg J$). This is called $(n + 1)$ rule. The number of lines in a coupling pattern is given, in general, by **2nI + 1** for coupling with n number of nuclei spin I.
4. Coupling constants (scalar) are dependent on the number of bonds between the interacting nuclei. Coupling is almost negligible at distances greater than three bonds. However, when there is charge or electron

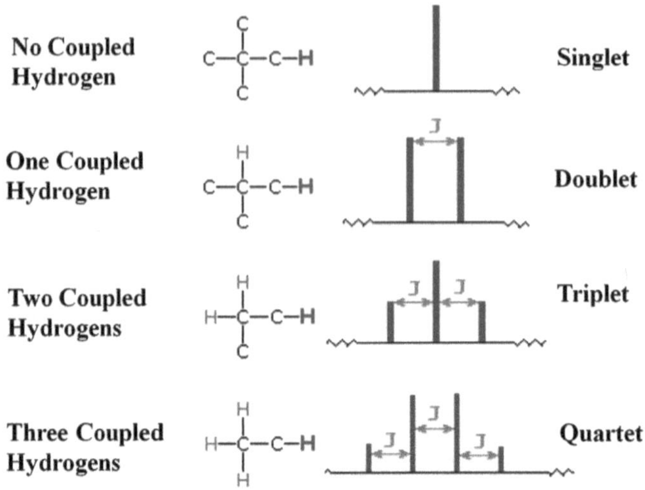

Figure 1.5 Spin–spin splitting and multiplicity.

delocalization as in aromatic systems, interaction beyond three bonds is possible.

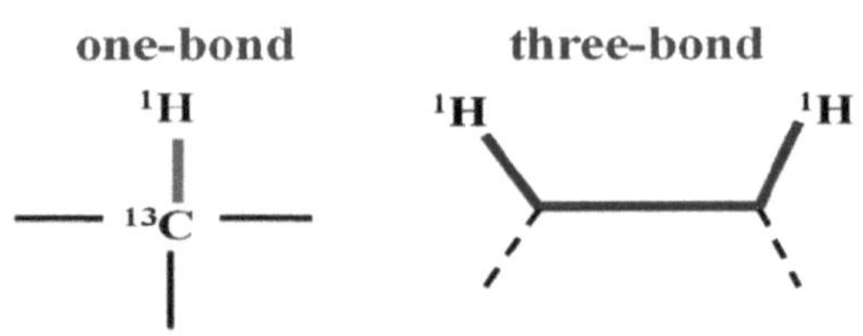

5. The difference between any two adjacent components of a multiplet is the same and yields the value of the spin–spin coupling constant J (in hertz) (Fig. 1.5).
6. The intensities of multiplets are symmetric about the center with relative intensities given by the coefficients of the terms in the expansion of $(a+b)^n$ (which is more conveniently called Pascal's triangle), where n is the number of equivalent nuclei causing the splitting. (Fig. 1.6).

To a first approximation, the relative intensities of the multiplets are given by binomial coefficients: 1:1 for a *doublet*, 1:2:1 for a *triplet*, and 1:3:3:1 for a *quartet*.

1.8.2 Mechanism of Nuclear Spin–Spin Interaction

There are two pathways of magnetic interaction (coupling) occurring between nuclei (A and X) with nonzero spins.

1. Direct interaction (dipole–dipole coupling D)
2. Indirect or scalar coupling (spin–spin splitting J)

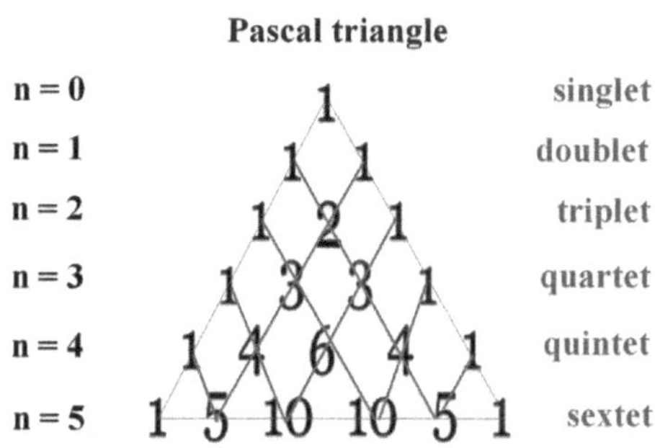

Figure 1.6 Pascal triangle.

1.8.2.1 Direct Interaction (Dipole–Dipole Coupling D)

Based on the theory of electrodynamics, the energy of interaction through space between two nuclear dipole moments μ_1 and μ_2 (associated with the two nuclear spins) is given by

$$E = k\,\mu_1\mu_2\left(3\cos^2\theta - 1\right)/r^3 \quad (1.8)$$

k is a proportionality constant, r is the distance between two magnetic nuclei with μ_1 and μ_2, and θ is the angle between them. In a solid, this direct dipole-dipole interaction energy is found to be considerable. However, in fluids (i.e., liquids and gases) because of random tumbling motion of the molecules, the average value of $(3\cos^2\theta-1)$ becomes zero and the value E vanishes. So, in liquids and gases, there is no direct nuclear spin–spin interaction.

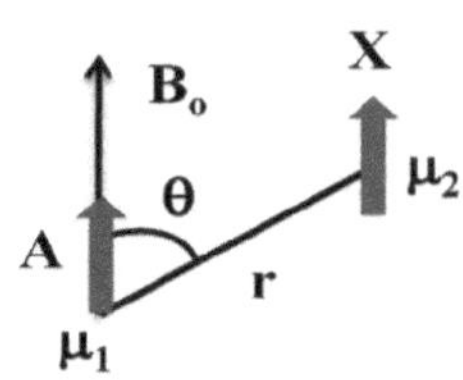

The magnitude of direct spin–spin interaction is about 1000 times as large as the scalar coupling (e.g., at 2 Å distance H...H dipolar coupling is c.30,000 Hz). These direct couplings make the observation of high-resolution NMR spectra in solids and very viscous liquids difficult. In liquids, the random motion of molecules completely averages the dipolar coupling, so no direct interactions are seen and become isotropic spectrum only due to scalar coupling as explained below.

1.8.2.2 Indirect or Scalar Coupling (Spin–Spin Splitting J)

N.F. Ramsey and E.M. Purcell (1951) [12a, 12b] proposed that in liquids and gases, the interaction between two nuclear spins is transmitted through the electrons of the chemical bonds that intervene between them. This is called "*indirect electron-coupled nuclear spin–spin interaction.*" In other words, the magnetic interaction between individual nuclear spins transmitted by the bonding electrons through which the nuclear spins are indirectly connected; this is called indirect spin–spin coupling (J). Consider, for example, two nuclear spins C (carbon) and H (hydrogen) separated by a chemical bond formed between electrons with spins ↑ and ↓ as follows:

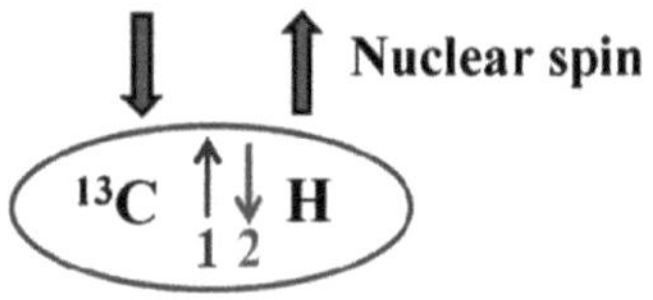

If electron ↑ has a finite probability at the nucleus with nuclear spin C, then the electron spin tends to be aligned antiparallel to the nuclear spin C. The electron spin ↑ now influences the alignment of spin ↓ of H. This in turn affects the alignment of nuclear spin H. Thus, the energy levels of nucleus H depend on the spin orientation of nucleus C. It is thus evident that the nuclear spins C and H interact through the intervention of the electron spins ↑ and ↓.

This Ramsey–Purcell mechanism of nuclear spin–spin interaction is merely considered an extension of the so-called **Fermi contact interaction**, proposed by Fermi in 1930. The Fermi contact interaction is defined as the magnetic interaction between an electron and an atomic nucleus when the electron is found inside that nucleus. According to Fermi contact interaction, if an s-electron has a finite value of its wave function at the center of a nucleus, then the electron spin can be felt by the nuclear spin.

The magnitude of this nuclear spin–spin interaction depends on the following factors:

1. Increasing number of bonds intervening between the nuclei involved decrease in J. This is expected because now the nuclei are too far to interact.
2. A double bond, however, transmits spin–spin coupling more readily than a single bond, i.e., the spin–spin coupling constant J between

two atoms joined by a double bond is generally greater than that joined by a single bond.

3. The spin–spin coupling can also be transmitted through lone pair of electron between nuclei that are sufficiently close together, although not directly bonded.

1.9 CLASSIFICATION OF NUCLEAR MAGNETIC RESONANCE SYSTEM

NMR spectra are classified based on Pople notation as follows:

1. Each chemically different type of proton is given a capital letter A, B, C...M, N...X, Y, Z.
2. The number of protons in each group protons is indicated as a subscript A_2B_3....
3. Protons with similar δ values are assigned letters close to one another in the alphabet A, B, C; M, N; X, Y, Z.
4. Two protons with same chemical shift but magnetically nonequivalent is assigned, for example, AA′ XX′.
5. AX, AM are weakly coupled $\Delta\delta/J > 8$. For three protons very different from each other, AMX, ABX are weakly coupled.
6. AB, XY are strongly coupled pair of protons $\Delta\delta/J < 8$. ABC, XYZ are strongly coupled.

Accordingly, NMR spectra are classified into (1) AB, (2) AX, (3) AM, (4) ABC, (5) AMX, etc. If the chemical shift values of two or three nuclei are close together, the spin system is called an AB or ABC systems, respectively. However, the chemical shift values of two or three nuclei are far apart (as in HF molecule); the spin system is called an AM or AX or AMX systems, respectively. For example, the spin system A_2BXY_2 has two A nuclei, one B nucleus with chemical shift close to that of the A nuclei, and one X and two Y nuclei with chemical shift values close to one another but far removed from those of A and B.

1.9.1 Chemically and Magnetically Equivalent Nuclei

Based on their chemical shift (δ) and spin–spin coupling (J), the magnetic nuclei (protons) are classified into two types (1) chemically equivalent and (2) magnetically equivalent.

1. *Chemically equivalent*: If the two protons in a molecule are chemically equivalent, then these protons will have identical chemical shift (due to symmetry present in the molecule) and may or may not have identical

coupling between them. For example, in 1,2-dichloroethane, the two methylene protons are identical and have the same chemical shift values.

$$Cl-CH_2-CH_2-Cl$$

2. *Magnetic equivalent*: If the two protons in a molecule are magnetically equivalent, then they have both identical chemical shift and coupling constant. For example, the two protons are chemically equivalent but magnetically inequivalent.

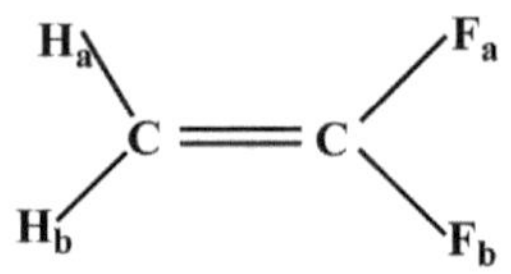

1.10 TYPES OF SPIN—SPIN SPLITTING

Based on the magnitude of chemical shift (δ) and spin—spin coupling constant (J), there are three types of splitting explained as follows:

1. Zero-order splitting ($J = 0$)
2. First-order splitting ($\Delta\delta/J > 8$)
3. Second- or higher-order splitting ($\Delta\delta/J < 8$)

1.10.1 First-Order Splitting

The appearance of NMR spectra depends on $\Delta\delta/J$, i.e., the ratio of the change in chemical shift to the spin—spin coupling constant. The spectra are called first-order spectra if $\Delta\delta/J$ is very large. The high-resolution NMR spectrum of ethanol is a first-order spectrum. However, when the ratio $\Delta\delta/J$ is not large, several peaks appear overlapping to give a complex NMR spectrum (second-order splitting).

If the chemical shift difference between coupled protons in hertz is much larger than the J coupling constant, $\Delta\delta/J > 8$, then the NMR systems (for example, viz., AX, AMX, AX_2) exhibit first-order splitting. In other words the proton is only weakly coupled to other protons that are far away in chemical shift (Fig. 1.7). The spectrum is called *first-order spectrum*. The splitting pattern also depends on the magnetic field. The second-order splitting at the lower field (frequency) can be resolved into first-order splitting pattern at the high field (frequency). The first-order splitting pattern follows multiplicity rule ($n + 1$) and Pascal's triangle to determine splitting pattern and intensity distribution.

For first-order spectrum, $J_{ab} = J_{ac}$.

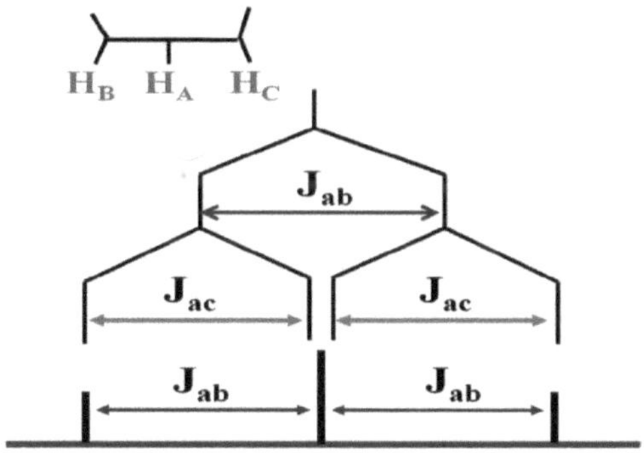

For example, consider $A_3M_2X_2$ NMR system as follows, H_A and H_X have the triplet pattern split by H_M using n + 1 rule. The signal of H_M is split into six peaks by H_X and H_A. The first-order pattern is easily interpreted because of separation with equal splitting pattern.

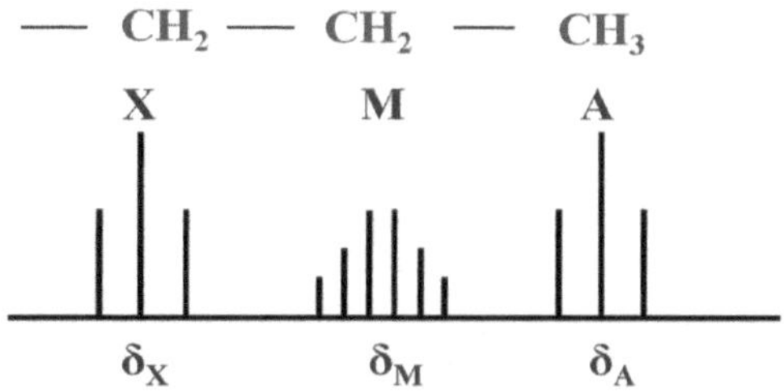

1.10.2 Second-Order Splitting

The n + 1 rule cannot predict the exact splitting patterns of non—first-order splitting. First-order splitting produces "normal" splitting patterns that have equal J values. Non—first-order splitting occurs when a nuclear spin couples with two or more sets of nearby nuclei that have different J values.

For non—first-order splitting, $J_{ab} \neq J_{ac}$.

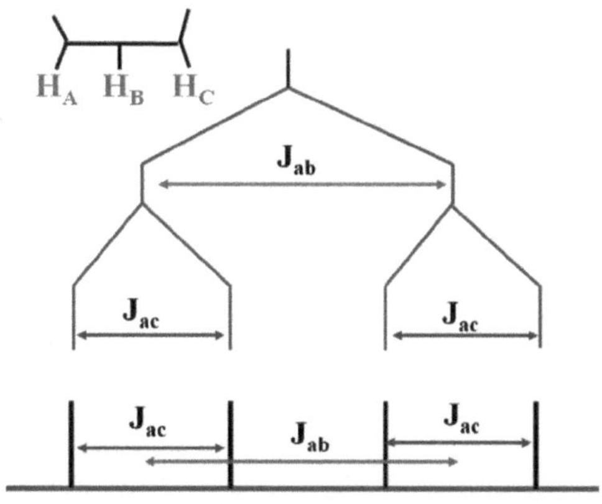

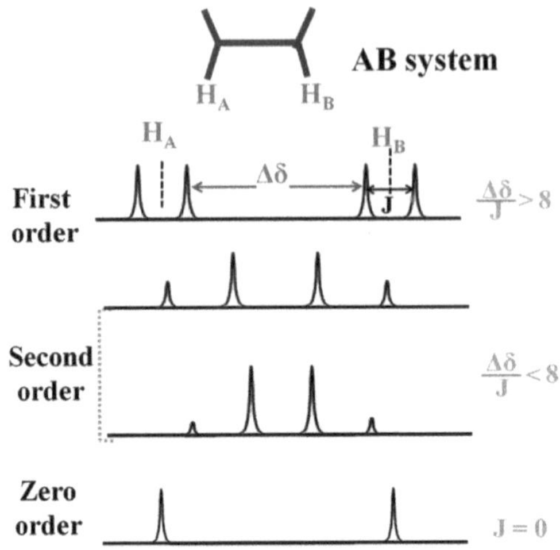

Figure 1.7 Nuclear magnetic resonance spectra for different spin–spin splitting for AB system.

High-order splitting pattern takes place when chemical shift difference ($\Delta\delta$) is much less or the same order of magnitude as the J coupling.

$$\Delta\delta/J \leq 8$$

For example, NMR systems, viz., AB, ABC, AB_2 ZY, etc., exhibit second-order splitting. The second- or higher-order pattern is observed as leaning of a classical pattern. The inner peaks are taller and the outer peaks are shorter in case of AB system (Fig. 1.7). This is called the "roof effect."

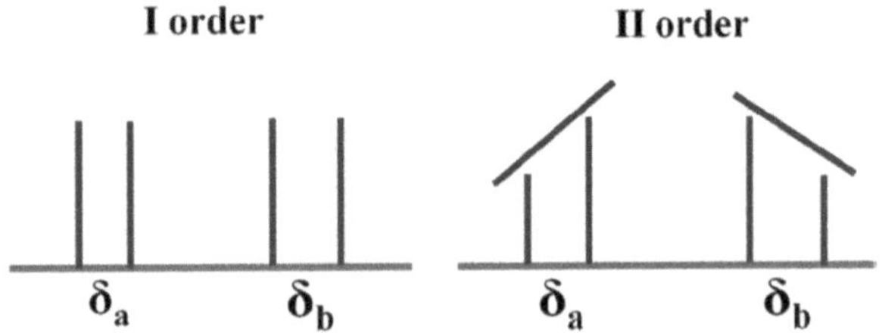

1.10.3 Effect of J on Energy Levels of AB System

The Hamiltonian equation and the energy of the spin–spin interaction (J) in AB system is given by

$$\widehat{H}_{Nuclear} = \widehat{H}_{NZ} + \widehat{H}_J$$

$$E = -(\nu_A m_A + \nu_B m_B) + J_{AB} m_A m_B \text{ (in frequency units)} \tag{1.9}$$

$\widehat{H}_{NZ}$ is the nuclear Zeeman interaction, $\widehat{H}_J$ the spin–spin splitting, $m_A = \pm 1/2$ (I = 1/2 for A), and $m_B = \pm 1/2$ (I = 1/2 for B). ν_A and ν_B are the resonance frequencies of A and B, respectively.

The effect of J on the energy levels of AB system is given in Fig. 1.8. The four energy levels obtained are given in Fig. 1.8.

$$\begin{aligned} E_1 &= -1/2(\nu_A + \nu_B) + (J/4) \\ E_2 &= -1/2(\nu_A - \nu_B) - (J/4) \\ E_3 &= +1/2(\nu_A - \nu_B) - (J/4) \\ E_4 &= 1/2(\nu_A + \nu_B) + (J/4) \end{aligned} \tag{1.10}$$

There are three cases depending on the magnitude of J values as follows:

1. **J = 0**: When there is no coupling between the A and B nuclei, the energy involved in either transition (A_1 or A_2) is equal. It gives one signal for A nucleus. Likewise, two transitions for B nucleus (B_1 and B_2), which are equal in energies, give a signal corresponding to chemical shift of it. This is also known as *zero-order splitting*.
2. **J > 0**: When J > 0 between the two nuclei, the energy levels of the spin system will be either stabilized or destabilized as shown in Fig. 1.8. So, the two transitions each for A_1 and A_2 for A nucleus ($A_2 > A_1$) and B_1 and B_2 ($B_2 > B_1$) for B nucleus, respectively, due to coupling are not equal. So, the NMR spectrum shows four lines.

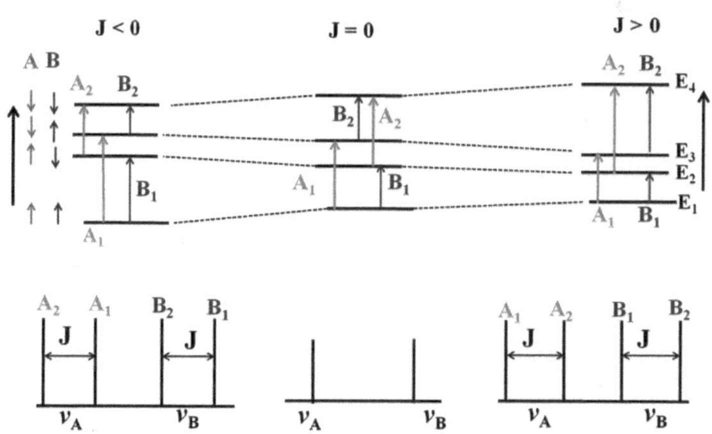

Figure 1.8 Energy levels of AB system due to different J values.

3. **J < 0**: Like above, when $J < 0$ between the two nuclei, the energy levels of the spin system will be either stabilized or destabilized as shown in Fig. 1.8. So, the two transitions A_1 and A_2 for A nucleus ($A_1 > A_2$) and B_1 and B_2 ($B_1 > B_2$) for B nucleus, respectively, again due to coupling are not equal. So, the NMR spectrum again shows four lines, two each for A and B, however, interchanged.
The transitions that we see in the spectrum are given by

$$\begin{aligned} A_1 &= E_3 - E_1 = \nu_A - 1/2\,J_{AB} \\ A_2 &= E_4 - E_2 = \nu_A + 1/2\,J_{AB} \\ B_1 &= E_2 - E_1 = \nu_B - 1/2\,J_{AB} \\ B_2 &= E_4 - E_3 = \nu_B + 1/2\,J_{AB} \end{aligned} \tag{1.11}$$

Thus, the observed four lines are quantitatively explained.

1.11 LOW- AND HIGH-RESOLUTION NUCLEAR MAGNETIC RESONANCE

1.11.1 Low-Resolution Nuclear Magnetic Resonance

It refers to low-frequency NMR (below 60 MHz) where spin–spin splitting is not observed. So, it looks much simpler but it cannot distinguish between the individual peaks in the various groups of peaks except its position (chemical shift). However, it provides the following:

1. The number of peaks tells the number of different environments for the hydrogen atoms, i.e., the types of protons.
2. The ratio of the areas under the peaks tells the ratio of the number of hydrogen atoms in each of these environments.
3. The chemical shifts give you important information about the sort of environment of the hydrogen atoms, i.e., shielded or deshielded.

1.11.2 High-Resolution Nuclear Magnetic Resonance

In high-resolution NMR spectra (frequency above 60 MHz), single peaks in the low-resolution spectrum are split into clusters of peaks [13].

One peak: singlet
Two peaks in the cluster: doublet
Three peaks in the cluster: triplet
Four peaks in the cluster: quartet

We can get exactly the same information from a high-resolution NMR spectrum as from a low-resolution spectrum. But in addition, the amount of spin–spin splitting (J) of the peaks gives valuable electronic, stereochemical, and connectivity information.

1.12 SIMPLIFICATION OF COMPLEX NUCLEAR MAGNETIC RESONANCE SPECTRA

The complete NMR spectral analysis of a compound is frequently made difficult, when signals overlap. Consequently, useful information is often buried because of complexity of the spectrum. In such a spectrum, many signals may overlap as is the case of closely related methylene groups in a molecule. In such a situation an intense and broad unresolved signal may appear between δ 1 and 2 ppm for methylene protons. Another cause of complexity is where a coupling constant is comparable with the chemical shift difference between the coupled protons because of higher order of splitting. Complex NMR spectra can be simplified by the following ways.

1.12.1 Using Shift Reagents

Spectral simplifications can be obtained by means of the so-called **shift reagents**, such as the complexes of the rare-earth metals, particularly those of europium and praseodymium (Section 3.7). A small amount of the shift reagent added to the sample under study causes a marked increase in the chemical shift, thereby reducing the complex pattern of the spectrum. The magnetic field associated with the metal ion, which is a paramagnetic moiety, causes marked changes in the observed shifts of the protons in the substrate. Shift reagents provide a useful technique for spreading out ^{1}H NMR absorption patterns, which normally overlap, without increasing the strength of the applied magnetic field.

For instance, in the ^{1}H NMR spectrum of n-hexanol, the high-field triplet is distorted, which represents the absorption of a methyl group adjacent to a $-CH_2-$ group. The low-field broad multiplet (presumably not first order) is due to the methylene group adjacent to the hydroxyl group. The protons of the remaining methylene groups (four) are all buried in the methylene envelope between δ 1.2 and 1.8 ppm. When the same spectrum is recorded after the addition of a soluble shift reagent Europium(III) complex, $Eu(DMP)_3$, the spectrum is spread out over a wider range of frequencies so that it is now simplified almost to first order (see Fig. 3.23).

1.12.2 High-Field Nuclear Magnetic Resonance

Complex NMR spectra can be simplified by methods that make chemical shift differences larger than spin–spin coupling constant. Because δ is field-dependent, while J is field-independent, it is evident that a complex proton NMR spectrum, obtained at say 1.4 T (60 MHz), will be simplified if it is obtained at a higher field, say 2.34 T (100 MHz). The NMR spectra run at 100 MHz yield chemical shifts difference almost twice as large as those run at 60 MHz.

Therefore, the magnitude of coupling constants J is not dependent on the strength of the applied field but their appearance in the spectrum does change as the magnetic field changes. One major cause of complexity in the ^1H NMR spectra is due to the situation where the chemical shift difference between the coupled protons becomes comparable to the coupling constant $\Delta\delta \approx J$. At a high field, all spectra essentially reduce to the simple first-order type because $\Delta\delta \gg J$. For example, an AB coupling pattern in a 60 MHz spectrum tends to become an AX type in 300 MHz spectrum. The signals become easily recognizable, and also unresolved signals now spread out. Moreover, too closely spaced signals are also resolved better at larger field strengths.

1.12.3 Deuterium Exchange and Labeling

Deuterium ^2H or D, the heavy isotope of hydrogen ($I = 1$), has been used extensively in ^1H NMR spectroscopy for two reasons. First, it can be easily introduced into a molecule. Second, the presence of deuterium in a molecule is not detected in a ^1H NMR spectrum. Deuterium has a much small magnetic dipole moment than hydrogen, and therefore it absorbs at different field strengths. Although it broadens the peak of a neighboring proton, it does not split it because it couples only slightly with a proton. For example, in ethyl bromide ($CH_3{-}CH_2{-}Br$), deuterium replaces the methyl hydrogens and the multiplicity of the $-CH_2-$peaks changes from a quartet ($3 + 1$ peaks) to a triplet (in $CH_2D{-}CH_2{-}Br$) to a doublet (in $CHD_2{-}CH_2{-}Br$) and finally to a singlet ($CD_3{-}CH_2{-}Br$) as the hydrogens on the $-CH_3$ group are continuously replaced by deuterium. Carboxylic acid and aldehyde protons appear between δ 12 and 9 ppm. However, acidic protons can be differentiated from aldehyde because protons in carboxylic acid attached to a hetero atom exchange readily with D_2O.

1.12.4 Spin Decoupling/Double Resonance Method

This method employs two different RF radiations also called double irradiation or spin decoupling technique very frequently used in the simplification and easy interpretation of complex NMR spectra. It is based on the principle that one group of equivalent protons in a molecule is irradiated, i.e., decoupled by applying very strong stationary RF field at the Larmor (resonance) frequency of the protons. Thus, the irradiated protons could no longer undergo spin–spin coupling with the neighboring group protons. As a result of this spin decoupling, the spectrum is simplified and easily interpreted (see Section 2.6.1).

1.12.5 Multidimensional Nuclear Magnetic Resonance

Multidimensional NMR techniques, viz., 2D and 3D, have been largely used for structure elucidation of large complex molecules where overlapping of signals occur (see Section 2.1.7). The main advantages are dispersion of overlapping signals into two or three dimensions and thereby eliminating overlapping and also obtaining correlations to establish the structure [14].

1.12.6 Magic Angle Spinning

In solid-state NMR, the broad lines are observed because of significant dipolar couplings and chemical shift anisotropy dependent on term $(1-\cos^2\theta)$ (see Section 3.5). By spinning the sample at about 54.74°C (called magic angle), the broad lines collapse into sharp isotropic peaks. Thus, the spectrum in solid state is simplified for interpretation and analysis.

1.13 SENSITIVITY AND BOLTZMANN STATISTICS

When a group of spins is placed in a magnetic field, each spin aligns in one of the two possible orientations. At any particular temperature, the number of spins in the lower energy level is N_{low} (N_α) and the number of spins in the higher energy level is N_{high} (N_β). In the presence of an external magnetic field, different nuclear spin states (with different values of m_I) have different energies. The energy difference is proportional to $\mathbf{B_0}$ (Fig. 1.9). At thermal equilibrium, these states will have different populations. Their ratio is given by the Boltzmann equation

$$N_{high}/N_{low} = e^{-(\Delta E/kT)} \tag{1.12}$$

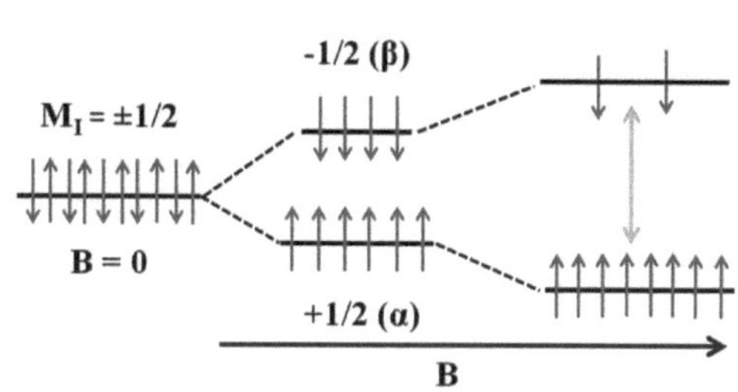

Figure 1.9 Splitting of nuclear spin states and their population difference.

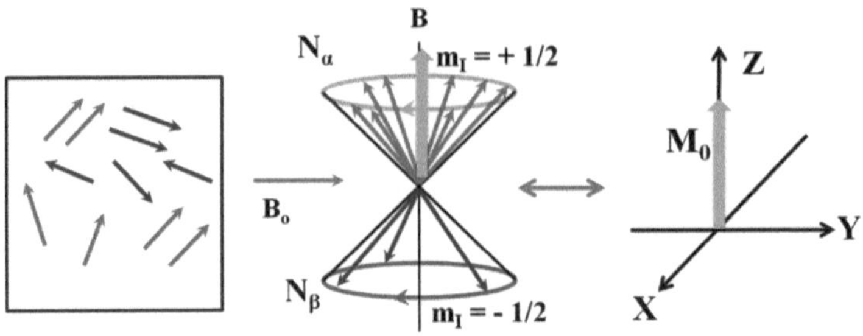

Figure 1.10 Representation of net magnetization vector (M_0).

N_{high} and N_{low} are the respective populations of the upper and lower spin states, $\Delta E = E_{high} - E_{low}$ the energy difference between the two states, k is Boltzmann's constant, 1.3805×10^{-23} J/K, and T is the temperature in Kelvin. As the temperature decreases, so does the ratio N_{high}/N_{low}. As the temperature increases, the ratio increases and approaches 1.

In currently achievable magnetic fields (B), the difference between nuclear spin energy levels ΔE is much smaller than kT, implying that N_{low} is only very slightly in excess of N_{high}. For ^1H in a 9.4 T field (400 MHz) and 300 K, one obtains a population ratio $N(\alpha)/N(\beta)$ of 1.000064, i.e., for one million spins in the upper β state, there are one million and sixty-four in the lower energy α state (Fig. 1.9). It is this excess 64 spins that respond to the NMR experiment and create the net magnetization $\mathbf{M_0}$ (Fig. 1.10). To summarize, the larger the B_0 (at high field) the greater the energy difference ΔE between the levels, and the larger the ΔE the more excess population exists in the lower energy state leading to higher sensitivity.

1.14 PULSED/FOURIER TRANSFORM NUCLEAR MAGNETIC RESONANCE

To carry out a normal continuous wave (CW) NMR experiment, either the strength of the magnetic field or the frequency must be varied. This is a time-consuming process, taking several minutes to scan over the required range. Furthermore, to acquire signals with low intensity, it is necessary to run the experiment repeatedly to average and remove the noise associated with the measurement. Instead of individually scanning the frequency domain, in pulse NMR, the sample will be excited by a single pulse or multiple pulses of strong radio waves of range of frequencies and then the free induction decay (FID) is picked up on a receiver. By Fourier transforming the FID, a multifrequency-based NMR spectrum is observed. This can be completed in a few seconds or less, making it far less time-consuming to perform than the multiple scans needed for the CW method. Modern NMR spectrometers use the pulsed Fourier transform (FT) method [15–22].

1.14.1 Nuclear Magnetic Resonance Pulse

The B_1 RF field is applied as a pulse of duration t_p, which usually lasts for a few microseconds. The angle θ_t (called the pulse angle) through which the net magnetization (M_0) is tipped from the z-axis increases with the amplitude of B_1 and with the length of time, t_p, of the pulse applied (Fig. 1.11):

$$\theta_t\ (\text{rad}) = \gamma B_1\ t_p \tag{1.13}$$

1. In the presence of the $\mathbf{B_0}$ magnetic field, the spins in α and β states are precessing about the z-axis. Their processional frequency ω_0 is given by the Larmor equation. The vector sum of the individual nuclear spin magnetic moment $\boldsymbol{\mu}$ vectors yields a net equilibrium magnetization M_0 along the positive z-axis (Fig. 1.10). Due to the precession about the z-axis, the x and y components of the individual μ vectors sum to zero, leaving only the z component.
2. Excitation of the nuclei from the lower energy α-spin state to the higher energy β-spin state is achieved (Fig. 1.12) with an oscillating RF magnetic field $\mathbf{B_1}$ applied with a transmitter coil as a short duration pulse along the x-axis (RF pulse). This oscillating magnetic field can be regarded as a rotating magnetic field. When the rotating frequency of $\mathbf{B_1}$ is equal to the precession frequency of the nuclear moments (ω_0), $\mathbf{B_1}$ excites nuclei in α-spin state to the β-spin state. This causes the net magnetization M_0 tipping it from the z-axis into the yz-plane (Fig. 1.13). One can completely transform the z-magnetization into y-magnetization if the duration of the pulse is the length of a $\pi/2$ pulse, a 90° pulse. The component of the magnetization is in the xy-plane, initially along

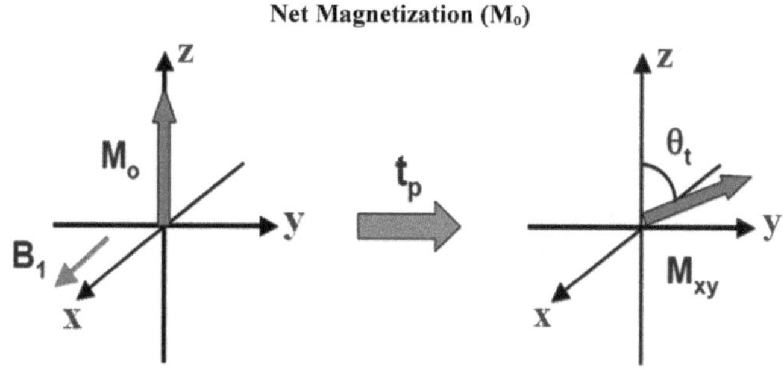

Figure 1.11 Pulse angle.

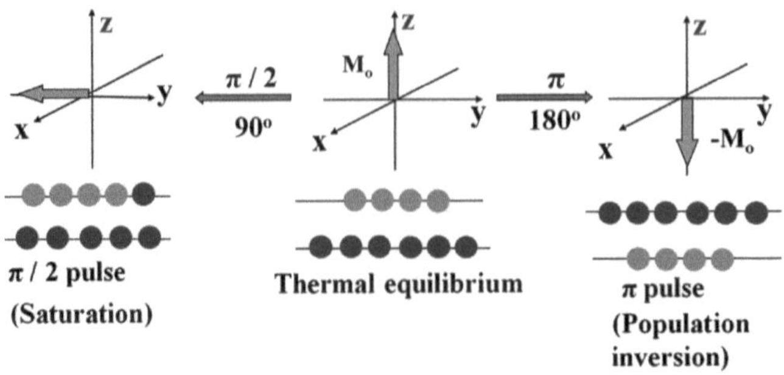

Figure 1.12 Effect of nuclear magnetic resonance pulse on population difference.

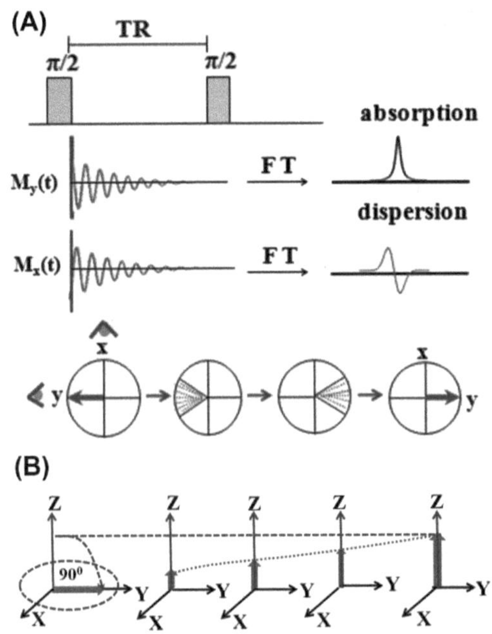

Figure 1.13 (A) Free induction decay and (B) saturation recovery.

the y-axis (M_y). The net magnetization (M_y) along the y-axis is detected with respect to time (t) using an antenna coil illustrated with an eye in Fig. 1.13.

3. In the pulse method of acquiring an NMR spectra, all the nuclei of one species in the sample are quickly excited simultaneously by a "hard" RF pulse of energy for a very short (~ μs) duration of time. The power of this pulse is of the order of several watts at a range of frequencies so that all the nuclei absorb the energy. This pulse of energy is applied perpendicular to the z-direction of the applied magnetic field along the x-direction. As a result, the net magnetization of the sample (M_0) is itself turned 90 degrees and is thus "tipped" along y-direction and then spread into the xy-plane as shown in Fig. 1.13.

This type of pulse is often called a $\pi/2$ pulse (as $\pi/2 = 90°$). Now that the magnetization vector has turned on its side into the xy-plane, the magnetization begins to precess, which is detected by coil of wires along both the x- and y-direction called quadrature detection. The signal will oscillate and eventually decay as the nuclear spins begin to precess incoherently from each other and relax to their ground state. The signal is called FID (Fig. 1.13) called time-domain NMR.

When viewed along the y-axis, the magnetization falls from initially high value (M_0) and decays in a sinusoidal manner as the vector precesses about the z-axis. Similarly, when viewed along the x-axis, the magnetization increases from zero value and falls in a sinusoidal manner. The amplitude of the signal decays with time as nuclei in the higher energy β-spin state return back to the lower energy α state in a process known as nuclear spin relaxation. If the molecule has only a single chemical shift (Fig. 1.14A),

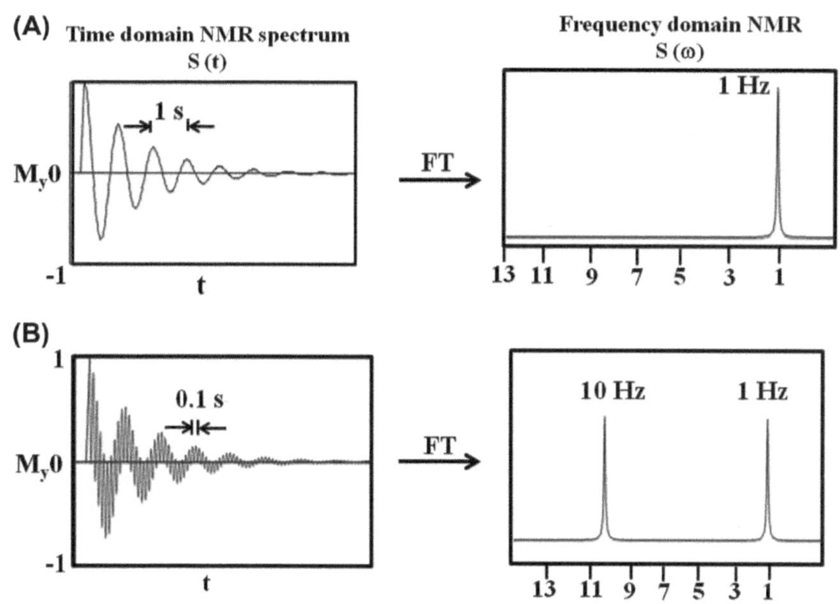

Figure 1.14 Time-domain and frequency-domain nuclear magnetic resonance (NMR) spectra for (A) one and (B) two chemical shifts.

the signal appears as a simple decaying sine wave and the chemical shift in hertz is the frequency of the sine wave relative to a reference frequency. Most molecules have many nuclei with many different chemical shifts and correspondingly many different precession frequencies (Fig. 1.14B).

This FID signal is not very useful in its present form. To change it into an actual NMR spectrum, we have to take the FT of the signal, which is accomplished by the following mathematical relationship:

$$S(\omega) = \int S(t)e^{-i\omega t}dt \tag{1.14}$$

$$\begin{aligned} e^{i\omega t} &= -\cos \omega t + i \sin \omega t \\ &= -M_y + iM_x \end{aligned}$$

In FID, the frequency of the oscillations (ω) changes the position of the peak, i.e., chemical shift. If the oscillating signal decays more quickly, the peak becomes broader. Furthermore, if there are clearly two frequencies overlapping (Fig. 1.14B), which appear as "beats" in the FID, the FT shows the presence of two or more frequencies in it.

1.14.2 Quadrature Detection

In time domain, the plot of M_y as a function of time is a cosine wave (Fig. 1.13A). Fourier transforming this gives peaks at both $+\upsilon$ and $-\upsilon$ because the Fourier transformation cannot distinguish between $+\upsilon$ and $-\upsilon$ rotation of the vector from the data supplied.

Similarly, a plot of M_x as a function of time is a sine wave. Fourier transforming this gives peaks at $+\upsilon$ and $-\upsilon$ because the FT cannot distinguish between a positive vector rotating at $+\upsilon$ and a negative vector rotating at $-\upsilon$ from the data supplied.

The solution is to input both the M_x and M_y into the Fourier transformation and is called linear detection. The FT is designed to handle two orthogonal input functions called the real and imaginary components. Detecting the M_x and M_y components for input into the FT is called linear detection. This was the detection scheme on many modern NMR spectrometers and some magnetic resonance imagers. It required the computer to discard half of the frequency domain data.

Simultaneous detection ofboth M_x and M_y is also called quadrature detection. Using the quadrature detection, the FT can distinguish between $+\upsilon$ and $-\upsilon$.

1.15 RELAXATION PROCESS AND RELAXATION TIMES

Consider absorption of electromagnetic RF radiation by a system whose nuclei possess spin $I = 1/2$. (1) Let N_α and N_β be the number of spins occupying the lower and upper energy levels, respectively. In the resonance condition, the sample absorbs energy (Fig. 1.15). The population difference must decrease with time. (2) Consequently, the energy absorbed by the sample should also decrease due to saturation and ultimately becomes zero. (3) But, the intensity of NMR spectrum is usually independent of time indicating a constant population difference. (4) So, there must be some mechanism in the system, which allows the spins occupying the upper state to lose energy by nonradiative transitions and return to the lower state.(5) The transfer of excess energy from the excited nuclear spin system to the other degrees of freedom and thereby return to thermal equilibrium after the absorption of RF pulsed energy is known as **relaxation process**.

There are two different relaxation processes (times)

1. Spin-lattice relaxation and
2. Spin—spin relaxation for nuclei.

1.15.1 Spin-Lattice Relaxation Time (T_1)

Initially, the spins in the upper energy state transfer the excess energy to the surroundings (lattice). This phenomenon is called spin-lattice relaxation (Fig. 1.15). The time required for the population difference to become (37%) $1/e$ times the population difference at $t = 0$ is referred to as the **spin-lattice relaxation time** or **longitudinal relaxation time** T_1 (Fig. 1.16). In other words, T_1 relaxation describes the return of the z component of longitudinal (z-) magnetization to its equilibrium value. This process is called saturation recovery (Fig. 1.13B). T_1 varies over a large range, being 10^{-2}—10^{-4} s (shorter) for solids and 10^{-4}—10 s (longer) for liquids.

1.15.2 Spin—Spin Relaxation Time (T_2)

In addition to the spin-lattice relaxation, relaxation of spin can also occur by a direct interaction of two spin states known as spin—spin relaxation time.

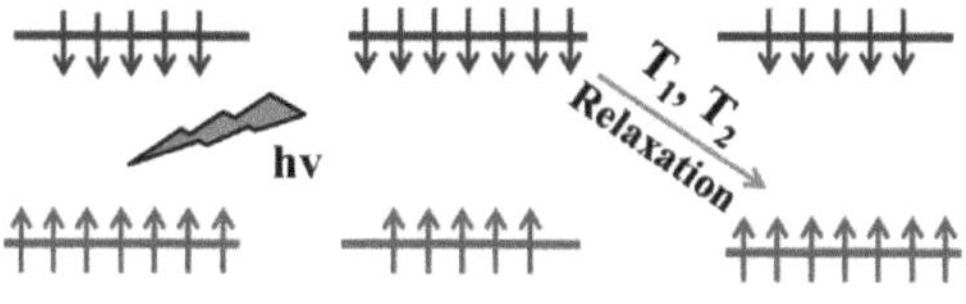

Figure 1.15 Relaxation process and nuclear magnetic resonance absorption.

In other words, T_2 relaxation describes the decay of transverse (x, y) magnetization called transverse relaxation time T_2. It is also due to loss of phase coherence among nuclei. T_2 is less than or equal to T_1; when relaxation is very fast, NMR lines are broad and J-coupling may not be resolved.

1.15.3 Mechanism of T_1 and T_2

1. *Paramagnetic effect*: There are different mechanisms that decrease T_1 and result in line broadening. However, the mechanism particularly effective is due to paramagnetic substances such as impurities, oxygen, etc. This effect is due to the fact that the magnetic moment of unpaired electron is about 10^3 times as large as the moment of a nucleus, resulting in very low values for T_1. Consequently, very broad line results when paramagnetic species are present.
2. *Nuclear quadrupole effect*: The nuclei with $I \geq 1$ possess nuclear quadrupole moment that can interact with the local fluctuating electric field gradients (which arise from the motion of polar solvent molecules and solute molecular vibrations) and undergo rapid spin-lattice relaxation T_1 and spin—spin relaxation T_2. This is a very effective process, and as a result deuterium (spin: $I = 1$) undergoes spin-lattice relaxation T_1 much more rapidly than a proton. It causes severe line broadening of neighboring protons.
3. *Nuclear dipole—dipole interaction*: It arises due to the dipole-dipole effect of one nuclei on the other nearby nuclei and dissipates the excess energy to the adjacent nuclei resulting in T_1 and T_2 shortening. In other words, the static component of the close nuclei causes a slight variation in the effective field of the nuclei being examined. The contribution to the effective magnetic field varies with the angle, distance, and magnitude of the field of this second nucleus (see Section 1.8.2). Since the effect decreases sharply with distance, only neighboring nucleus will have an appreciable effect on the effective field.
4. *Inhomogeneity in applied field*: Another cause of shortening of T_2 and broadening results from the variation in the applied field B_0 over the sample dimensions because of field inhomogeneity. So, the observed spectrum is a superposition of several spectra corresponding to the different values of B_0 in the different parts of the sample.
5. *Transverse magnetization*: A 90 degrees RF pulse rotates the magnetization from $M_{0(z)}$ to xy-plane. The rotating component of the nuclei can induce a transition of the neighboring nuclei, which has a frequency equivalent to the Larmor frequency of the nucleus undergoing the

transition. Thus, this rotating component can induce a transition of the neighboring nucleus to a different spin state. The nucleus causing the transition also simultaneously undergoes a change in its spin state. This decreases the spin—spin relaxation time (T_2) and also the lifetime of the excited state results in line broadening.

6. *In solid state*: The molecules are rigid, tightly packed, and close together. As a consequence, the NMR spectra of solids are very broad because T_2 is short. The fixed positions of the atoms in the solid give rise to some very large, oriented local fields. The spin—spin relaxation processes are more efficient in solid state and T_2 is smaller. Because of the very broad lines usually obtained when solid spectra are examined, most NMR experiments have been carried out on gases, liquids, or solutions. A sample-spinning procedure has been reported to increase resolution and reduce line widths in solids (see Section 3.5).

1.16 MEASUREMENT OF T_1 AND T_2

There are four common methods to obtain the NMR signals and they are used to measure T_1 and T_2 (Table 1.3).

1. FID method and saturation recovery
2. Spin echo (SE) method (or) Hahn echo method
3. Stimulated echo (STE)
4. Inversion recovery method (IR)

1.16.1 Free Induction Decay Method and Saturation Recovery

Several different pulse sequences have been utilized to measure T_1 and T_2. The simplest common and best technique for determining T_1 is to apply a 90 degrees pulse. This creates transverse magnetization components as shown in Fig. 1.13B. After complete relaxation time τ (TR), a next 90 degrees pulse is applied. The FID following the 90 degrees pulse is acquired.

Table 1.3 Different types of nuclear magnetic resonance signals

Type of MR signal	Method of formation	Sequence
1. Free induction decay	One radiofrequency (RF) pulse	90°
2. Spin echo	Two RF pulses	90°–180°
3. Stimulated echo	Three or more RF pulses	90°–90°–90°
4. Inversion recovery	Three or more RF pulses	180°–90°

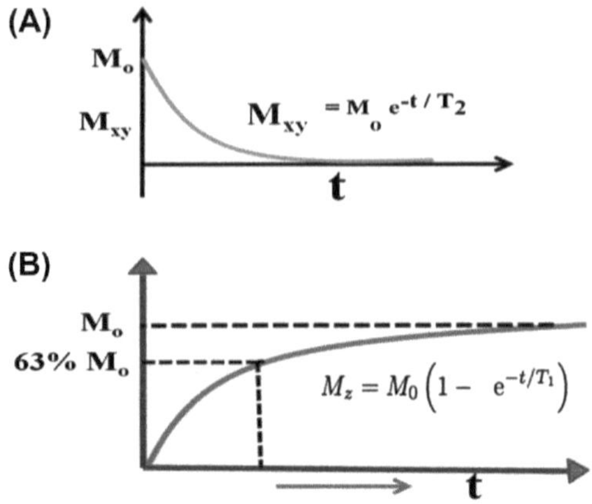

Figure 1.16 (A) Free induction decay and (B) saturation recovery.

A 90 degrees pulse is applied to detect any magnetization only in the transverse plane (XY) (see eye in Fig. 1.13). The value of T_2 may be determined by measuring the magnitude M_x and M_y of the FID as a function of time (t) using the following expression (Fig. 1.16A):

$$M_{x,y} = M_0 e^{-t/T_2} \tag{1.15}$$

where M_0 is the magnitude of the FID, when $\tau =$ infinity (i.e., at thermal equilibrium).

However, the value of T_1 may be determined by measuring the recovery of M_z of the FID along the z-axis until it reaches its equilibrium value M_0 as a function of time using the below expression (Fig. 1.16A). This method is called saturation recovery.

$$M_z = M_0\left(1 - e^{-t/T_1}\right) \tag{1.16}$$

Their relaxation rates $R_{1(2)} = 1/T_{1(2)}$ are also used. Both the relaxation times and processes are assumed to be first-order rate processes and are obtained by fitting the curve.

T_1 is called the longitudinal relaxation time because it is a measure of how fast the magnetization relaxes back along the z-axis. Likewise, T_2 may be called the transverse relaxation time because it measures how fast the spins exchange energy in the transverse (xy) plane with other nuclear spins.

1.16.2 Spin Echo Method (or) Hahn Echo Method

Echo phenomena are important features of coherent spectroscopy, which have been used in NMR, laser spectroscopy, and neutron scattering. Echoes

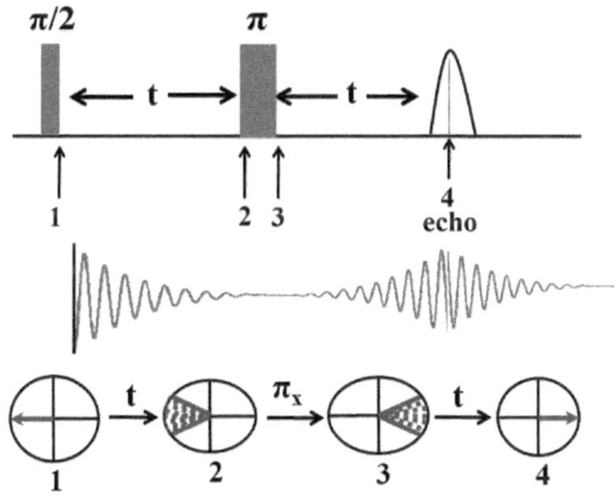

Figure 1.17 Pulse sequence for spin echo formation and detection.

were first detected in NMR by Erwin Hahn in 1950 and therefore SEs are referred to as Hahn echoes.

An SE is formed after application of two pulses: an excitation pulse (90 degrees) and a refocusing pulse (180 degrees), commonly a 90−180 degrees sequence (Fig. 1.17). One microwave pulse produces a signal that decays away called FID. The advantage of this SE method is that NMR/ESR spectrum is inhomogeneously broadened. So, we can recover the disappeared or buried weak signal using another microwave pulse (180 degrees) to produce Hahn echo.

A sample is irradiated with two or more RF pulses and observes an echo signal at some time after the RF pulses. More specifically, started with a 90 degrees excitation pulse, waited a short time period t, then 180 degrees refocusing pulse was applied and waited for another equal time period t for the peak of the SE signal to occur (Fig. 1.17 (4)). In NMR, an SE is the refocusing of spin magnetization by a pulse of resonant electromagnetic radiation. Modern NMR/EPR and magnetic resonance imaging use widely this SE method.

1.16.2.1 Significance of Spin Echo Method

The NMR signal observed following an initial excitation pulse ($\pi/2$ or 90°) decays with time because of both spin−spin relaxation (T_2) and any inhomogeneous effects (T_2*), which cause different spins in the sample to process at different rates. Relaxation leads to an irreversible loss of magnetization. However, the inhomogeneous dephasing in FID can be removed by applying a 180 degrees inversion pulse that inverts the magnetization vectors. The inhomogeneous effects are due to magnetic field gradient and a

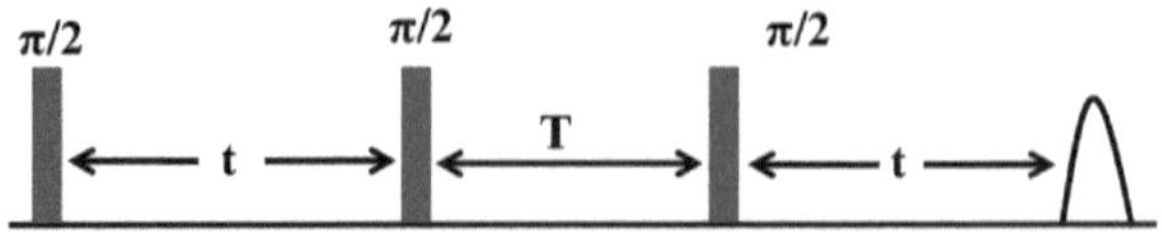

Figure 1.18 Stimulated echo pulse sequence.

distribution of chemical shifts. If the inversion pulse is applied after a period t of dephasing, the inhomogeneous evolution will rephase to form an echo at time 2t. In simple cases, the intensity of the echo relative to the initial signal is given by e^{-2t/T_2}, where T_2 is the time constant for spin—spin relaxation, i.e., $M_{xy} = M_0 e^{-2\tau/T_2}$ and similarly the saturation recovery followed (at position 4) is fitted using the expression $M_z = M_0\left(1 - 2e^{-2/T_1}\right)$ to obtain T_1.

1.16.3 Stimulated Echo

An echo that is formed by the application of three or more RF pulses, such as the classic STE sequence [90-90-90-echo], is as follows (Fig. 1.18).

For example, the three-pulse sequence, $\pi/2 - t - \pi/2 - T - \pi/2 - t$, creates a STE at time t, i.e., increase after the third $\pi/2$ pulse. The stimulate echo is an example of pancake-like echo.

1.16.4 Inversion Recovery Method

The inversion recovery experiment can also be used to measure T_1 times. The pulse sequence used are $180° - \tau - 90°$ as follows.

At time $t = 0$, the 180 degrees pulse inverts the magnetization vector $M_z = -M_z$. After this, the magnetization lies along the negative z-axis. The spin-lattice relaxation makes the M_z magnetization increase during τ delay time until it is back to original value, $M_z = M_0$ (Fig. 1.20). If at some time t following the 180 degrees pulse, the 90 degrees pulse is applied, which rotates the magnetization around the x-axis and will then lie along the y-axis, one can find the value M_z at any time of t using the following equation.

The T_1 times can be calculated by fitting the curve (Fig. 1.20). Thus, the IR experiment can be used to measure T_1 relaxation times.

However, saturation recovery is fast and simple to measure T_1 times more quickly than an IR pulse sequence (Figs 1.16 and 1.19). SR consists of 90°degrees RF pulse at relatively short repetition times (Fig. 1.13).

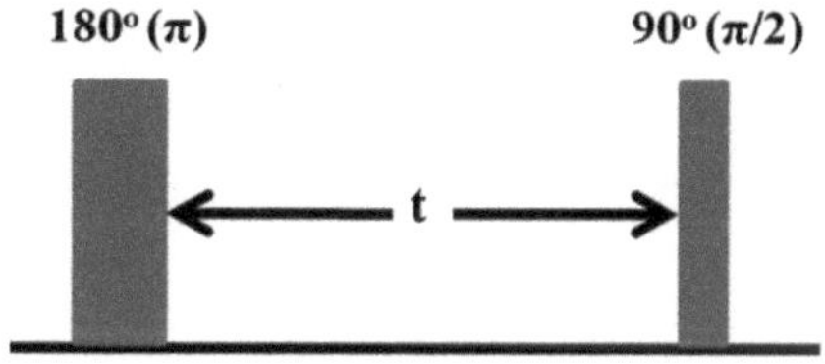

Figure 1.19 Inversion recovery pulse sequence.

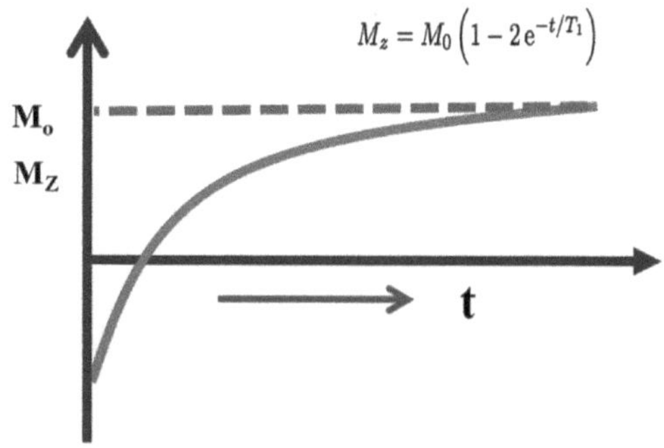

Figure 1.20 Inversion recovery for T_1 measurement.

1.17 BLOCH EQUATION

The Bloch equations (BEs) are a set of differential equations, which can be used to describe the behavior of a magnetization vector under any conditions. When properly integrated, the Bloch's equations will yield the X′, Y′, and Z′ components of magnetization as a function of time. The Bloch's equations thus describe the fate of magnetization in the presence of a magnetic field and RF field.

Consider an ensemble of nuclei, for example, protons. The total magnetic moment is given by

$$M = \sum_i \mu_i \tag{1.17}$$

where μ_i is the magnetic moment of individual spin due to its spin angular momentum. In terms of total angular momentum (L) of a sample, it is expressed as

$$M = \gamma L \tag{1.18}$$

In the presence of external magnetic field (B), there is an interaction between magnetic moment with B. As a result, a force called torque (T) changes the angular momentum of the system ($T = M \times B$).

$$T = dL/dt \tag{1.19}$$

Differentiating Eq. (1.18)

$$\begin{aligned} dM/dt &= \gamma dL/dt \\ &= \gamma(M \times B) \end{aligned} \tag{1.20}$$

$$M \times B = \begin{vmatrix} i & j & k \\ M_x & M_y & M_z \\ B_x & B_y & B_z \end{vmatrix}$$
$$= (M_yB_z - M_zB_y)i + (M_zB_x - M_xB_z)j + (M_xB_y - M_yB_x)k$$

In terms of individual components,

$$\begin{aligned} \frac{dM_x}{dt} &= \gamma(M_yB_z - M_zB_y) \\ \frac{dM_y}{dt} &= \gamma(M_zB_x - M_xB_z) \\ \frac{dM_z}{dt} &= \gamma(M_xB_y - M_yB_x) \end{aligned} \tag{1.21}$$

When the static magnetic field is applied along z-direction, B_z is nonzero and $B_x = B_y = 0$. The spin processes around B_z. In the absence of any relaxation and RF radiation (i.e., for free precession), the BEs are:

$$\begin{aligned} \frac{dM_x(t)}{dt} &= \gamma M_y(t)B_z \\ \frac{dM_y(t)}{dt} &= -\gamma M_x(t)B_z \\ \frac{dM_z(t)}{dt} &= 0 \end{aligned} \tag{1.22}$$

However, including relaxation terms considering the spin-lattice (T_1) and spin—spin relaxation (T_2) as first-order processes, the modified BEs are given by

$$\begin{aligned}
\frac{dM_x}{dt} &= \gamma(M_yB_z - M_zB_y) - \frac{M_x}{T_2} \\
\frac{dM_y}{dt} &= \gamma(M_zB_x - M_xB_z) - \frac{M_y}{T_2} \\
\frac{dM_z}{dt} &= \gamma(M_xB_y - M_yB_x) - \frac{(M_z - M_0)}{T_1}
\end{aligned} \tag{1.23}$$

Applying again the condition that the field B_0 is applied along z-axis, therefore $B_0 = B_z$, $B_x = B_y = 0$, then the BEs become

$$\begin{aligned}
\frac{dM_x}{dt} &= \gamma M_yB_0 - \frac{M_x}{T_2} \\
\frac{dM_y}{dt} &= -\gamma M_xB_0 - \frac{M_y}{T_2} \\
\frac{dM_z}{dt} &= -\frac{(M_z - M_0)}{T_1}
\end{aligned} \tag{1.24}$$

Furthermore, by applying the RF field, it transforms the magnetization in laboratory frame (x, y, z) into rotating frame (x′, y′, z) (Fig. 1.21). The BEs under this condition are given below.

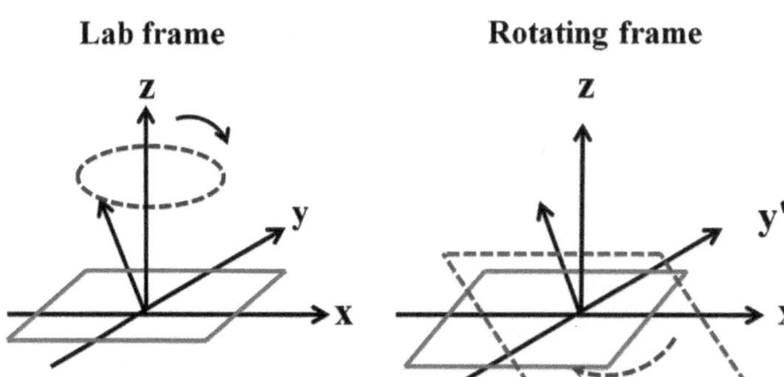

Figure 1.21 Representation of lab and rotating frames.

$$\frac{dM_x}{dt} = \gamma(M_yB_0 + M_zB_1\sin(\omega t)) - \frac{M_x}{T_2}$$
$$\frac{dM_y}{dt} = -\gamma M_zB_0(\cos(\omega t) - M_xB_0) - \frac{M_y}{T_2} \quad (1.25)$$
$$\frac{dM_z}{dt} = -\gamma(M_xB_0\sin(\omega t) + M_yB_1\cos(\omega t)) - \frac{(M_z - M_0)}{T_1}$$

The BEs are further integrated to obtain the magnetization components along x-, y-, and z-axis.

$$M_x(t) = [M_x(0)\cos\omega t - M_y(0)\sin\omega t]e^{(-t/T_2)}$$
$$M_y(t) = [M_x(0)\sin\omega t + M_y(0)\cos\omega t]e^{(-t/T_2)} \quad (1.26)$$
$$M_z(t) = M_{eq} + [M_z(0) - M_{eq}]e^{(-t/T_1)}$$

1.18 LINESHAPE AND LINE WIDTH ANALYSIS

1.18.1 Lineshape

Spectral lineshape is a characteristic feature observed in spectroscopy, corresponding to an energy change in an atom, molecule, or ion. When the resonant transition energy is measured by means of NMR/ESR or any other spectroscopic technique, the spectroscopic line is not infinitely sharp but has a particular lineshape. The shape of lines in an NMR/EPR spectrum is determined by the process of FID. Therefore, the ideal lineshapes in spectroscopy include mostly and sometimes Lorentzian, Gaussian, and Voigt functions. They are defined by the three parameters viz., line position, maximum height (amplitude), and half-width. Furthermore, the observed spectral lineshape and line width are also affected by instrumental factors and parameters settings.

1.18.1.1 Lorentzian Lineshape

It is also called homogenous lineshape because all the molecules behave in the same way. It is represented mathematically by the equation as follows and shown in Fig. 1.22A.

$$Y(B) = \frac{A}{1 + \left[(B - B_0)^2/(w/2)\right]} \quad (1.27)$$

where A = amplitude, w = full width at half maximum ($LW_{1/2}$) in Hz (FWHM), x_0 = peak position in Hz.

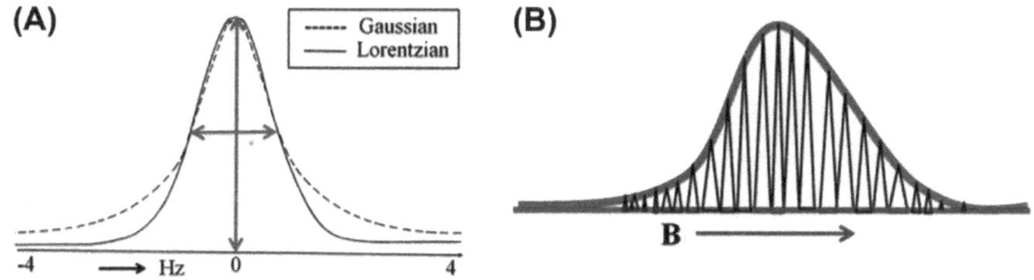

Figure 1.22 (A) Comparison of Lorentzian and Gaussian lineshapes. (B) Gaussian inhomogenous lineshape.

However, in EPR, the lines are broad because of efficient relaxation process and hence its lineshape is either Lorentzian or Gaussian.

The minimum obtainable peak width at half-height is directly related to the resolution of an instrument, i.e., how close two peaks can be and still be distinguishable. Resolution is usually measured using *o*-dichlorobenzene as standard, which has very narrow lines in its ^{1}H NMR spectrum. The manufacturers' resolution specification is usually 0.20 Hz.

It arises largely because of pressure/concentration broadening in gaseous state and natural lifetime broadening. Other factors that influence lineshape include the quality of the NMR tube, sample spinning, sample concentration, dissolved oxygen, and paramagnetic impurities. The latter three will lead to an overall broadening of the lines.

1.18.1.2 Gaussian Lineshape

Inhomogeneous lineshape is called Gaussian because all molecules behave differently (distribution). The expected lineshape of the EPR resonance line is Lorentzian, i.e., for isotropic spectrum, if there is no hyperfine broadening and g-strain. The presence of g-value distribution (g-strain) is inferred from the Gaussian part in the ESR lineshape. This normally results in a change from a Lorentzian to Gaussian lineshape. We know from folding studies and from structural NMR and X-ray studies that samples of proteins come with a distribution of conformations. For EPR, this means that the paramagnet in each molecule has a slightly different structural surrounding and thus a slightly different g-value. This structural inhomogeneity or g-strain is reflected in the spectroscopy in the form of an inhomogeneous lineshape. An important consequence of this g-value anisotropy is that the line width, w, is in general, large and broad.

A Gaussian lineshape also arises from the statistical distribution of the spin magnetic moments, resulting in a dipolar (inhomogeneous) broadening. Hence, the Gaussian lineshape is a superposition of many individual

(Lorentzian) components (Fig. 1.22B). One usually refers to such composites as inhomogeneous broadened. It also arises because of Doppler broadening and power broadening in gaseous state. The mathematic representation of it is given as follows and its plot is shown in Fig. 1.22.

$$Y(B) = Ae^{[(B-B_0)/(w/2)]^2} \ln 2 \tag{1.28}$$

Thus, it occurs because of the following factors

- An inhomogeneous external magnetic field
- Unresolved hyperfine structure
- Anisotropic interactions
- Dipolar interactions

1.18.1.3 Voigt Lineshape

The Voigt lineshape is a convolution of inhomogenous Gaussian lineshape function with a homogenous Lorentzian lineshape function.

1.18.2 Line Width

In spectroscopy, several factors contribute to the broadening, i.e., line width of spectral lines. The principal sources affecting the line width are as follows

1. *Lifetime broadening.* According to the uncertainty principle, the uncertainty in energy, ΔE, and the lifetime, Δt, of the excited state are related by

$$\Delta E\ \Delta t > \hbar/2 \tag{1.29}$$

This relation determines the minimum possible line width. As the excited state decays exponentially in time, this effect produces a line with Lorentzian shape. The natural width of a spectra line, w, is proportional to the reciprocal of the average time, t, the system spends in the excited state.

$$w \propto 1/t \tag{1.30}$$

This important statement can be understood by a consideration of the uncertainty principle. As the lifetime of the excited state becomes very short (i.e., Δt is small), the uncertainty in the energy of this state, ΔE, becomes great. The lifetime of a nucleus in a particular spin state will be determined by the relaxation processes (and times) already discussed. According to the equation, when ΔE is large, a wide range of frequencies is absorbed in transitions, resulting in a broad line.

In infrared, visible, and ultraviolet spectroscopy, the limit of resolution of commonly available instruments is not sufficient to allow measurement of the natural line width. In NMR/EPR spectroscopy, the natural line width often can be obtained, and as a result information about the lifetime of a state can be deduced from the observed line width.

The lifetime of the state will depend on the efficiency of the spin-lattice T_1 and spin–spin relaxation T_2 processes. A large value for T_1 and T_2 indicates an inefficient relaxation process and a long lifetime for the excited state. As a result, a sharp Lorentzian line results as predicted by the equation.

2. *Doppler broadening.* This is caused by the fact that the velocity of atoms or molecules relative to the observer follows a Maxwell distribution, so the effect is dependent on temperature. If this was the only effect, then the lineshape would be Gaussian.
3. *Pressure broadening* (collision broadening). Collisions between atoms or molecules reduce the lifetime of the upper state, Δt, increasing the uncertainty ΔE. This effect depends on both the density (that is, pressure or concentration for a gas) and the temperature, which affects the rate of collisions. This broadening effect is described by a Lorentzian profile in most cases.
4. *Proximity broadening.* The presence of other molecules close to the molecule involved affects both the line width as well as line position. It is the dominant process for liquids and solids. An extreme example of this effect is the influence of hydrogen bonding on the spectra of protic liquids.
5. *Chemical exchange.* A certain chemical and conformational exchange processes are found to have a very pronounced effect on the lineshape and width and appearance of NMR spectra. As both the number of peaks and the fine splitting structure for a given compound depend on the exchange phenomena, it is necessary to understand these effects to interpret spectra obtained for many systems. Furthermore, it will be seen that line broadening caused by rapid chemical exchange provides kinetic data on very fast exchange reactions and their activation parameters (see Sections 3.2 and 3.3).

1.19 NUCLEAR MAGNETIC RESONANCE AND ELECTRON PARAMAGNETIC RESONANCE PROBLEMS

1. Calculate the proton NMR resonance frequency in magnetic field of 1.5 T. ($g_N = 5.585$ and $\mu_N = 5.05 \times 10^{-27}\ JT^{-1}$)

We know that $h\upsilon = g_N\beta_N B$ Given $B = 1.5\ T(\beta_N = \mu_N)$

$$\upsilon = \frac{5.585 \times 5.05 \times 10^{-27}\ JT^{-1} \times 1.5\ T}{6.626 \times 10^{-34}\ Js}$$

^1H NMR resonance frequency $= 63.8 \times 10^7$ MHz

2. What is the magnetic field required for proton magnetic resonance at 220 MHz? ($g_N = 5.585$ and $\mu_N = 5.05 \times 10^{-27}\ JT^{-1}$)

$$h\upsilon = g_N \beta_N B \quad \text{Given } \upsilon = 220 \times 10^6 \text{ Hz}$$

$$B = \frac{6.626 \times 10^{-34}\ \text{Js} \times 220 \times 10^6\ \text{s}^{-1}}{5.585 \times 5.05 \times 10^{-27}\ \text{JT}^{-1}}$$

Magnetic field, $B = 5.17 \times 10^{-1}$ T

3. Calculate the magnetogyric ratio of ^{19}F nucleus with NMR frequency 40 MHz in a magnetic field of 1 T.
The Larmor frequency is given by

$$\omega = \gamma B \quad \text{Given } \upsilon = 40 \times 10^6 \text{ Hz, } B = 1 \text{ T}$$

$$\therefore \gamma = \omega / B \quad \text{where } \omega = 2\pi\upsilon$$

$$\gamma = 2\pi\upsilon / B$$

$$\gamma = \frac{2 \times 3.14 \times 40 \times 10^6\ \text{s}^{-1}}{1\ \text{T}} = 2.51 \times 10^{-8}\ \text{T}^{-1}\ \text{s}^{-1}$$

4. Calculate the g_e value of an unpaired electron placed in a magnetic field, 0.33 T at ESR resonance frequency 9.0 GHz. ($\mu_B = 9.273 \times 10^{-24}\ JT^{-1}$)

$$h\upsilon = g_e \beta_e B \quad \text{Given } B = 0.33 \text{ T (Bohr mageton, } \beta_e = \mu_B)$$

$$g_e = \frac{6.626 \times 10^{-34}\ \text{Js} \times 9.0 \times 10^9 \text{s}^{-1}}{0.33\ \text{T} \times 9.273 \times 10^{-24}\ \text{JT}^{-1}} = 1.95$$

5. Calculate the ESR resonance frequency of the free electron when placed in a magnetic field of 0.34 T.

$$h\upsilon = g_e \beta_e B \quad \text{Given } B = 0.34 \text{ T (Bohr mageton, } \beta_e = \mu_B)$$

$$\upsilon = \frac{2.0023 \times 9.273 \times 10^{-24}\ \text{JT}^{-1} \times 0.34\ \text{T}}{6.626 \times 10^{-34}\ \text{Js}} = 9.5 \times 10^4\ \text{GHz}$$

REFERENCES

[1] R. Freeman, A Handbook of Magnetic Resonance, John Wiley & sons, New York, 1987;
(a) E.K. Zavoisky, Paramagnetic relaxation of liquid solutions for perpendicular fields, Zhur. Eksperiment. i Theoret. Fiz. 15 (1945) 344–350;
(b) C.J. Gorter, L.J.F. Broer, Negative result of an attempt to observe nuclear magnetic resonance in solids, Physica 9 (1942) 591.

[2] C.P. Slichter, Principles of Magnetic Resonance, third ed., Springer-Verlag, New York, 1990.

[3] J.K.M. Sanders, B.K. Hunter, Modern NMR Spectroscopy: A Guide for Chemists, Oxford University Press, Oxford (Oxfordshire), New York, 1987.

[4] A.E. Derome, Modern NMR Techniques for Chemistry Research, first ed., Pergamon Press, 1987.

[5] R. Freeman, A Handbook of Nuclear Magnetic Resonance, second ed., Longman Scientific & Technical (Wiley), 1997.

[6] R.K. Harris, Nuclear Magnetic Resonance Spectroscopy, A physicochemical view Longman Scientific & Technical (Wiley), 1986.

[7] J.N.S. Evans, Biomolecular NMR Spectroscopy, 1995.

[8] K. Wuthrich, NMR of Proteins and Nucleic Acids, 1986.

[9] F.A. Bovey, L. Jelinski, P.A. Mirau, Nuclear Magnetic Resonance Spectroscopy, second ed., Academic Press, San Diego, 1988.

[10] A. Carrington, A.D. McLachlan, Introduction to Magnetic Resonance with Applications to Chemistry and Chemical Physics, Harper & Row, 1967.

[11] J. Cavanagh, W.J. Fairbrother, A.G. Palmer III, N.J. Skelton, Protein NMR Spectroscopy, Principles and Practice, Academic Press, 1996.

[12] J.P. Hornak. The Basics of NMR, 1996. An interactive web based textbook found at: http://www.cis.rit.edu/htbooks/nmr/nmr-main.htm;
(a) N.F. Ramsey, R.V. Pound, Nuclear audiofrequency spectroscopy by resonant heating of the nuclear spin system, Phys. Rev. 81 (1951) 278–279;
(b) E.M. Purcell, R.V. Pound, A nuclear spin system at negative temperature, Phys. Rev. 81 (1951) 279–280.

[13] T.D.W. Claridge, High Resolution NMR Techniques in Organic Chemistry, Pergamon Press, 1999.

[14] R.R. Ernst, G. Bodenhausen, A. Wokaun, Principles of Nuclear Magnetic Resonance in One and Two Dimensions, Oxford Science Publications, 1987.

[15] E. Fukushima, S.B.W. Roeder, Experimental Pulse NMR: A Nuts and Bolts Approach, Addison-Wesley Pub. Co., Advanced Book Program, 1981.

[16] G.D. Mateescu, A. Valeriu, 2D NMR, Density Matrix and Product Operator Treatment, Prentice Hall, 1993.

[17] Farrar, T. Clark, E.D. Becker, Pulse and Fourier Transform NMR; Introduction to Theory and Methods, Academic Press, 1971.

[18] M. Munowitz, Coherence and NMR, John Wiley & Sons, 1988.

[19] R.S. Macomber, A Complete Introduction to Modern NMR Spectroscopy, 2002.

[20] J.K.M. Sanders, B.K. Hunter, Modern NMR Spectroscopy: A Guide for Chemists, Oxford University Press, 1993.

[21] J.C. Hoch, A.S. Stern, NMR Data Processing, John Wiley & Sons, New York, 1996.

[22] P.J. Hore, J.A. Jones, S. Wimperis, NMR: The Toolkit, Oxford University Press, 2000.

CHAPTER TWO

^{1}H and ^{13}C Nuclear Magnetic Resonance Spectroscopy

Chandran Karunakaran, Paulraj Santharaman, Murugesan Balamurugan
VHNSN College (Autonomous), Virudhunagar, India

Contents

Spin Resonance Spectroscopy
ISBN: 978-0-12-813608-9
https://doi.org/10.1016/B978-0-12-813608-9.00002-2

Abstract

This chapter focuses on interpretation, assignment of chemical shifts including factors affecting it, additive rule for its calculation, spin–spin coupling, and comparison of ^{1}H and ^{13}C nuclear magnetic resonance (NMR) spectroscopy. It also highlights symmetry in NMR systems and the effect of chirality on NMR. Spin decoupling and nuclear double resonance spectroscopy including single frequency off-resonance decoupling (SFORD), distortionless enhancement by polarization transfer (DEPT), insensitive nuclei enhanced by polarization transfer (INEPT), and chemically induced dynamic nuclear polarization for sensitivity enhancement (CIDNP) and spectral simplification are discussed. Nuclear Overhauser effect (NOE) enhances the signal intensity through space effect, and its mechanism and types are discussed. Multidimensional NMR viz., 2D and 3D for structure elucidation are also presented. The principles and advantages of 2D techniques viz., homonuclear (^{1}H—^{1}H viz., correlations spectroscopy, total correlation spectroscopy (TOCSY), nuclear Overhauser effect spectroscopy (NOESY), rotating frame nuclear Overhauser effect spectroscopy, exchange spectroscopy, ^{13}C—^{13}C: incredible natural-abundance double-quantum transfer experiment) and heteronuclear (^{1}H—^{13}C viz., heteronuclear multiple quantum correlation, heteronuclear single quantum coherence (HSQC), heteronuclear multibond connectivity are illustrated using examples. 3D NMR experiments constructed by combining the two 2D experiments viz., NOESY—HSQC, TOCSY—HSQC and triple resonance experiments using ^{1}H, ^{13}C, ^{15}N nuclei are also discussed.

Keywords: ^{13}C and ^{1}H NMR assignment; 2D and 3D NMR; INEPT; NOE; Organic NMR; Symmetry in NMR.

2.1 INTERPRETATION OF ^{1}H NUCLEAR MAGNETIC RESONANCE

2.1.1 What Can Be Obtained From a ^{1}H Nuclear Magnetic Resonance Spectrum?

Analysis and interpretation of nuclear magnetic resonance (NMR) spectrum provided the following useful informations to understand the structure of a molecule and compound [1—3].

1. The number of peaks (in low-resolution NMR without spin—spin splitting) tells us about how many different kinds or types of proton present in the molecule.
2. The chemical shift values, i.e., the positions of the peaks tell us about the electronic environment of each kind of proton and the functional group.
3. The relative intensities of the peaks (or area of the peaks) tell us about the number of protons of each kind, which are present. The area of the peak is obtained by integration.

4. The spin–spin splitting of a peak into several peaks (in high-resolution NMR) identifies the environment of a proton with respect to other nearby protons in the molecule. In other words, it tells us about adjacent functional groups and therefore connectivity.

For example, $-CH_3$ protons connected to $-CH_2-$ protons show triplet signal.

2.2 ASSIGNMENT OF ^{1}H NUCLEAR MAGNETIC RESONANCE

The different positions of NMR absorptions are described as **chemical shifts** (δ). A chemical shift is defined as the difference in parts per million (ppm) between the resonance frequency of the observed proton and that of the tetramethylsilane (TMS) hydrogens. TMS is the most common reference compound used in NMR. It is set at $\delta = 0$ ppm.

$$\text{Chemical shift, ppm } \delta = \frac{(\text{Frequency of signal} - \text{frequency of reference}) \times 10^6}{\text{Spectrometer frequency (in Hz)}}$$

The assignment of chemical shift values of each peak to protons in the functional group of molecules is given in Table 2.1.

The proton chemical shifts exhibit a range from −20 (for metal hydrides *viz.*, $[H\text{-}Rh(CN)_5]^{3-}$) to 40 ppm (for naked protons). But the majority of the organic compounds appear in the range 0–10 ppm.

Table 2.1 Typical selected ^{1}H chemical shift values

Proton	δ (ppm)	Proton	δ (ppm)
Aldehyde	9.5–10.5	CH_3 attached to = bonds	1.8–2.5
Aromatic	6.5–8.2	CH_3 attached to aromatic	1.8–2.5
Alkene	4.5–6.1	Methyl (CH_3–CO–)	1.8–2.7
Alkyne	2.0–3.2	Methylene (–CH_2–O–)	~3.6–4.7
Acetal	4.5–6.0	Methylene (CH_2–R_1R_2)	~1.3
Alkoxy	3.4–4.8	Methine (CH–$R_1R_2R_3$)	~1.5
Methyl (CH_3–R)	~0.9	Cyclopropane	0.22
N-methyl	3.0–3.5	Me_4Si (tetramethylsilane)	0.0
Methoxy	3.3–3.8	Metal hydride	−5 to −20

2.3 FACTORS INFLUENCING THE CHEMICAL SHIFT OF PROTONS

2.3.1 Shielding in Nuclear Magnetic Resonance

Magnetic nuclei are shielded by valence electrons surrounding them, which circulate in an applied magnetic field. This generates a local diamagnetic current in the opposite direction (Fig. 2.1). This diamagnetic shielding will affect the frequency of radiofrequency (RF) radiation necessary to cause a nucleus to spin flip (the resonance frequency). Therefore, the nuclei will absorb radiation of slightly different frequencies compared with bare nuclei depending on their local magnetic environment, which is determined by the structure of the compound. Because the magnetic field strength dictates the energy separation of the nuclear spin states and, hence, the resonance frequency, different types of nuclei will occur at different chemical shifts. This makes NMR so useful and powerful for structure determination. Otherwise, all nuclei would have the same chemical shift. Thus, the chemical shifts of NMR provide a wealth of structural information about molecules [4].

The following factors affect the chemical shift values of protons:

1. Inductive effects by electronegative groups
2. Magnetic anisotropy
3. The ring current effect in cyclic π-systems
4. H-bonding
5. Mesomeric effect
6. Hybridization

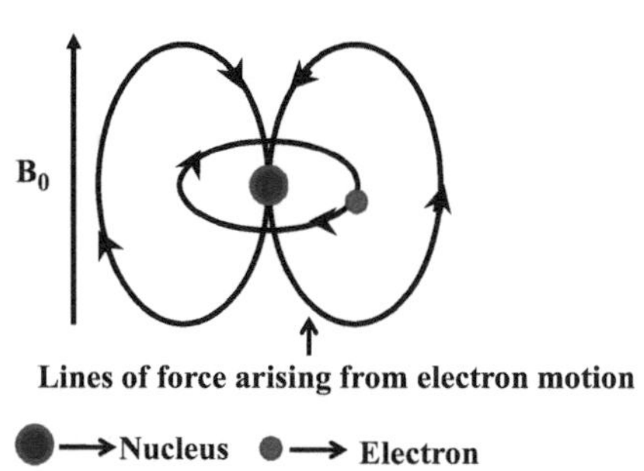

Figure 2.1 Diamagnetic shielding of nucleus by electron.

2.3.1.1 Electronegativity

As described above, the electrons around the nucleus create a magnetic field that opposes the applied field. This reduces the magnetic field experienced at the nucleus. Because the induced field opposes the applied field, the electrons are said to be diamagnetic and the effect on the nucleus is referred to as diamagnetic **shielding**. Because the field experienced by the nucleus affects the energy difference between the different spin states, the frequency (and hence the chemical shift δ) will change depending on the electron density around the nucleus. Electronegative groups decrease the electron density around the nucleus, and there is less shielding (i.e., **deshielding**) so the chemical shift value increases.

The applied magnetic field, B_0, induces circulations in the electron cloud surrounding the nucleus such that its magnetic moment μ, opposed to B_0, is produced (Lenz's law). Nuclei in a region of high electron density are more shielded from the applied field than those in regions of lower electron density. If inductive effects present in a molecule reduce the electron density in the hydrogen 1s orbital, deshielding (shift to higher frequencies) is observed. For example, the increasing electronegativity from left to right in Table 2.2 decreases the shielding effect and hence the increase in chemical shifts [4].

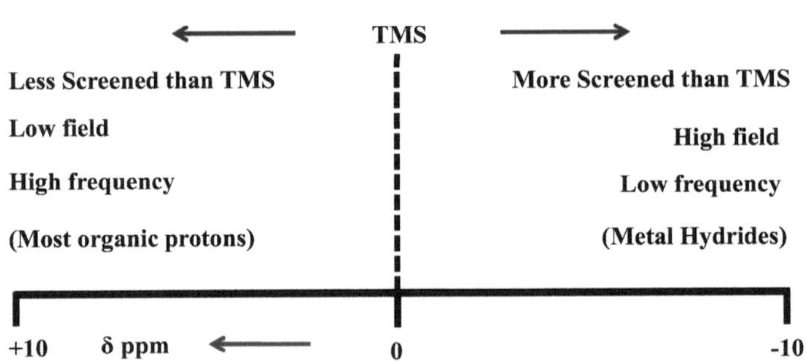

Table 2.2 Effect of electronegative substituent on chemical shift

	CH_3H	CH_3I	CH_3Br	CH_3Cl	CH_3F
δ (CH_3) ppm	0.13	1.98	2.45	2.84	4.13
Electronegativity (Pauling)	2.1	2.5	2.8	3.0	4.0

2.3.1.2 Magnetic Anisotropy

Magnetic anisotropy means that there is a nonuniform magnetic field. Electrons in π-systems (e.g., aromatics, alkenes, alkynes, carbonyls, etc.) interact with the applied magnetic field, which induces a magnetic field that causes the anisotropy (Fig. 2.2). As a result, the nearby nuclei will thus experience three different fields: (1) the applied field, (2) the shielding field of the valence electrons, and (3) the field due to the π system. Depending on the position of the nucleus in this third field, it can be either shielded (smaller δ) or deshielded (larger δ), which implies that the energy required for the frequency of the absorption will change.

From the above figure (Fig. 2.2), the aromatic, ethylenic, and aldehydic protons are present in the deshielded region and hence are deshielded. On the other hand, the acetylenic protons are in the shielded core and hence are shielded [4].

2.3.1.3 Ring Current Effect in Cyclic π-Systems

An aromatic molecule consists of π electrons free to move on a circle formed by the σ framework. This induces a ring current under the influence of the external magnetic field B_0. The secondary magnetic field arising from this ring current effect supports the external magnetic field outside the ring or opposes the magnetic field inside the ring (Fig. 2.3). As a consequence,

1. Protons in the molecular plane and outside the ring resonate at higher frequencies (deshielded) (Fig. 2.3);
2. Protons in the middle, above, or below the ring resonate at lower frequencies (shielded).

Hence, higher frequency chemical shifts occur outside the ring and lower frequency chemical shifts inside the ring.

Also, in large unsaturated ring systems where the number of π-electrons satisfies the Hückel rules (4n + 2), the ring current effects are observed. However, the opposite ring current behavior is observed for antiaromatic ring systems. In other words, if the cyclic conjugated antiaromatic system

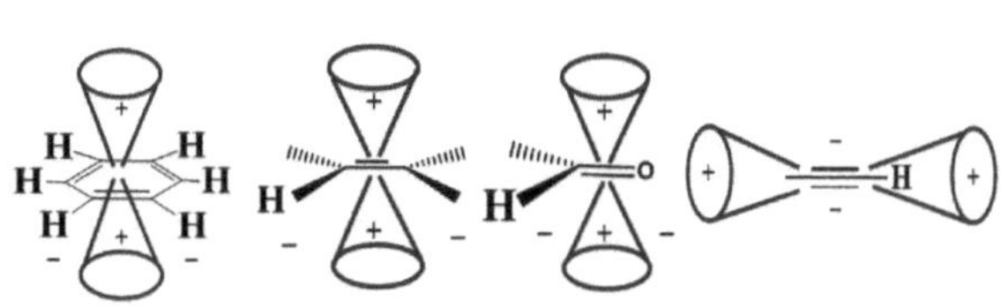

δ = 7 - 8 ppm   δ = 5 - 7 ppm   δ = 9 - 10 ppm   δ = 2 - 3 ppm

Figure 2.2 Magnetic anisotropy effects.

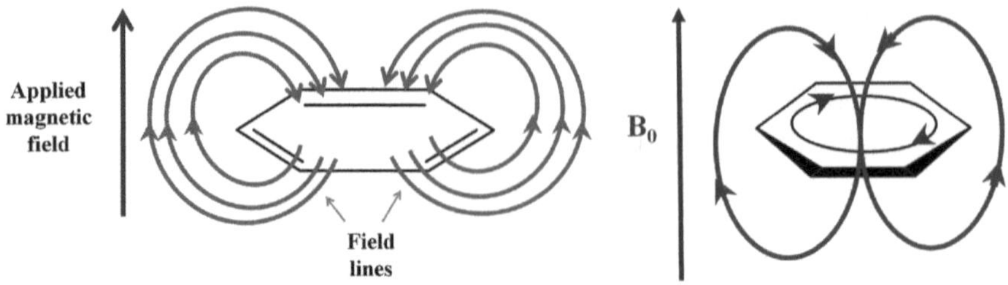

Figure 2.3 Representation of ring current effect.

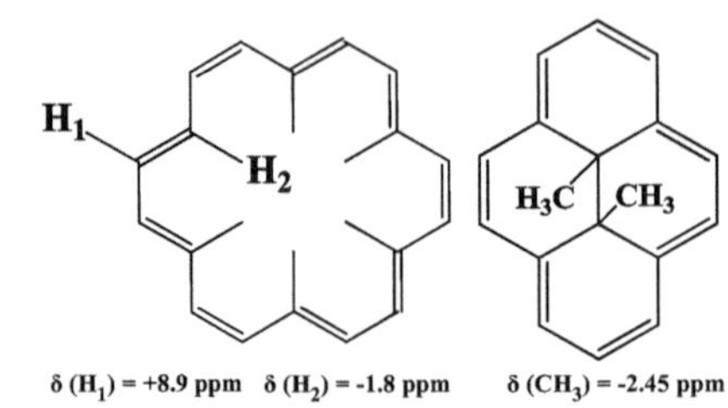

Figure 2.4 Effect of ring current on large unsaturated ring systems.

is planar and has 4nπ electrons, the chemical shift effects are opposite, i.e., the protons outside the ring are shielded (low field) and inside the ring (high field) are deshielded (Fig. 2.4).

2.3.1.4 Hydrogen Bonding

Hydrogen bonding results from the presence of electronegative atoms in the neighborhood of protons. The more the hydrogen bonding the more the proton is deshielded, and its chemical shift will be higher. However, because the amount of hydrogen bonding is susceptible to factors such as solvation, acidity, concentration, and temperature, it can often be difficult to predict. Protons that are involved in hydrogen bonding (especially **−OH** or **−NH**) are typically observed over a large range of chemical shift values (Table 2.3). For instance, the protons in a very dilute solution of H_2O in $CDCl_3$ resonate at 1.5−1.6 ppm (at 0.3−0.4 ppm in C_6D_6), whereas the ^1H resonance in H_2O is usually observed around 4.8 ppm due to hydrogen bonding. The H−D exchange using D_2O can cause signals from H-bonds, viz. OH/NH/SH protons, to disappear from the ^1H NMR spectrum, thereby detecting and helping with assignments.

Examples of strong hydrogen bonds and their effects on chemical shifts (Fig. 2.5).

Table 2.3 Typical ^1H chemical shift ranges for H-bonding OH and NH protons

Proton center	Compound	δ/ppm
−OH	Alcohols	1−5
	Phenols	4−10
	Acids	9−13
	Enols	10−17
−NH	Amines	1−5
	Amides	5−6.5
	Amino groups in peptides	7−10

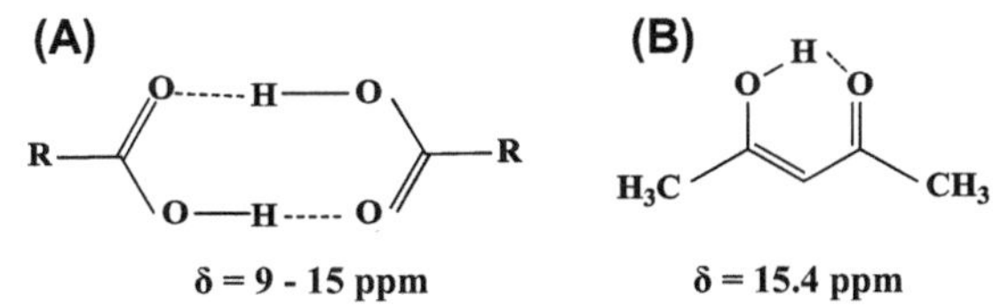

Figure 2.5 Intermolecular (A) and intramolecular (B) hydrogen bonding.

2.3.1.5 Mesomeric Effect

Substituents, such as OR or NR_2, can act both as an electron withdrawing group (due to electronegativity, thus causing a shift to higher frequencies) and as a donor of lone pairs on O or N (if double bonds are present, thus causing a shift to lower frequency). Mesomeric effects can be thus used to explain the shielding observed in aromatic compounds, such as aniline, in which the protons in the *ortho* and *para* positions are more strongly shielded than in the *meta* position (Fig. 2.6).

2.3.1.6 Hybridization

As *s*-character of carbon atom increases, the electronic cloud is held closer to the carbon and provides less electron density for shielding of protons, and thus the chemical shift, δ, increases (shifted downfield) (Fig. 2.7). But, we have the observed order of $sp^2 > sp > sp^3$. The chemical shift value actually observed at 2−3 ppm for acetylene. This suggested that factors other than

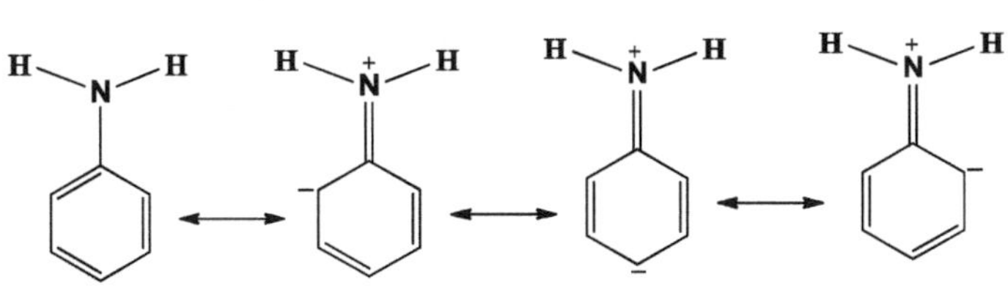

Figure 2.6 Mesomeric effect increases electron density on *ortho* and *para* positions.

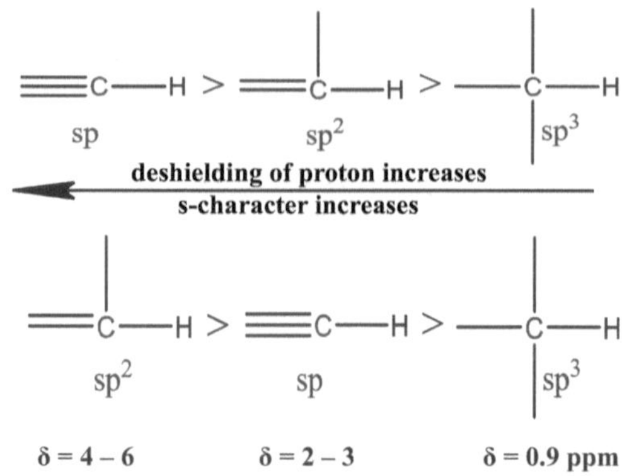

Figure 2.7 Effect of sp character and magnetic anisotropy on chemical shift values.

the *sp* character of carbon might affect the chemical shift in this case. The above discrepancy can be readily explained by what is called magnetic anisotropic effect.

2.3.1.7 Other Factors

1. Solvent
   The changes in ^{1}H resonance frequency induced by the solvent are usually less than 1 ppm. Complications because of solvent effects can be avoided by using "inert" solvents, viz. carbon tetrachloride, cyclohexane, etc. On the other hand, strong solvent effects can be used to "simplify" the spectrum. For example, an aromatic solvent such as benzene is used. This will solvate areas of low electron density, and also due to ring current effects, this can change chemical shifts by as much as 1 ppm.
2. Temperature
   The chemical shift difference between the resonance signals in the ^{1}H NMR spectrum of a sample of liquid methanol is usually temperature dependent (see Section 5.2). It is used for the purpose of accurate temperature calibration. On increasing the temperature, the extent of H-bonding diminishes, and hence the resonance for the hydroxyl proton moves to low frequency, toward the resonance signal of the methyl protons.
3. Shift reagents
   In electron spin resonance (ESR), the interaction between nuclear and electron spins manifests itself as the hyperfine splitting of the signal. On the other hand, in NMR experiments, this interaction leads to a shift of the NMR signal to higher or lower frequency relative to the signal of the same proton in the ligand (or) diamagnetic compound called *contact*

shift (see Section 3.6). Paramagnetic impurities also cause severe line broadening by significantly accelerating the spin-lattice relaxation of protons.

The paramagnetic *lanthanide shift reagents* were used to simplify the ^{1}H NMR spectra. A lanthanide shift reagent is an octahedral complex of lanthanide ions, viz. Eu, Dy, Pr, and Yb, with a ligand chosen such that the complex is soluble in NMR solvents. This is dissolved in a solution of the compound. These lanthanides are capable of forming higher coordination numbers than 6. If the organic molecule possesses a suitable coordination site (generally, any atom with an available lone pair), it can interact through inner sphere or outer sphere complex. This causes a shift of the protons of the organic molecule, thereby spreading and dispersing the overlapping signals.

2.4 COUPLING CONSTANT (J)

Simply, the coupling constant, J (usually in frequency units, Hz), is a measure of the interaction between a pair of protons. The scalar coupling J is a through-bond interaction, in which the spin of one nucleus perturbs (polarizes) the spins of the intervening electrons, and the energy levels of neighboring magnetic nuclei are in turn perturbed by the polarized electrons. This leads to a lowering of the energy of the neighboring nucleus when the perturbing nucleus has one spin and a rising of the energy when it has the other spin.

2.4.1 Characteristics of Coupling Constant

1. The coupling constant, J (*always* in Hz), is *field-independent* (i.e., J *is constant at different external magnetic field strengths*) and is mutual (i.e., $J_{AX} = J_{XA}$).
2. Because the effect is usually transmitted through the bonding electrons, the magnitude of J falls off rapidly as the number of intervening bonds increases.
3. Coupling of one (^{1}J), two (^{2}J), and three (^{3}J) bonds usually dominates the fine structure of NMR spectra.

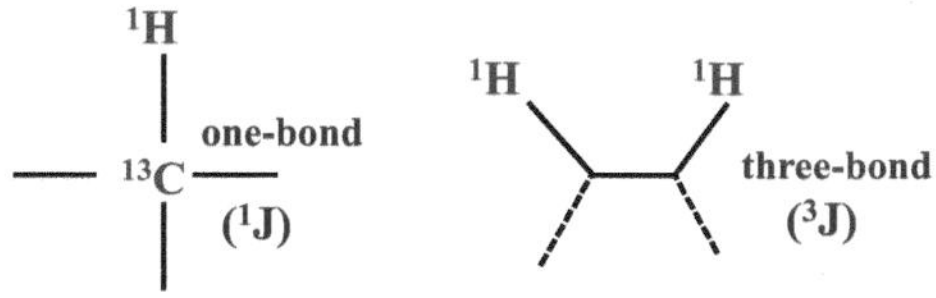

Table 2.4 H–H coupling

Environment	Coupling	Value (Hz)
Aliphatic, H–C–H (*geminal*)	$^{2}J_{HH}$	−10–16
Aliphatic, H–C–C–H (*vicinal*)	$^{3}J_{HH}$	6–8
Aldehyde, H–C–OC–H	$^{3}J_{HH}$	2–3
Alkene, H–C–H, *geminal*	$^{2}J_{HH}$	0–3
Alkene, H–C=C–H, *trans*	$^{3}J_{HH}$	12–18
Alkene, H–C=C–H, *cis*	$^{3}J_{HH}$	6–12
Aromatic, H–C∼C–H	$^{3}J_{HH}$	6–10
Aromatic, H–C∼C∼C–H	$^{4}J_{HH}$	1–3

Table 2.5 C–H coupling

Environment	Coupling	Value (Hz)
Aliphatic, C–H	$^{1}J_{CH}$	125–135
Aliphatic, CX–H (X=N,O,S)	$^{1}J_{CH}$	135–155
Alkene, =C–H	$^{1}J_{CH}$	155–170
Alkyne, ≡C–H	$^{1}J_{CH}$	240–250
Aromatic, ∼C–H	$^{1}J_{CH}$	155–165

4. But coupling of four (^{4}J) and five (^{5}J) bonds is often seen, especially through π-bonds (double and triple bonds, aromatic carbons).
5. Coupling constants can be either positive or negative, defined as follows: coupling constants are positive if the energy of A is lower when X has the opposite spin as A (αβ or βα) and negative if the energy of A is lower when X has the same spin as A (αα or ββ).
6. Correlation of $^{1}H-^{1}H$ coupling constants with structure: Interproton spin–spin coupling constants are useful in obtaining structural and geometrical data about a molecule and also about the connectivity (Tables 2.4 and 2.5).

2.4.2 Types of Coupling

There are several types of coupling depending on the number of intervening bonds.

1. *Vicinal*—It arises from protons on adjacent carbon atoms.

vicinal **6 to 8 Hz** **three bond** **^{3}J**

2. *Geminal*—It arises from two protons attached to the same carbon atom. In saturated systems, the magnitude of the geminal coupling constant $^2J_{H-C-H}$ (two protons attached to the same carbon atoms) is typically between 10 and 16 Hz, but, however, the values between 0 and 22 Hz have also been reported in some unusual structure.

geminal 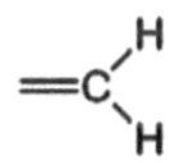 **0 to 5 Hz** **two bond** ^2J

3. *Cis*—It means that *functional groups* or protons are on the same side of the carbon chain in alkene.

cis 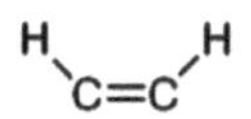 **6 to 15 Hz** **three bond** ^3J

4. *Trans*—It means that *functional groups* are on the opposite side of the carbon chain in alkene.

trans 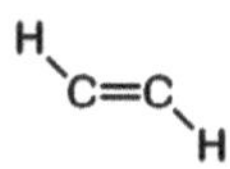 **11 to 18 Hz** **three bond** ^3J

5. *Ortho*, *para*, and *meta* coupling constants.

2.4.2.1 Ortho, Meta, *and* Para *Coupling Constants*

In aromatic systems, the coupling constants between protons attached to an aromatic ring are characteristic of the relative position of the coupled protons, i.e., *ortho*, *meta*, or *para*.

Analogously, in condensed polynuclear aromatic compounds and other heterocyclic compounds, the magnitude of the coupling constants between protons in the aromatic rings reveals the relative position of the coupled protons.

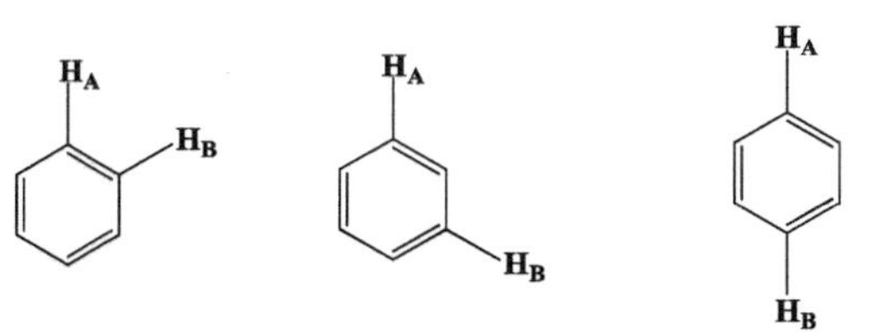

$^3J_{AB\,(ortho)} = 6–10$ Hz $^4J_{AB\,(meta)} = 1–3$ Hz $^5J_{AB\,(para)} = 0–1.5$ Hz

2.4.2.2 *Effect of Dihedral Angle (ϕ) on Vicinal Couplings (^3J)—Karplus Relation*

Vicinal couplings exhibit a characteristic dependence on the dihedral angle as shown in Fig. 2.8 for saturated nonaromatic systems.

It (^3J) can be related to the dihedral angle for protons in nonaromatic systems

$$^3J = 7 - 1.0\cos\phi + 5\cos^2\phi$$

Similarly, for ^1H, ^{13}C couplings, the equation has been used

$$^3J = 3.81 - 0.9\cos\phi + 3.83\cos^2\phi$$

This relationship, also known as the **Karplus relation**, has been successfully employed to determine the stereochemistry of compounds [5].

The vicinal coupling (protons on adjacent carbon atoms) $^3J_{H-C-C-H}$ can have values in the range 0−16 Hz depending mainly on the dihedral angle ϕ as shown in Fig. 2.8.

The Karplus relation gives the angular dependence of the vicinal coupling constant as

$$^3J_{H\text{-}C\text{-}C\text{-}H} = 10\cos^2\phi \quad \text{for } 0 < \phi < 90^\circ \text{ and}$$

$$^3J_{H\text{-}C\text{-}C\text{-}H} = 15\cos^2\phi \quad \text{for } 90 < \phi < 180^\circ$$

It follows from these equations that if the dihedral angle ϕ between two vicinal protons is near 90°, then the coupling constant will be very small. Conversely, if the dihedral angle ϕ between two vicinal protons is between

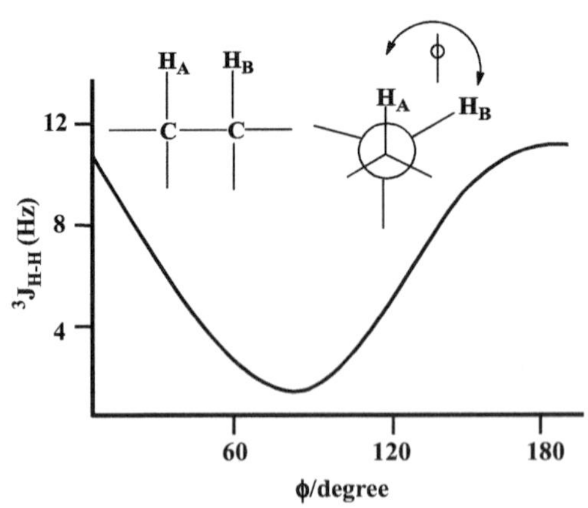

Figure 2.8 Effect of dihedral angle (ϕ) on ^3J (Karplus curve).

90° and 180°, then the coupling constant will be relatively large. Hence, Karplus relationship is very useful in determining the stereochemistry of organic molecules. But this must be treated with caution because the vicinal coupling constants also depend markedly on the nature of substituents. In systems that assume an average conformation, due to a flexible hydrocarbon chain, $^3J_{H-H}$ value lies between 6 and 8 Hz.

For unsaturated (olefinic) systems, the vicinal coupling constants depend on the nature of the substituents attached to the C=C, but for vast majority of substituents, the ranges for $^3J_{H-C=C-H}$ (*cis*) and $^3J_{H-C=C-H}$ (*trans*) do not overlap. This reveals that the stereochemistry of the double bond can be determined by measuring the coupling constant between the vinylic protons. When the C=C bond is in a ring, the J_{H-C-C} coupling indicates the ring size.

Also, the magnitude of the long-range allylic coupling is controlled by the dihedral angle between the C–H_A bond and the plane of the double bond in a relationship similar to the Karplus relation.

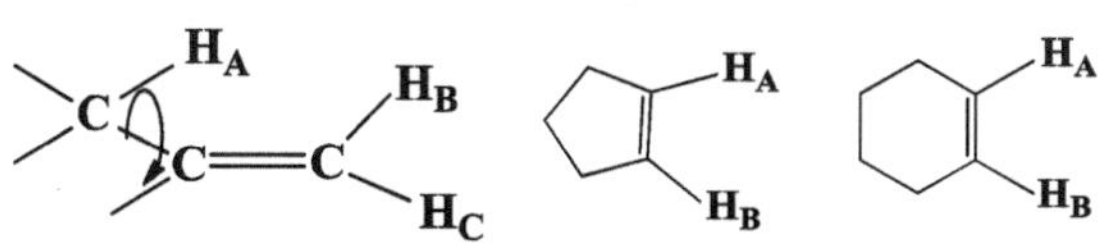

5 – 7 Hz 9 – 11 Hz

2.5 SYMMETRY IN NUCLEAR MAGNETIC RESONANCE

Based on symmetry in NMR spectra, the protons and other nuclei can be classified as [6]:

1. Homotopic,
2. Enantiotopic,
3. Diastereotopic, and
4. Heterotopic.

- Both the homotopic and enantiotopic protons will have identical chemical shifts. They may or may not have identical coupling to other nuclei.
- However, diastereotopic and heterotopic will have different chemical shifts and coupling to neighboring magnetic nuclei.

1. Homotopic

When two hydrogens are attached to a carbon, substitution or replacement of each of these two hydrogens in turn with a different atom gives rise to the same product. For example, in propane replace one and then the other hydrogen with chlorine atom. If they yield identical products, the protons are homotopic.

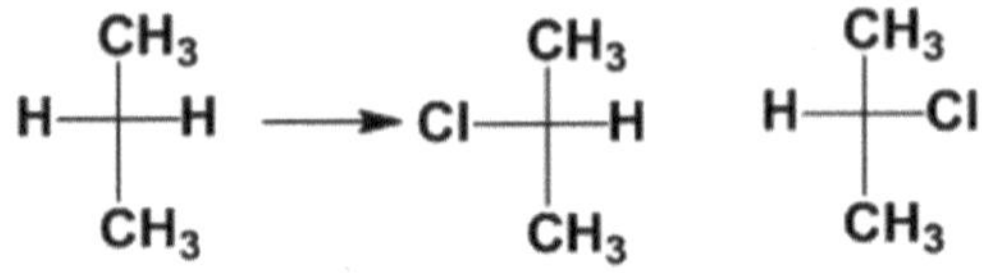

2. Enantiotopic
When two hydrogens are attached to a carbon, substitution or replacement of each of these two hydrogen gives rise to different enantiomers called enantiotopic hydrogens. For example, in propanol replace one and then the other hydrogen with chlorine atom. If they are nonidentical, the protons are enantiotopic.

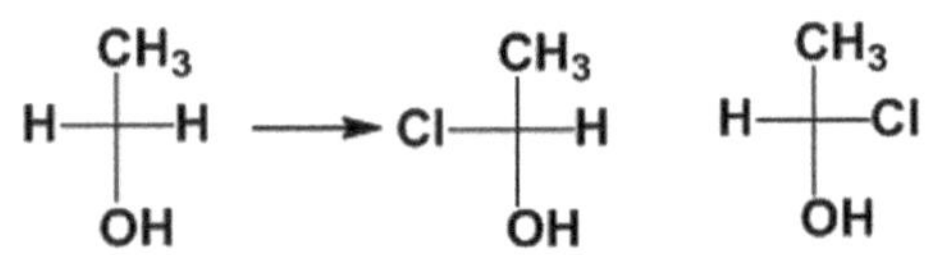

3. Diastereotopic
When two hydrogens are attached to a carbon, substitution or replacement of each of these two hydrogens gives rise to different diastereomers called diastereotopic protons. So, the two protons are said to be diastereotopic.
For example,

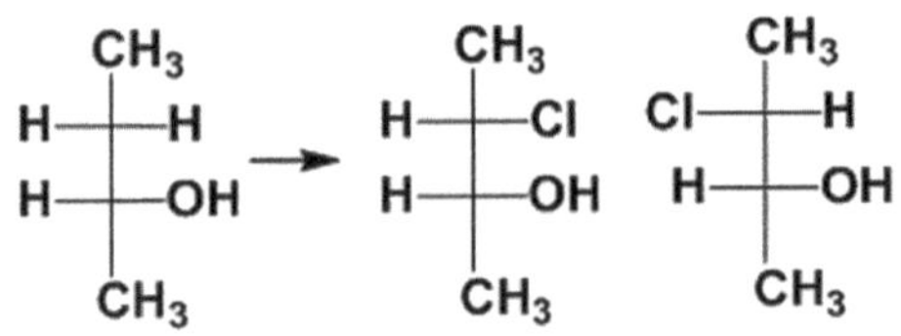

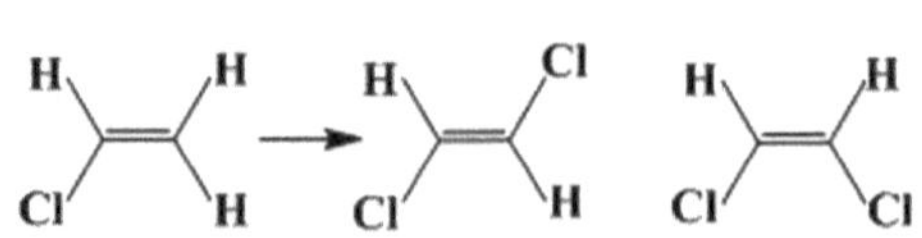

4. Heterotopic
When two hydrogens are attached to a carbon, substitution or replacement of each of these two hydrogens by chlorine atom gives rise to

different structural isomers called heterotopic hydrogens. So, the two protons are said to be heterotopic.
For example,

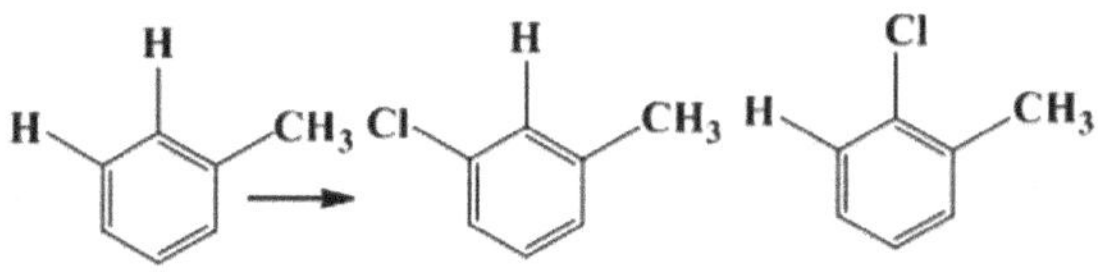

2.5.1 Effect of Chirality on Nuclear Magnetic Resonance

In an achiral solvent, enantiomers will give identical NMR spectra. However, in a chiral solvent or in the presence of chiral reagents to the NMR solvent, enantiomers will have different spectra. This is often used to establish the enantiomeric purity of compounds [1].

For those molecules containing a stereogenic center, the NMR spectra will sometimes be more complex than expected. For example, groups such as $-CH_2-$ groups require particular attention in molecules, which contain an asymmetric center. The carbon atom of a $-CX_2-$ group is called a prochiral carbon if there is a chiral center elsewhere in the molecule. A prochiral carbon atom is defined as a carbon in a molecule, which would become chiral if one of its substituents was replaced by a different substituent. The protons of a $-CH_2-$ group in an acyclic aliphatic compound are normally expected to be equivalent and resonate at the same frequency in the ^1H NMR. However, when there is a stereogenic center in the molecule, each of the protons of the $-CH_2-$ group will now appear at different chemical shifts. Also, because they are nonequivalent, the protons will couple to each other giving a large coupling of ~15 Hz.

The effect of chirality plays an important role in the characterization of natural products including amino acids, proteins, or peptides using NMR spectra. Most molecules derived from natural sources consist of chiral and prochiral centers. Further, they are present as a single pure enantiomer. In these molecules, the two protons of the methylene groups will be nonequivalent. Fig. 2.9 shows the aliphatic protons in the ^1H NMR spectrum of the cysteine $\left(HSCH_2CHNH_3{}^+COO^-\right)$ amino acid. Cysteine has a stereogenic, i.e., prochiral center and the signals of the methylene group appear separately at δ 3.18 and δ 2.92 ppm. Each of the methylene protons is split into a doublet of doublets due to coupling firstly to the other methylene proton and secondly to the proton on the α-carbon (Ha) [4].

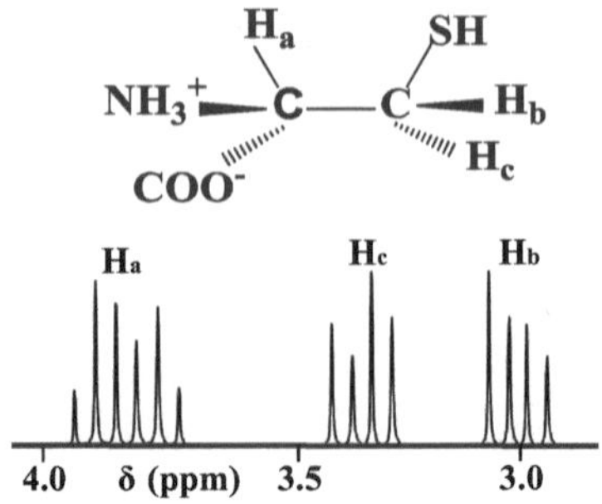

Figure 2.9 ^{1}H Nuclear magnetic resonance of cysteine indicating nonequivalence of the methylene protons.

2.6 ^{13}C NUCLEAR MAGNETIC RESONANCE SPECTROSCOPY

Organic compounds are mainly made up of carbon and hydrogen. However, the most abundant isotope of carbon (^{12}C) cannot be observed by NMR. On the other hand, ^{13}C is a rare nucleus (with $I = 1/2$ like ^{1}H) but low abundance (1.1% natural abundance) coupled with the relative low resonance frequency lead to its relative insensitivity as an NMR-active nucleus. However, with the increasing availability of pulsed FT NMR spectrometers, it is now common to acquire many spectra and add them together. So, ^{13}C NMR spectra of good quality can be readily obtained [1,3,7]. The following Table 2.6 shows the major differences between ^{1}H NMR and ^{13}C NMR spectroscopy.

2.6.1 Coupling and Decoupling in ^{13}C Nuclear Magnetic Resonance Spectra

Because the ^{13}C nucleus is isotopically less abundant; it is extremely unlikely that any two adjacent carbon atoms in a molecule will both be ^{13}C. As a result, ^{13}C—^{13}C coupling is not observed in ^{13}C NMR spectra, i.e., there is no signal multiplicity or splitting in a ^{13}C NMR spectrum arising from ^{13}C—^{13}C coupling. However, ^{13}C couples strongly to any protons that is directly attached ($^1J_{CH}$ is typically about 125 Hz for saturated carbon atoms in organic molecules) [1].

2.6.2 Broadband Decoupling

It is the usual practice to irradiate the ^{1}H nuclei during ^{13}C acquisition so that all ^{1}H are fully decoupled from the ^{13}C nuclei (called *broadband*

Table 2.6 Comparison of ^1H and ^{13}C nuclear magnetic resonance (NMR)

S.No	Properties	^1H NMR	^{13}C NMR
1	Natural abundance	99.98%	1.1%
2	Gyromagnetic ratio (γ)	2.6753×10^8 rad s^{-1} T^{-1}	0.6729×10^8 rad s^{-1} T^{-1}
3	Magnetic moment μ_p $\mu_N = e\,\hbar/2m_p$	2.7928 μ_N	0.7023 μ_N
4	Sensitivity	More sensitive	400 times less sensitive than proton NMR
5	Method	Either continuous or Fourier transform (FT)	FT only
6	Chemical shift range	0–14 ppm	0–240 ppm. So, there is less overlapping of signals
7	Coupling constant range	0–15 Hz ($^1H-^1H$)	125–250 Hz ($^{13}C-^1H$)
8	Solvent peak (for example, $CDCl_3$)	Not observed	Observed
9	Area under the peak	Considered	Not considered
10	Tetramethylsilane peak	Singlet	Quartet
11	Effect of substituent on adjacent carbon atom	Can change chemical shift	Cannot affect chemical shift
12	Applications	**1.** Types of protons present **2.** Identification of structural isomers **3.** Detection of hydrogen bonding	**1.** Types of carbon present **2.** Detection of aromaticity **3.** Detection of some double bond character due to resonance **4.** Detection of cyclic compounds

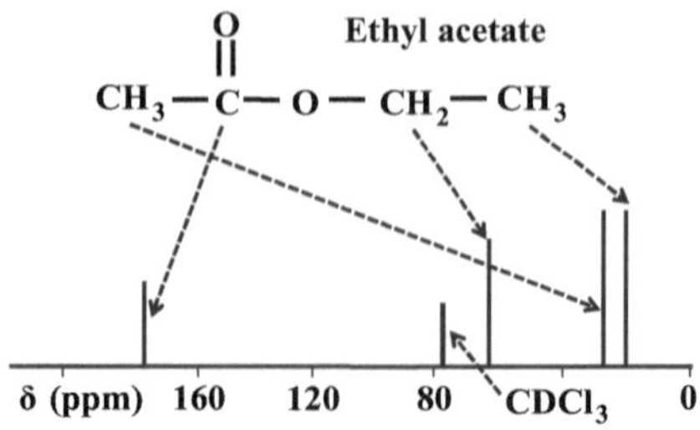

Figure 2.10 Proton-decoupled ^{13}C nuclear magnetic resonance spectrum of ethyl acetate in $CDCl_3$.

decoupling or noise decoupling). Thus, ^{13}C NMR spectra usually appear as a series of singlets (when ^1H is fully decoupled). Furthermore, each distinct ^{13}C environment in the molecule gives rise to a separate signal. For example, the proton-decoupled ^{13}C NMR spectrum of ethyl acetate in $CDCl_3$ as solvent is shown in Fig. 2.10.

On the other hand, if ^1H is not decoupled, the signals in the ^{13}C spectrum appear as multiplets due to the $^1J_{C-H}$ couplings.

1. ~125 Hz for sp^3-hybridized carbon atoms, ~160 Hz for sp^2-hybridized carbon atoms, and ~250 Hz for sp-hybridized carbon atoms.
2. CH_3- signals appear as quartets, $-CH_2-$ signals appear as triplets, $-CH-$ groups appear as doublets, and quaternary C (no attached H) appears as singlet.

This multiplicity information, taken together with chemical shift data, is useful in identifying and assigning the topology/functional groups in organic molecules. In ^{13}C spectra acquired without proton decoupling, there is usually much more "long-range" coupling information visible in the fine structure of each multiplet. The fine structure arises from coupling between the carbon and protons that are not directly bonded to it (e.g., from $^2J_{C-C-H}$, $^3J_{C-C-C-H}$). The magnitude of long-range C—H coupling is typically <10 Hz and this is much smaller than $^1J_{C-H}$. Hence, more detailed analysis of the long-range C—H couplings can be used to provide additional information about the structure of the molecule.

In most organic compound, ^{13}C nuclei, which have directly attached protons, receive a significant signal enhancement (but not easily predictable) when the protons are decoupled as a result of the nuclear Overhauser effect (NOE). As a consequence, peak intensity does not necessarily represent the number of ^{13}C nuclei giving rise to the signal. That is, it is not quite possible to integrate the routine ^{13}C spectra directly (unless specific precautions have been taken). However, with proper controls, ^{13}C NMR could be used for quantitative analysis and also valuable technique for the analysis of mixtures. For measuring the relative signal intensity reliably, the ^{13}C NMR spectra must be recorded with techniques to suppress the NOE and with a long delay between the acquisitions of successive spectra to ensure that all of the carbons in the molecule are completely relaxed between spectral acquisitions.

2.7 ^{13}C CHEMICAL SHIFTS AND THEIR ASSIGNMENT

Because the organic chemists are interested primarily in the molecular carbon skeleton, a ^{13}C NMR spectrum gives structural information much more directly than a ^{1}H NMR spectrum. Here, quaternary carbons, as those of many functional groups (CN, C=O, C=NR), exhibit signals but not in ^{1}H NMR. In certain cases, ^{13}C NMR readily distinguishes between groups that might otherwise be confused in proton NMR (Tables 2.7 and 2.8). For example, cyanide (110−120 ppm) and isocyanide (135−150 ppm) could be readily differentiated.

A schematic of ^{13}C assignment of signals of different functional carbon atoms is shown in Fig. 2.11.

Table 2.7 Typical ^{13}C chemical shift values (in ppm) of hydrocarbons

CH_3-	10−30	$=CR_2$	130−150
R_3C-, R_2CH, RCH_2	25−50	Alkyne	70−85
$=CH_2$	105−120	Ar−H	115−130
$=CH-$	110−140	Ar−C	130−150

Table 2.8 Typical ^{13}C chemical shift values (in ppm) of carbons bonded to oxygen

Ketones	200−210	Acetals	90−110
Aldehydes	190−200	R_3C-O	70−85
Conjugated C=O	180−200	R_2HC-O	60−80
Carboxylic acids	170−180	RH_2C-O	45−65
Carboxylic esters	160−170	H_3C-O	50−60
Phenols (C_1)	150−160	Epoxides	40−60
Furans (C_2)	140−150		

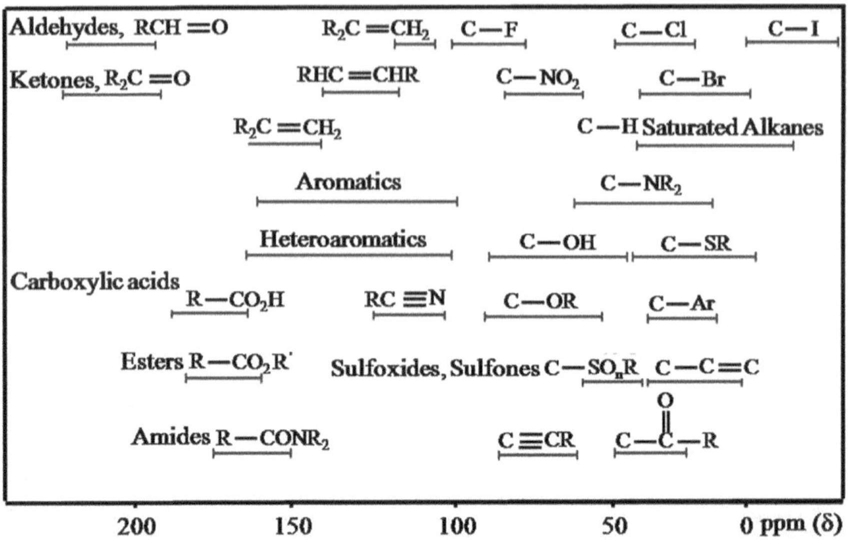

Figure 2.11 Schematic of ^{13}C assignment of signals of different functional carbon atoms.

2.8 ADDITIVITY RULE

The influence of neighboring groups (either shielding or deshielding) on ^{1}H and ^{13}C chemical shifts is additive (to an extent). Shoolery has predicted the additivity rules for calculating the chemical shift of protons of the unknown type from the known chemical shifts.

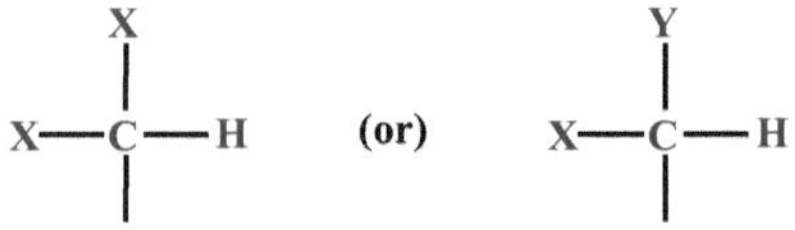

For protons on sp^3 carbons, the chemical shift is given by $\delta(\text{ppm}) = 0.233 + \sum \sigma_i$.

2.8.1 Effect on Chemical Shifts by Two Directly Attached Functional Groups

Shoolery's rule states that the sum of the constants (σ) for the attached functional groups is added to δ 0.23 as follows. For

$$\mathrm{Y - CH_2 - X}$$

$$\delta\,(\mathrm{Y - CH_2 - X}) = 0.23 + \sigma_y + \sigma_x$$

Table 2.9 Substituent constants for alkyl methylene (and methyl) protons

Y or Z	Substituent constants (σ)	Y or Z	Substituent constants (σ)
−H	0.34	−OC(=O)R	3.01
$-CH_3$	0.68	−OC(=O)Ph	3.27
−C=C−	1.32	−C(=O)R	1.50
−C≡C−	1.44	−C(=O)Ph	1.90
−Ph	1.83	−C(=O)OR	1.46
$-CF_2$	1.12	$-C(=O)NR_2(H_2)$	1.47
$-CF_3$	1.14	−C≡N	1.59
−F	3.30	$-NR_2(H_2)$	1.57
−Cl	2.53	−NHPh	2.04
−Br	2.33	−NHC(=O)R	2.27
−I	2.19	$-N_3$	1.97
−OH	2.56	$-NO_2$	3.36
−OR	2.36	−SR(H)	1.64
−OPh	2.94	$-OSO_2R$	3.13

For example, the chemical shift of a methylene group protons attached to two functional groups can be calculated by means of the following substituent constants (σ values) in Table 2.9.

The chemical shift for the methylene protons of $C_6H_5CH_2Br$, for example, is calculated from the σ value in Table 2.9.

$$\begin{aligned} & 0.23 \\ \sigma_{Ph} &= 1.85 \\ \sigma_{Br} &= 2.33 \\ & \text{--------} \\ \text{(calculated)}\ \delta &= 4.41 \qquad \text{observed } \delta = 4.43 \text{ ppm} \\ & \text{--------} \end{aligned}$$

2.8.2 Effect on Chemical Shifts by Three Directly Attached Functional Groups

For

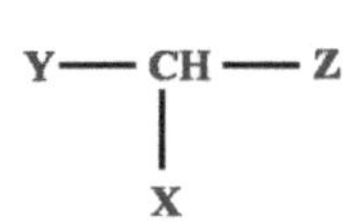

$$\delta\ \mathrm{CHXYZ} = 2.50 + \sigma_x + \sigma_y + \sigma_z$$

Table 2.10 Substituent constants for methine protons

Group	(σ)	Group	(σ)
−F	1.59	−Ar	0.99
−Cl	1.56	−C=C	0.46
−Br	1.53	−C≡C	0.79
$-NO_2$	1.84	−C≡N	0.66
$-NH_2$	0.64	−COR, −COOR, −COOH	0.47
$-NH_3^+$	1.34	$-CONH_2$	0.60
−NHCOR	1.80	−COAr	1.22
−OH, −OR	1.14	−SH, −SR	0.61
−Oar	1.79	$-SO_2R$	0.94
−OCOR	2.07	−R	0

This relation is satisfactory if at least two of the substituents are electron withdrawing groups. In other words, only a single substituent may be an alkyl group. Within these limits, the standard error of estimate is 0.20 ppm [4].

For example, the chemical shift of the methine proton is calculated from Table 2.10 as follows:

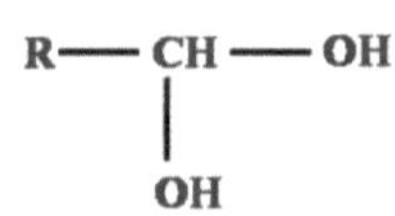

$$\delta = 2.50 + 1.14 + 1.14 + 0.00 = 4.78 \text{ (calculated)}, \quad \text{and observed } \delta = 4.72.$$

2.8.3 Calculation of Chemical Shifts in Unsaturated Systems

The proton chemical shift for unsaturated systems *viz*., substituted alkenes as shown below is calculated from the σ-values given in Table 2.11.

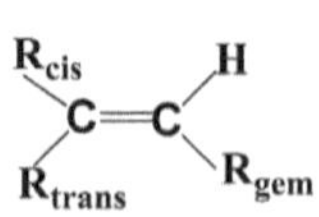

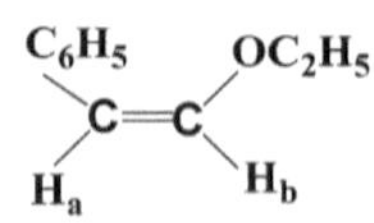

Table 2.11 Substituent constants (Z) for chemical shifts of substituted ethylenes

Substituent R	Z: *gem*	*cis*	*trans*
−H	0.0	0.0	0.0
−Alkyl	0.44	−0.26	−0.29
−Alkyl-ring	0.71	−0.33	−0.30
$-CH_2O$, $-CH_2I$	0.67	−0.02	−0.07
$-CH_2S$	0.53	−0.15	−0.15
$-CH_2Cl$, $-CH_2Br$	0.72	0.12	0.07
$-CH_2N$	0.66	−0.05	−0.23
−C≡C	0.50	0.35	0.10
−C≡N	0.23	0.78	0.58
−C=C	0.98	−0.04	−0.21
−C=C conj	1.26	0.08	−0.01
−C=O	0.10	1.13	0.81
−C=O conj	1.06	1.01	0.95
−COOH	1.00	1.35	0.74
−COOH conj	0.69	0.97	0.39
−COOR	0.84	1.15	0.56
−COOR conj	0.68	1.02	0.33
−CHO	1.03	0.97	1.21
$-CONH_2$	1.37	0.93	0.35
−COCl	1.10	1.41	0.99
−OR, R:aliph	1.18	−1.06	−1.28
−OR, R:conj	1.14	−0.65	−1.05
−OCOR	2.09	−0.40	−0.67
−Aromatic	1.35	0.37	−0.10
−Cl	1.00	0.19	0.03
−Br	1.04	0.40	0.55
$-NR_2$ (R:aliph)	0.69	−1.19	−1.31
$-NR_2$ (R:conj)	2.30	−0.73	−0.81
−SR	1.00	−0.24	−0.04
$-SO_2$	1.58	1.15	0.95

conj, conjuagated

$$\delta_H = 5.25 + Z_{gem} + Z_{cis} + Z_{trans}$$

$H_a = C_6H_{5gem}$ 1.35

OR_{trans} -1.28

0.07 $\delta_{Ha} = 5.25+0.07 = 5.32$

$H_b =$ OR_{gem} 1.18

C_6H_{5trans} -0.10

1.08 $\delta_{Hb} = 5.25+1.08 = 6.33$

2.8.4 Effect of Substituents on Aromatic Ring Systems

Inductive substituent effects are approximately additive in aromatic compounds. So, the chemical shift values of protons and carbon-13 can be calculated using the additivity rule using Tables 2.12 and 2.13.

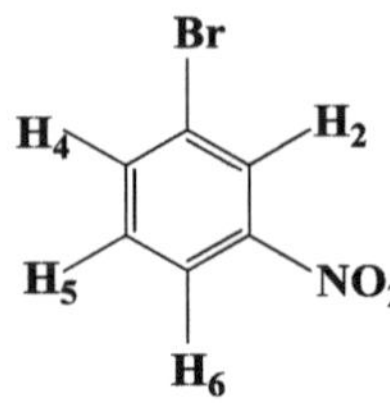

Table 2.12 Chemical shifts increments (in ppm) for substituted benzenes

Substituent	*Ortho*	*Meta*	*Para*
NO_2	0.95	0.17	0.33
CHO	0.58	0.21	0.27
COCl	0.83	0.16	0.3
COOH	0.8	0.14	0.2
$COOCH_3$	0.74	0.07	0.2
$COCH_3$	0.64	0.09	0.3
CN	0.27	0.11	0.3
C_6H_5	0.18	0	0.08
CCl_3	0.8	0.2	0.2
$CHCl_2$	0.1	0.06	0.1
CH_2Cl	0	0.01	0
CH_3	−0.17	−0.09	−0.18
CH_2CH_3	−0.15	−0.06	−0.18
$CH(CH_3)_2$	−0.14	−0.09	−0.18
$C(CH_3)_3$	0.01	−0.1	−0.24
CH_2OH	−0.1	−0.1	−0.1
F	−0.3	−0.02	−0.22
Cl	0.02	−0.06	−0.04
Br	0.22	−0.13	−0.03
I	0.4	−0.26	−0.03
OCH_3	−0.43	−0.09	−0.37
$OCOCH_3$	−0.21	−0.02	–
OH	−0.5	−0.14	−0.4
$p\text{-}CH_3C_6H_4SO_3$	−0.26	−0.05	–
NH_2	−0.75	−0.24	−0.63
SCH_3	−0.03	0	–
$N(CH_3)_2$	−0.6	−0.1	−0.62

Table 2.13 ^{13}C chemical shifts (δ) for aromatic carbon in benzene derivatives Ph—X in ppm relative to benzene at δ 128.5 ppm (positive sign denotes a downfield shift)

X	*Ipso*	*Ortho*	*Meta*	*Para*
—H	0.0	0.0	0.0	0.0
$-NO_2$	19.9	−4.9	0.9	6.1
$-CO-OCH_3$	2.0	1.2	−0.1	4.3
$-CO-NH_2$	5.0	−1.2	0.1	3.4
$-CO-CH_3$	8.9	0.1	−0.1	4.4
—C≡N	−16.0	3.5	0.7	4.3
—Br	−5.4	3.3	2.2	−1.0
$-CH{=}CH_2$	8.9	−2.3	−0.1	−0.8
—Cl	5.3	0.4	1.4	−1.9
$-CH_3$	9.2	0.7	−0.1	−3.0
$-OCO-CH_3$	22.4	−7.1	0.4	−3.2
$-OCH_3$	33.5	−14.4	1.0	−7.7
$-NH_2$	18.2	−13.4	0.8	−10.0

Calculation of chemical shifts in aromatic system

$$\delta\ (H_2) = 7.27 + 0.22 + 0.95 = 8.44$$

$$\delta\ (H_4) = 7.27 + 0.22 + 0.33 = 7.82$$

$$\delta\ (H_5) = 7.27 - 0.13 + 0.17 = 7.31$$

$$\delta\ (H_6) = 7.27 - 0.13 + 0.95 = 8.19$$

2.9 STERIC EFFECTS

^{13}C chemical shifts are highly sensitive to molecular geometry. Carbon atoms separated by several bonds strongly influence each other if they are spatially close enough (Fig. 2.12) [4].

Upfield shifts (i.e., low δ values) are observed in all cases where the stereochemistry leads to Van der Waals interactions of the type indicated above ($X{=}CH_3$). In rigid molecules, this effect can be as large as 10 ppm. Smaller γ effects are observed if there is free rotation about the C—C bonds (Fig. 2.13).

$\Delta\delta_{st} = F_{HH}(r)\cos\theta$. Depending on θ, $\Delta\delta_{st}$ can have either positive or negative values ($F_{HH}(r)$—the repulsive force between the protons).

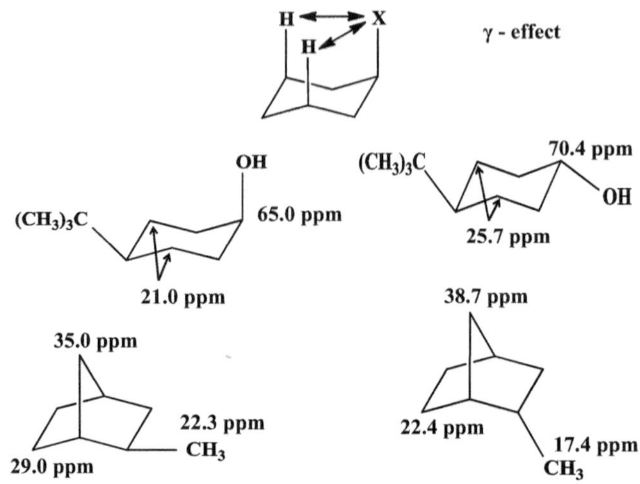

Figure 2.12 Influence of steric effect on chemical shift value of ^{13}C nuclear magnetic resonance.

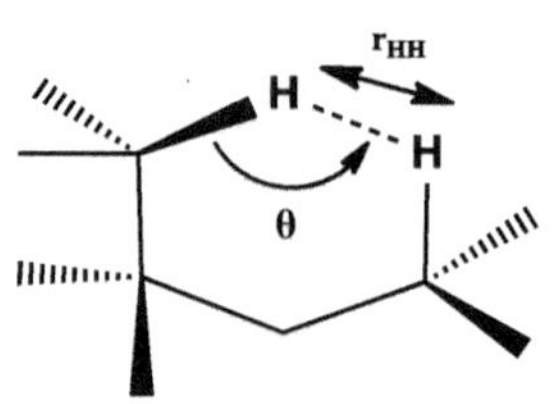

Figure 2.13 The repulsive force between the protons.

2.10 STEREOELECTRONIC EFFECTS ON $^{1}J_{C-H}$ (PERLIN EFFECTS)

There are well-defined stereoelectronic effects on the C—H coupling constants of axial and equatorial protons in cyclohexanes (called *Perlin effects*). These are most pronounced when the strong electronic interactions of n → σ* or σ → π* types cause perturbations of the C—H bond length and bond strength either by electron donation into the C—H σ* orbital or electron withdrawal from the σ hybridized orbital. In cyclohexane itself, the axial C—H coupling is slightly smaller than the equatorial ($\Delta J = 4.0$ Hz), attributable to a stronger donation by the axial H—Cσ bond compared with the equatorial C—H bond. In 1,3-dioxane, the difference is much larger ($\Delta J = 10.1$ Hz), indicating larger donation by the p lone pair on each oxygen.

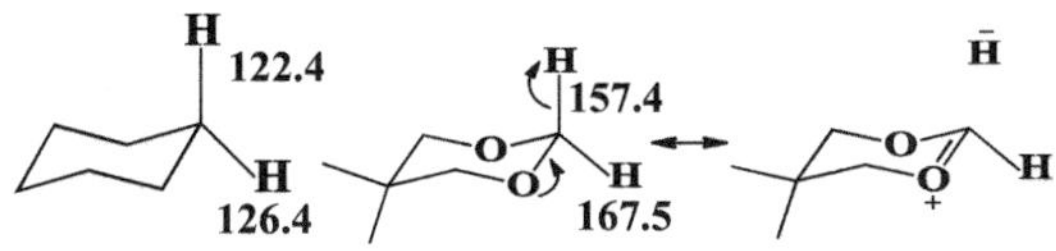

2.11 NUCLEAR MAGNETIC DOUBLE RESONANCE

It employs two different RF radiation frequencies also called double irradiation or spin decoupling technique very frequently used in the simplification and easy interpretation of complex NMR spectra. It is based on the principle that when one group of equivalent protons in a molecule is irradiated, i.e., decoupled by applying very strong stationary RF field at the Larmor (resonance) frequency of the protons, then the irradiated protons could no longer undergo spin−spin coupling with the neighboring group protons. As a result of this spin decoupling, the spectrum is simplified [1,8−10].

For example, Fig. 2.14 shows the ^{1}H NMR spectra of pure ethanol and acid added to it.

1. The CH_3 protons give triplet signal due to two $-CH_2-$ protons ($n + 1 = 2 + 1 = 3$).
2.
   a. The $-CH_2-$ protons give quartet signal due to spin−spin coupling with $-CH_3$ protons.
   b. Each signal further splits into doublet due to −OH single proton.

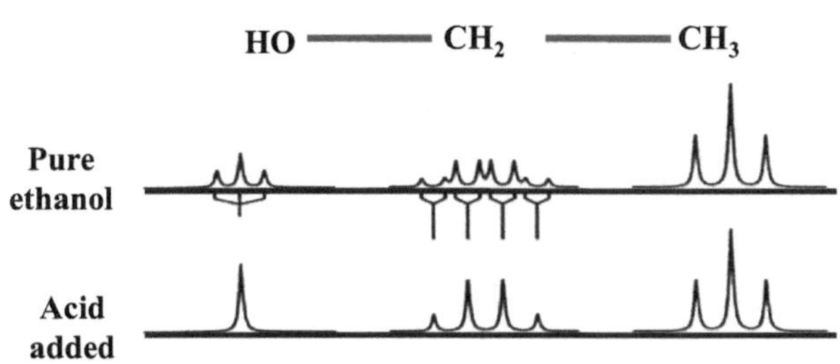

Figure 2.14 ^{1}H Nuclear magnetic resonance spectra of pure ethanol and acid added to it.

3. The —OH proton exhibits triplet signal due to splitting by $-CH_2-$ (two protons).
4. In the presence of acid, the —OH coupling is lost due to its catalytic fast exchange of —OH protons with acid protons.

Furthermore, a schematic ^1H NMR spectrum of pure ethanol and the double resonance spectra for pure ethanol are shown in Fig. 2.15.

1. When CH_3 group protons are irradiated, the resonance of the methylene protons collapses to a doublet because they are now coupled only with the OH proton; the coupling of CH_3 protons is disappeared.
2. Similarly, when CH_2 group protons are irradiated, the resonances of the OH and methyl protons both collapse to singlets.

In such cases, it is possible and useful to remove the splitting effects due to spin—spin coupling by irradiating the sample with an additional RF source at the exact resonance frequency of the proton giving the splitting. This additional RF radiation causes rapid spin flipping of the irradiated nuclei. As a result, the neighboring nuclei coupled to them cannot sense them as present in either an α or β state for long enough to give rise to splitting. So, the irradiated nuclei are **decoupled** from other nuclei in the spin system. This spin decoupling simplifies the complex multiplets by removing some of the splittings. Furthermore, decoupling is also a useful tool for assigning the complicated spectra. Now-a-days, the spectroscopist often uses a series of decoupling experiments to sequentially identify which nuclei are coupled to each other.

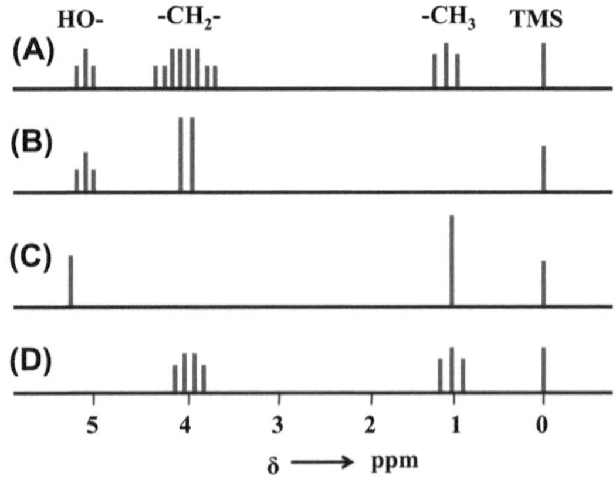

Figure 2.15 Double resonance spectra of pure ethanol (A) normal nuclear magnetic resonance, (B) CH_3 group irradiated, (C) CH_2 group irradiated, and (D) OH group irradiated. *TMS*, tetramethylsilane.

2.12 SINGLE FREQUENCY OFF-RESONANCE DECOUPLING (SFORD)

There is another double resonance method called single frequency off-resonance decoupling (SFORD) for obtaining ^{13}C NMR spectra (still retaining the multiplicity information). It involves a strong decoupling signal at a single frequency just outside the range of proton resonances being applied. This leads to incomplete or partial decoupling of protons from the ^{13}C nuclei. The spectrum thus obtained retains the multiplicity information. The technique is also known as off-resonance decoupling or SFORD. This multiplicity, due to the large one-bond C—H couplings, makes it possible to differentiate whether a carbon atom belongs to a methyl group (quartet), methylene group (triplet), a methine (CH) group (doublet), or a quaternary carbon (a singlet) similar to the fully proton-coupled spectrum. However, it was found that because the protons are only partially decoupled, the magnitude of the splitting measured from SFORD ^{13}C NMR spectrum is not the actual/exact C—H coupling constant. The splitting is always lower than the real coupling constants. SFORD spectra are thus useful only to establish signal multiplicity (Fig. 2.17).

2.13 DISTORTIONLESS ENHANCEMENT BY POLARIZATION TRANSFER (DEPT)

The distortionless enhancement by polarization transfer (DEPT) experiment is the most widely used method to determine the multiplicity of ^{13}C signals than SFORD. This method is a pulsed 1D NMR experiment. Here, a series of programmed RF pulses to both the ^1H and ^{13}C nuclei in a sample are excited. The acquired ^{13}C DEPT spectrum contains only signals arising from protonated carbons (nonprotonated carbons, i.e., quaternary carbons do not yield signals). There are three types of DEPT spectra that can be acquired. They are noted by numbers. DEPT-45 will give a positive signal for any carbon with an attached proton. DEPT-90 spectra will give a positive signal for methine carbons. DEPT-135 spectra will give positive signals for methine and methyl carbons (odd number of hydrogen) and negative signals for methylene carbons (even number of hydrogen). By a linear combination of these three spectra and the normal ^{13}C spectrum, the multiplicities of all of the carbons in a molecule may be determined. The example spectra in Fig. 2.16 show the appearance of each type of carbon in the different DEPT spectra.

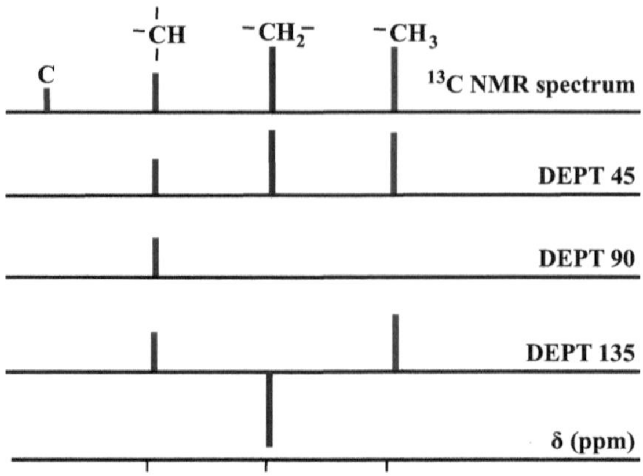

Figure 2.16 The appearance of each type of carbon in the different distortionless enhancement by polarization transfer (DEPT) spectra.

The signals obtained from carbons in CH_3 and CH groups (i.e., carbon nuclei those with an odd number of attached protons) are oppositely phased from those in CH_2 groups (i.e., those with an even number of attached protons). Hence, the signals from CH_3 and CH groups point upward, whereas the signals from CH_2 groups point downward (Fig. 2.17). So, DEPT

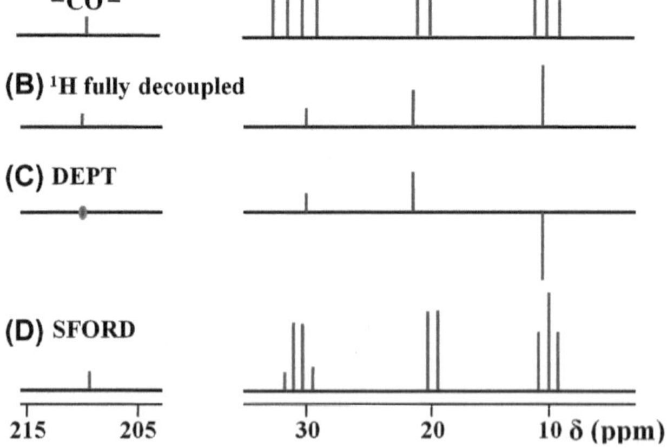

Figure 2.17 ^{13}C nuclear magnetic resonance (NMR) spectra of methyl cyclopropyl ketone.

experiment can be used to separate the overlapping signals due to carbons in CH_3, CH_2, and CH groups. This is called *spectral editing*.

1. In the case of proton-coupled ^{13}C NMR spectra of methyl cyclopropyl ketone (i.e., with no proton decoupling) (Fig. 2.17A), all of the resonances of protonated carbons appear as multiplets. The multiplet structure occurs because of coupling with the attached protons. The CH_3 (methyl) group appears as a quartet, the CH_2 (methylene) groups appear as a triplet, and the CH (methine) group appears as a doublet while the carbonyl carbon (with no attached protons) appears as a singlet. Therefore, all of the $^{1}J_{C-H}$ coupling constants could be measured directly from this spectrum.
2. The ^{13}C spectrum obtained with full proton decoupling for methyl cyclopropyl ketone (Fig. 2.17B) shows four singlet peaks, one for each of the four different carbon environments in the molecule.
3. The DEPT spectrum in Fig. 2.17C shows only the three resonances only for the protonated carbons. The carbon atoms that have an odd number of attached hydrogens (viz., CH and CH_3 groups) show pointing signals upward. Those with an even number of attached hydrogen atoms (i.e., CH_2 group) point downward. However, the carbonyl carbon does not appear in the DEPT spectrum because it has no attached protons.
4. The SFORD spectrum (Fig. 2.17D) shows again the expected multiplicity for all of the resonances similar to proton coupled. But the multiplets are narrower due to partial decoupling of the protons and the splitting is smaller than the true values of $^{1}J_{C-H}$.

2.14 INSENSITIVE NUCLEI ENHANCED BY POLARIZATION TRANSFER (INEPT)

The sensitivity of a low γ, spin I = ½ nucleus (especially ^{13}C) is determined by the difference in populations between the low- and high-energy states, governed by the Boltzmann distribution. If the low γ, spin I = ½ nucleus is coupled to a proton (e.g., $^{13}C-^{1}H$), the energy level diagram is further split as shown in Fig. 2.18 for a $^{13}C-^{1}H$ spin pair [11–13].

2.14.1 Mechanism of Insensitive Nuclei Enhanced by Polarization Transfer

The populations of the states involved in the ^{13}C transitions and hence the sensitivity of the ^{13}C signal can be altered by inverting the H_1 or H_2 transitions with 180 degrees pulses. This is illustrated in the figure below.

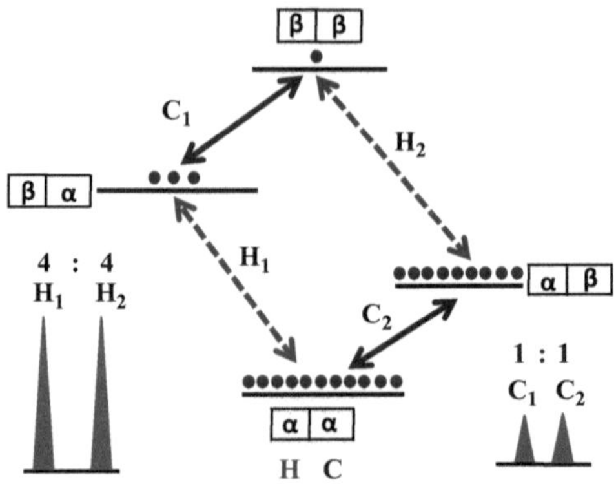

Figure 2.18 Equilibrium energy level diagram for a $^{13}C-^{1}H$ spin pair.

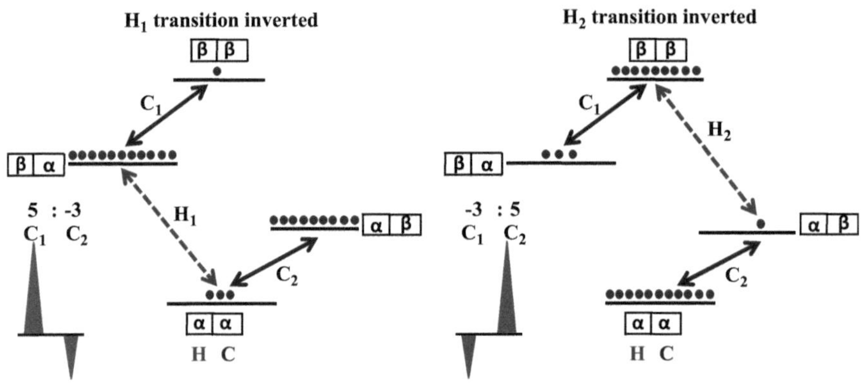

Figure 2.19 Polarization transfer and signal enhancement.

In Fig. 2.19 (left), the H_1 transition of a $^{13}C-^{1}H$ spin pair is inverted. This affects the populations of the energy levels involved in the C_1 and C_2 and hence ^{13}C transitions. After the inversion of H_1 transition, the intensities of the C_1 and C_2 of ^{13}C transitions have enhanced (5:−3) from their equilibrium value of (1:1). On the other hand, when the H_2 transition is inverted (Fig. 2.19, right), the intensity ratio of C_1:C_2 is −3:5. Hence, the sensitivity of the ^{13}C doublet has been enhanced compared to its equilibrium value. This phenomenon of enhancement is called insensitive nuclei enhanced by polarization transfer (INEPT). It is widely used sensitivity enhancement technique in NMR.

Furthermore, phase cycling can be employed to obtain a −4:4 antisymmetric doublet (as shown in Fig. 2.20), rather than doublets with

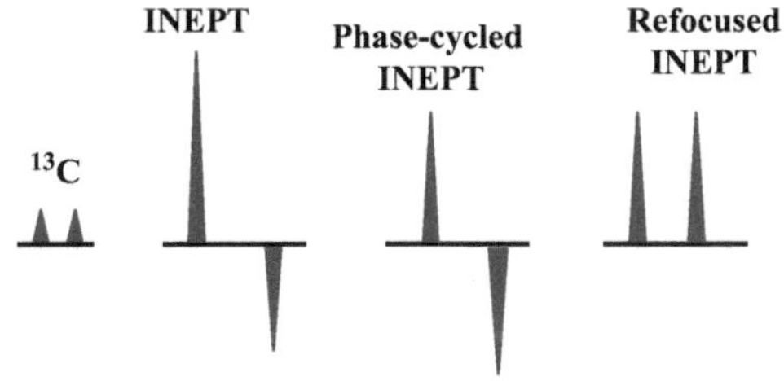

Figure 2.20 Polarization transfer and signal enhancement using insensitive nuclei enhanced by polarization transfer (INEPT) sequence.

components of unequal magnitude. On refocusing, it becomes 4:4 symmetric doublets with proton decoupling. As a result, each singlet (^{13}C signal) becomes four times (i.e., γ_H/γ_C) the intensity of the singlet (one would expect under equilibrium conditions without an NOE).

Similarly, for ^{15}N, a sensitivity gain of ~10 is obtained (see Section 2.16).

2.14.2 Advantage of Insensitive Nuclei Enhanced by Polarization Transfer

The additional advantage of INEPT is that its pulse sequence repetition rate is determined by the ^{1}H T_1 rather than the ^{13}C T_1. This results in additional sensitivity improvement when multiple scans are acquired because the ^{1}H T_1 is shorter than the ^{13}C T_1 by an order of magnitude. Hence, it is possible to collect ~10 times as many scans per unit time. This advantage is even more significant for ^{15}N.

2.15 CHEMICALLY INDUCED DYNAMIC NUCLEAR POLARIZATION (CIDNP)

If there is no saturation and polarization transfer, the spin populations of diamagnetic and paramagnetic species in the presence of the external magnetic field (NMR, ESR experiments) are usually in thermal equilibrium governed by Boltzmann distribution. However, during chemical reactions with the participation of free radicals, there is nonequilibrium nuclear and electron spin levels produced. This phenomenon is called chemically induced magnetic polarization. Furthermore, this is known as chemically induced dynamic nuclear polarization (CIDNP) (in the case of NMR studies) or chemically induced dynamic electron polarization (in the case

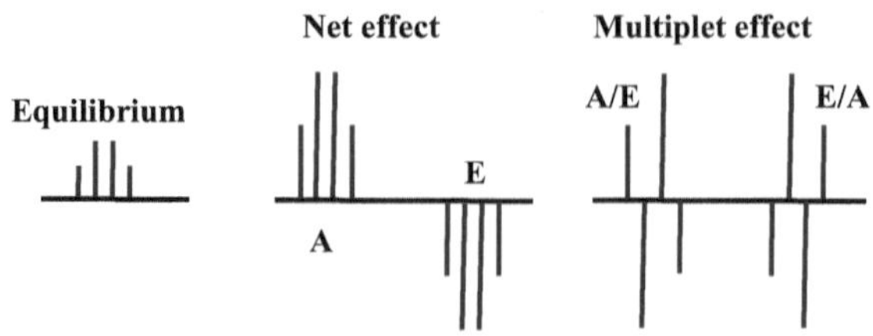

Figure 2.21 Chemically induced dynamic nuclear polarization effects on a quadruplet: thermal equilibrium, net effects, and multiplet effects.

of electron paramagnetic resonance (EPR) measurement). When such processes occur, the observed NMR spectra are not anymore the conventional spectra with respect to intensity of Zeeman transitions. Therefore, a strongly enhanced emission or absorption could be observed—but still the same with respect to chemical shifts of all observed NMR signals.

There are two types of CIDNP effects observed experimentally:

1. **Net effect**: When all of the resonances in multiplet show either enhanced emission (E) or enhanced absorption (A), then it is called net effect.
2. **Multiplet effect**: When different polarizations within the multiplet show either enhanced emission (E) or absorption (A), the patterns of absorption/emission (A/E) or emission/absorption (E/A) are observed in Fig. 2.21.

Sometimes both net and multiplet effects could be observed within the same multiplet. Fig. 2.21 shows possible CIDNP effects for nuclei A (quartet) in AX_3 spin system.

2.15.1 Mechanism of Chemically Induced Dynamic Nuclear Polarization

This experimentally observed CIDNP phenomenon has been explained using radical pair mechanism (RPM) theory. A vast majority of the CIDNP observations can be treated within this theory. According to this theory, the existence of a transient radical pairs in the reaction pathway is assumed (Fig. 2.22). These radicals could be directly observed by ESR in certain favorable cases. However, in the NMR the diamagnetic products of the reaction are detected. Hence, all information about radical pair intermediates is obtained indirectly by the changes in the NMR signal intensities of the products using CIDNP effect [14].

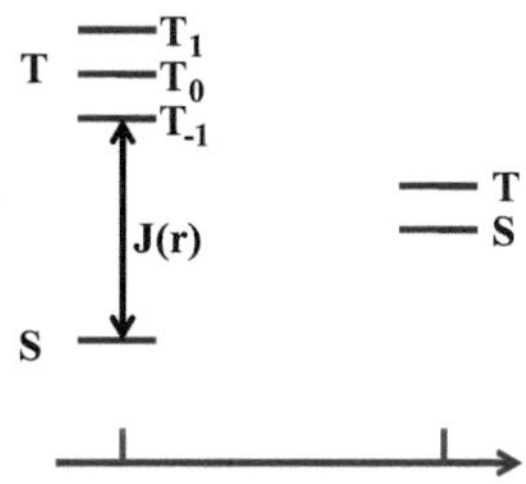

Figure 2.22 Energy level diagram of a separating radical pair in an external magnetic field (H_0) as a function of interradical distance (r).

2.15.2 Radical Pair Mechanism Theory

According to RPM, in the presence of external magnetic field, the CIDNP effect is mainly due to a nuclear spin-dependent recombination probability, i.e., the probability for two radicals to react (recombine, disproportionate) depends on the nuclear spin states of those two radicals. In the external magnetic field, the precursor molecule (for example, R_1-R_2) is cleaved into two radicals $R_1^{\bullet}$ and $^{\bullet}R_2$ with the correlated electron spins, and this is called radical pair. During the thermal cleavage singlet radicals pairs (S) are formed predominantly, whereas photochemical process leads to triplet radical pairs (T). In the magnetic field, triplet state ($m_s = -1, 0, +1$) is split into three sublevels T_1, T_0, and T_{-1}. When the two radicals are enough far apart, the singlet S and triplet T_0 states are degenerate. At the relatively small interradical distances (r), the singlet and triplet states possess different energies because of high exchange interaction J(r) (Fig. 2.22).

2.16 NUCLEAR OVERHAUSER EFFECT (NOE)

Irradiation of one nucleus while observing the resonance of another may result in a change in the amplitude of the observed resonance, i.e., an enhancement/decrease of the signal intensity. In other words, there is a change in the intensity of nucleus B when the spin states of nucleus A is perturbed (Fig. 2.23). This phenomenon is known as the NOE. The NOE is a "through space" effect and an important consequence of dipole—dipole (DD) relaxation and its magnitude is inversely proportional to the sixth power of the distance between the interacting nuclei. Because of the distance dependence of the NOE, it is an important method for establishing

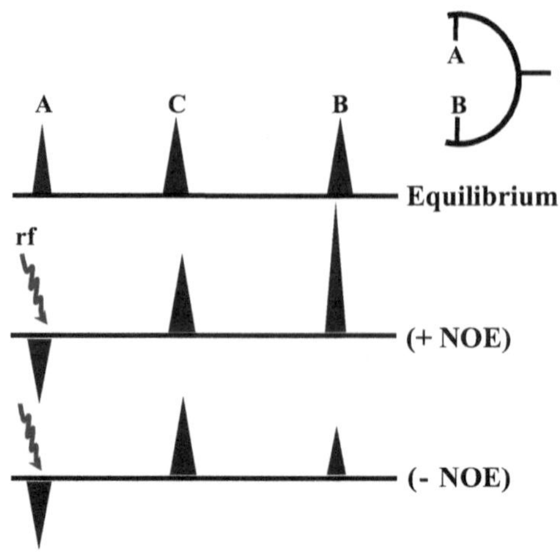

Figure 2.23 Nuclear Overhauser effect (NOE) for A and B spatially close together and A and C are far apart.

which groups are close together in space and can be used to determine intra-(and even inter-) molecular distances. Because, the NOE can be measured quite accurately, it is a very powerful means for determining the three dimensional structure (and stereochemistry) of organic compounds including proteins.

2.16.1 Nuclear Overhauser Effect (Heteronuclear)

Analogously, the intensity of ^{13}C resonances may be increased by up to 200% when H nuclei, which are directly bonded to the carbon atom, are irradiated. This effect is very important in increasing the intensity of ^{13}C spectra when they are proton decoupled. The efficiency of the proton/carbon NOE varies from carbon to carbon. This is a factor that contributes to the generally nonquantitative nature of ^{13}C NMR. While the intensity of protonated carbon atoms can be increased significantly by NOE, nonprotonated carbons (quaternary carbon atoms) receive little NOE and are usually the weakest signals in a ^{13}C NMR spectrum.

In a ^{13}C—^{1}H decoupling experiment, the ^{1}H transitions are irradiated while the ^{13}C resonance is observed, showing increase in its intensity called Overhauser effect. (1) When the irradiating field is strong, the homonuclear relaxation processes are not adequate to restore the equilibrium population of ^{1}H nuclei. Therefore, these nuclei transfer their energy to the ^{13}C nuclei via internuclear DD interaction. (2) These energized carbon nuclei relax to the lower energy level enhancing the intensity of the carbon signal.

Thus there is an increase in intensity of the ^{13}C signal larger than that expected due to spin decoupling experiment. (3) Such intensity enhancement in heteronulcear double resonance experiments are more pronounced when the nucleus with low μ/I is observed while the nucleus with high μ/I is decoupled. This enhancement in intensity due to DD relaxation mechanism is called the *NOE*. It can be used to enhance the intensity and get information about the relaxation processes.

NOE is thus useful to enhance the signal intensities (sensitivity) especially of low γ and low-abundant nuclei when they are coupled (scalar and dipolar) to high γ and high-abundant nuclei, for example, ^{13}C and ^1H. ^{13}C NMR acquired under broadband proton-decoupled conditions has the following advantages.

1. Signal intensity is increased because multiplets are removed and each carbon gives a single line (No C—H coupling).
2. NOE effect increases the signal intensities of those carbons that bear only protons. NOE effect increases with increasing number of attached protons on the carbon. $CH_3 > CH_2 > CH$ and quaternary carbons do not have NOE. When the ^{13}C relaxation operates only through DD mechanism, then the enhancement due to NOE is $\eta_C = \gamma_H/2\gamma_C = 1.988$. The maximum theoretical limit of NOE is about 200%.
3. Proton decoupling of nuclei with negative γ will result in lowering of intensity due to negative NOE effect. ^{15}N and ^{29}Si are common examples where in negative NOE is observed. Sometimes the negative NOE can cancel the natural signal.
   NOE can be negative in heteronuclear system if one of the γ is negative; for example, from ^1H to ^{15}N it is −494%.
4. The differential relaxation times and differential NOE of various carbons are responsible for the lack of quantitative correlation (integration of peaks) to the structure.
5. Steady state NOEs are those measured after continuous saturation of A spin, which leads to new steady-state spin population for B spin.

$$\text{Maximum possible NOE is } \gamma_A/2\gamma_B$$

Therefore, in homonuclear systems the maximum NOE is 50% and in heteronuclear systems, such as ^{13}C NOE under ^1H irradiation, it is maximum 200%. For example, in the E configuration, the CH_3 group protons are close to the indicated (Fig. 2.24). So, there is 31% positive NOE observed. However, no NOE is observed for Z configuration as the groups are far apart.

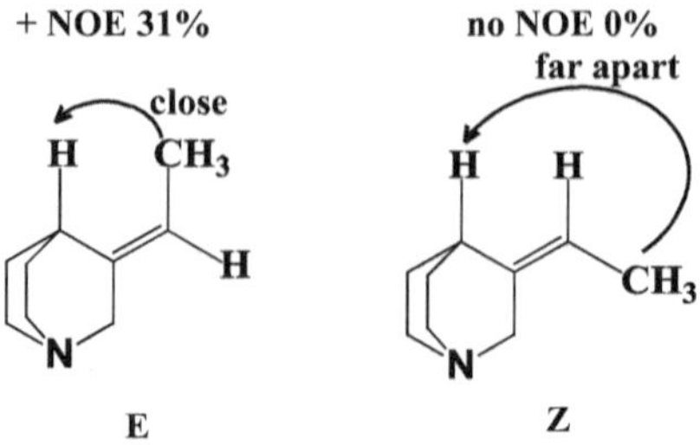

Figure 2.24 ^1H–^1H nuclear Overhauser effect (NOE) for E/Z stereochemistry assignment.

2.16.2 Distinction Between Decoupling and the Nuclear Overhauser Effect

In a decoupling experiment, the B_1 irradiation must be on during acquisition of the free induction decay (FID) (but not necessarily otherwise). On the other hand, in an NOE experiment, the decoupler is on during a delay period but may be turned off during the acquisition of the FID (Fig. 2.25).

2.16.3 Origin of the Nuclear Overhauser Effect

When a pair of protons is close in space, their magnetic dipoles interact (DD interaction). This interaction is distinct from J coupling, which is not a through space effect but is mediated by polarization of bonding electrons in the molecule. The effects of DD interactions on the appearance of NMR spectra are completely averaged by the normal tumbling of molecules in solution if the medium is isotropic and viscosity is low enough to allow sufficiently fast molecular motion (short enough correlation time, τ_c). The DD interactions between protons do, however, dominate the ^1H T_1 relaxation processes in most molecules that contain more than one proton [15].

To understand the NOE effect, consider a pair of protons **AX**, close in space, but not J coupled to each other. Such a system has four energy states, corresponding to the $\alpha\alpha$, $\alpha\beta$, $\beta\alpha$, and $\beta\beta$ spin states as shown in Fig. 2.26.

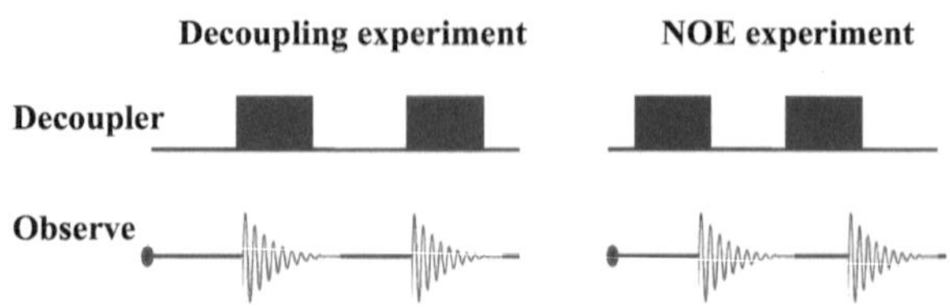

Figure 2.25 Decoupling and nuclear Overhauser effect (NOE) experiments.

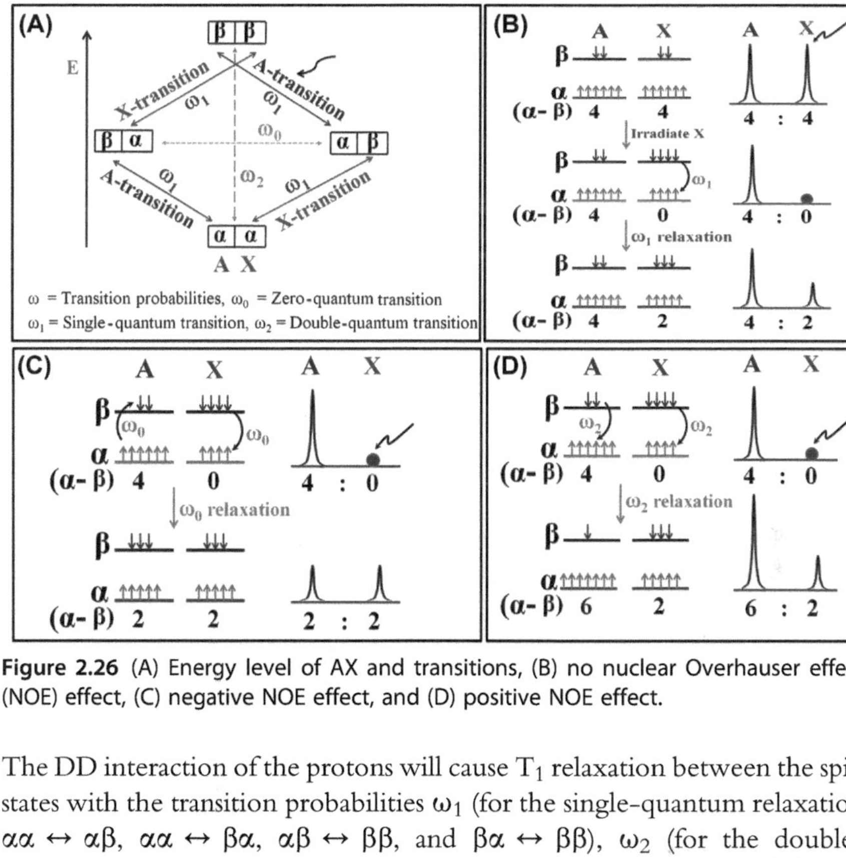

Figure 2.26 (A) Energy level of AX and transitions, (B) no nuclear Overhauser effect (NOE) effect, (C) negative NOE effect, and (D) positive NOE effect.

The DD interaction of the protons will cause T_1 relaxation between the spin states with the transition probabilities ω_1 (for the single-quantum relaxation $\alpha\alpha \leftrightarrow \alpha\beta$, $\alpha\alpha \leftrightarrow \beta\alpha$, $\alpha\beta \leftrightarrow \beta\beta$, and $\beta\alpha \leftrightarrow \beta\beta$), ω_2 (for the double-quantum relation $\alpha\alpha \leftrightarrow \beta\beta$), and ω_0 (for the zero-quantum relaxation $\alpha\beta \leftrightarrow \beta\alpha$). There will be an excess population in the $\alpha\alpha$ ground state and a less population in the $\beta\beta$ excited state.

When the X-transition is irradiated, the populations of the $\alpha\alpha$ and $\alpha\beta$ states become equalized (saturated), as do the $\beta\alpha$ and $\beta\beta$ states (Fig. 2.26). As relaxation occurs, the difference in these two populations depends crucially on which of the three relaxation processes dominates. Based on that, there are three cases as follows:

1. **No NOE (ω_1 dominates)**
   For AX spins (Fig. 2.26B), the peak intensities of A and X are in the ratio 4:4 in their natural equilibrium populations. On X-irradiation, it equalizes the populations of α- and β-spin states of X. So, there is no peak observed for X. Due to only ω_1 relaxation, the A nucleus is unaffected (i.e., the nuclei A and X are coupled) giving rise to no NOE effect on A nuclei.

2. **Negative NOE (ω_0 dominates)**
   Next consider the situation when either the ω_0 or the ω_2 processes are the only ones operative. In the ω_0 process, the dipolar interaction between A and X causes A nucleus to undergo $\alpha \rightarrow \beta$ transition when the X nucleus relaxes from $\beta \rightarrow \alpha$ ($\alpha\beta \rightarrow \beta\alpha$) (Fig. 2.26C). The net result is that as X returns to its normal population difference, it lowers the population difference for A. Thus, as the X intensity decreases, the A intensity decreases. If X is irradiated continuously then the signal for A will vanish (−100% NOE). **This is a negative NOE.**
3. **Positive NOE (ω_2 dominates)**
   For the ω_2 process, each time an X nucleus relaxes from β to α state, and A nucleus also undergoes a β to α transition ($\beta\beta \rightarrow \alpha\alpha$). This has the effect of increasing the population difference of A, i.e., an increase in intensity of A (Fig. 2.26D). **This is a positive NOE**. This phenomenon has sometimes been known as *spin pumping*.

2.16.4 Comparison of Different Nuclear Overhauser Effects

The reason we get NOE population changes is that the three dipolar relaxation pathways contribute to differing extents depending on a number of factors. A key one is that the balance between ω_2, ω_1, and ω_0 depends crucially on molecular motion (τ_c). In mobile solvents, molecular motion (for small molecules) is much faster than the Larmor precession frequency (ν_0) (Fig. 2.38). Under these conditions, the double-quantum relaxation ω_2 is more effective than ω_1 or ω_0. If ω_2 is the *dominant* relaxation process, then we get a positive NOE. This is opposite for negative NOE (largely observed for large molecules).

Thus when $\omega_2 > \omega_0$, the NOE will be positive. On the other hand, when $\omega_0 > \omega_2$, the NOE will be negative. However, when $\omega_2 = \omega_0$, then there is no NOE. The maximum NOE observable is reduced to the extent that T_1 relaxation pathways other than DD between X and A are operative. This includes intermolecular DD processes (for example, by solvent molecules or by dissolved dioxygen).

2.17 MULTIDIMENSIONAL NUCLEAR MAGNETIC RESONANCE SPECTROSCOPY

In multi- (or 2- or 3-) dimensional NMR spectroscopy, multipulse sequences are employed to provide additional information not easily

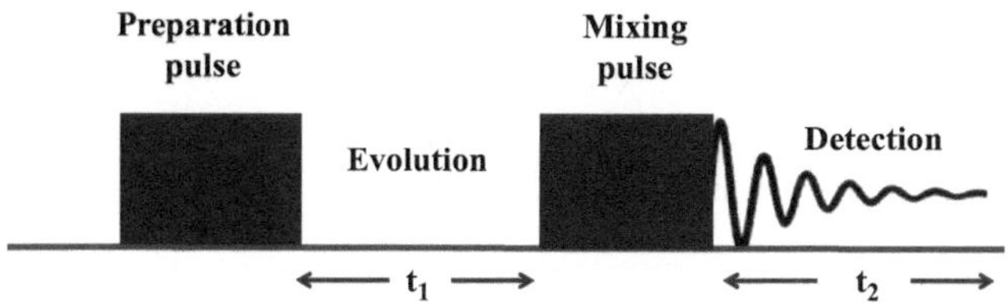

Figure 2.27 Schematic representation of pulse sequence of multidimensional nuclear magnetic resonance spectroscopy.

obtainable from normal one-dimensional spectra (Fig. 2.27). The following steps are commonly involved in the multidimensional especially 2D NMR spectroscopy.

1. **Preparation**: This is typically a 90 degrees pulse that generates transverse magnetization.
2. **Evolution**: Depending on the frequency of different protons, the individual magnetization spreads out at different speed during evolution time. This means that magnetization is "frequency-labeled" as a function of t_1.
3. **Mixing**: During mixing (pulse), the magnetization transfer between two nuclei through bond or through space or chemical exchanges occurs. It indicates that part of the magnetization from spin 1 was transferred to spin 2 in this mixing time.
4. **Detection**: The magnetization that did not get transferred during the mixing period will appear at the same frequency during the detection period. These are *diagonal peaks* of frequency (Ω_A, Ω_A). Magnetization that was at frequency A but moved to frequency B during the mixing period will precess at an off-diagonal frequency (Ω_A, Ω_B) (see Fig. 2.29).

2.18 2D NUCLEAR MAGNETIC RESONANCE SPECTROSCOPY

The most useful and commonly used forms of 2D NMR spectroscopy provide correlations between proton (or other NMR-active nuclei) signals based on some interaction between them, most frequently J-coupling (H—H, H—C, or even C—C), but it could also be dipolar coupling (the NOE interaction), rates of chemical exchange (DNMR swapping of spin environments), or relative diffusion rates (DOSY and related experiments). 1D NMR spectra are called "1D" because they have one *frequency* dimension, but actually have an additional dimension, intensity. 2D NMR means that there are **two frequency dimensions** [16–18]. The classification of 2D NMR is given in the following flow chart.

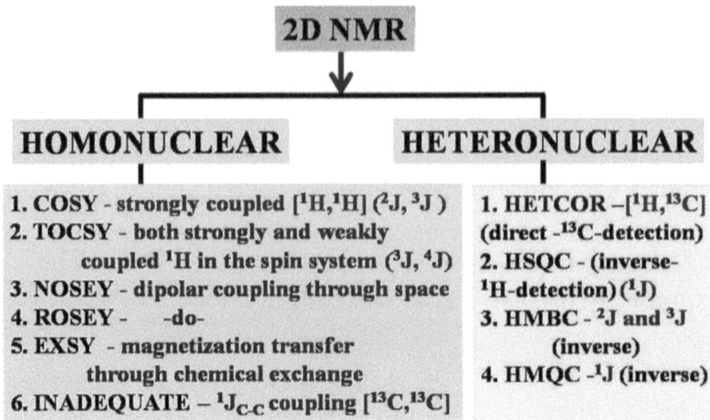

2.18.1 [^{1}H, ^{1}H]—Correlations Spectroscopy Nuclear Magnetic Resonance Spectra

COSY stands for correlations spectroscopy and is a method for finding **homonuclear, through-bond** (scalar couplings mainly ^{2}J) **correlations.** The COSY spectrum shows which pairs of protons in a molecule are directly and strongly coupled to each other. The correlations will only appear between the protons that possess a resolved coupling to each other. It is a 2D symmetrical spectrum that has the ^{1}H NMR spectrum of the substance on both of the chemical shift axis (F_1 and F_2). This is best illustrated by the basic **COSY-90 sequence** (Fig. 2.28):

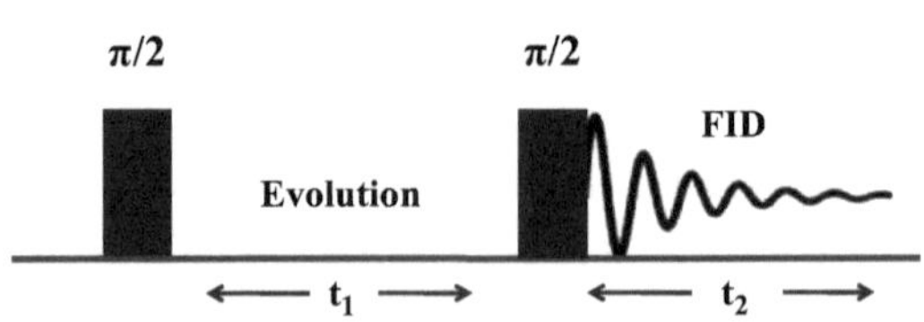

Figure 2.28 2D correlations spectroscopy-90 pulse sequence. *FID*, free induction decay.

2.18.1.1 Correlations Spectroscopy of Ethyl Acetate

The COSY-90 spectrum for ethyl acetate is shown in Fig. 2.29.

The important characteristics observed are as follows:

1. *Diagonal peaks*: It arises because of the fact that every proton is coupled to itself. These are corresponding peaks in 1D spectra.

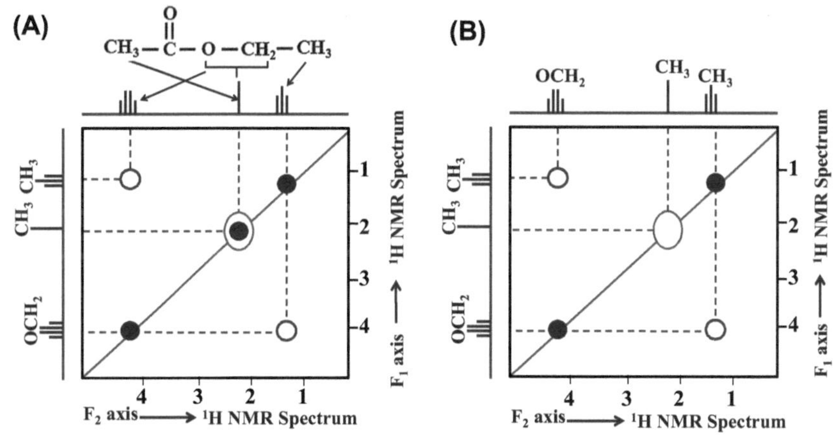

Figure 2.29 Correlations spectroscopy (COSY) (A) and double-quantum filtered COSY (B) of ethyl acetate.

2. *Cross-peaks*: In this case, there is an ethyl group, which means that a methyl and a methylene share a *vicinal* coupling (protons an adjacent carbon atoms). So, there is magnetization transfer due to this coupling, giving rise to cross-peaks or off-diagonal peaks (open circle). The antiphase character of COSY cross-peaks leads to cancellation of signal intensities for small couplings (Fig. 2.30).
3. *Isolated peaks*: Not every proton has large enough cross-couplings to give off-diagonal peaks. In general, COSY gives cross-peaks for ^2J and ^3J couplings. Therefore, methyl protons in ethyl acetate, which does not couple with any other protons, lead to isolated peaks.
4. *Shape*: The spectrum is in square shape because we are correlating a spectrum to itself.

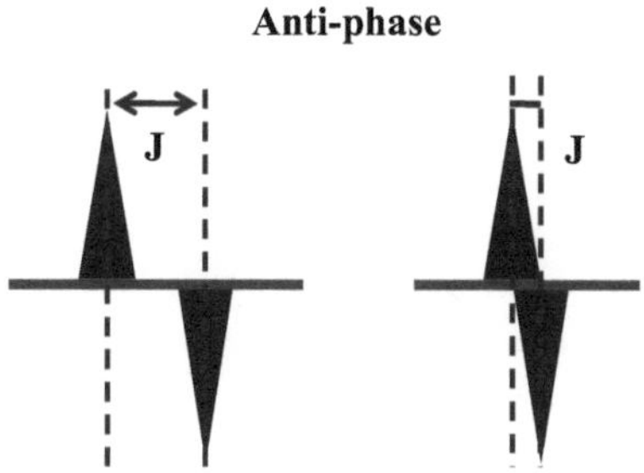

Figure 2.30 The antiphase character of correlations spectroscopy cross-peaks.

5. *1D Curves*: These are *not* part of the 2D experiment. They do not have to be drawn at all, but it is customary to add them for reference and interpretation.
6. *Symmetry*: Because A is coupled to B and then B is also coupled to A, there is reflection symmetry about the diagonal.

In addition to COSY-90, there are COSY-45 and COSY-60, where the second 90 degrees pulse is replaced by 45 or 60 degrees, respectively (Fig. 2.28). Although COSY-45 is slightly less sensitive than COSY-90 (by about 15%), COSY-45 gives more simplified spectrum than COSY-60 and COSY-90 and is therefore sometimes preferred [19].

2.18.1.2 Double-Quantum Filtered Correlations Spectroscopy

There are two disadvantages of COSY experiment as follows:

1. When the coupling constant becomes comparable with the line width, the positive and negative peaks in the cross-peak multiplet begin to overlap and cancel one another out (Fig. 2.29B). This leads to an overall reduction in the intensity of the cross-peak multiplet, and ultimately the cross-peak disappears into noise in the spectrum.
2. In addition, the dispersive tails of these diagonal peaks spread far into the spectrum. The result is a broad intense diagonal, which can obscure nearby cross-peaks.

These difficulties are overcome by a modification called double-quantum filtered COSY (DQF COSY). The pulse sequence is shown as follows (Fig. 2.31).

Up to the second pulse, the sequence is the same as COSY. However, it is arranged that only double-quantum coherence present during the (very short usually a few microseconds) delay between the second and third pulses is ultimately allowed to contribute to the spectrum. Hence the name "double-quantum filtered," as all the observed signals are filtered through double-quantum coherence. The final pulse is needed to convert the double-quantum coherence back into observable magnetization. Here, the uncoupled, i.e., isolated peaks are filtered and disappeared (Fig. 2.29B).

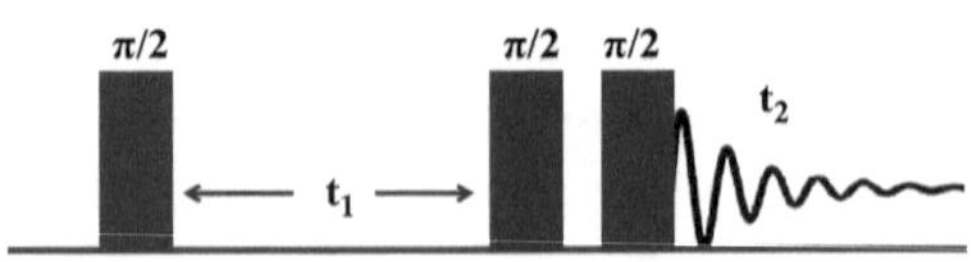

Figure 2.31 Double-quantum filtered correlations spectroscopy pulse sequence.

2.18.2 [^1H, ^1H]—Total Correlation Spectroscopy

Similar to the COSY, total correlation spectroscopy (TOCSY) is also a homonuclear, shift-correlated 2D NMR experiment, in which coherence transfer takes place via scalar couplings (through bond). The TOCSY spectrum is useful in identifying all of the protons strongly and weakly coupled including ^3J and ^4J, which belong to an isolated spin system. It uses a spin-lock pulse for coherence transfer (Fig. 2.32). Like the COSY (and nuclear Overhauser effect spectroscopy (NOESY)) spectra, the TOCSY spectrum also has peaks along a diagonal at the frequencies of all of the resonances in the 1D spectrum. The experiment relies on all spin—spin coupling rather than showing pairs of nuclei, which are only directly and strongly coupled together (COSY). The TOCSY shows a cross-peak (off-diagonal peak) for every nucleus, which is a part of the spin system not just those that are directly coupled.

During the spin-lock, all the protons including weakly coupled in a coupled spin system become "strongly coupled," leading to cross-peaks between all resonances of a coupled system. It has advantages over COSY in that even small couplings can be detected (in COSY small couplings give very weak cross-peaks because the peaks are antiphase). In contrast to the COSY, correlation between a spin and all other spins from the same spin system may be observed.

The TOCSY spectrum has a diagonal set of peaks (filled circles) as well as peaks that are off the diagonal (open circles). The schematic TOCSY spectrum is shown in Fig. 2.33 for two isolated different spin systems (H_{A1}, H_{A2}) and (H_{X1}, H_{X2}). The cross-peaks clearly indicate which resonances belong to each spin system. For example, correlations from the amide proton will, under favorable conditions, include all side-chain protons from the same amino acid (e.g., for alanine shown in Fig. 2.34). Correlations from the anomeric proton in sugars could display correlations to all other protons from the same sugar unit.

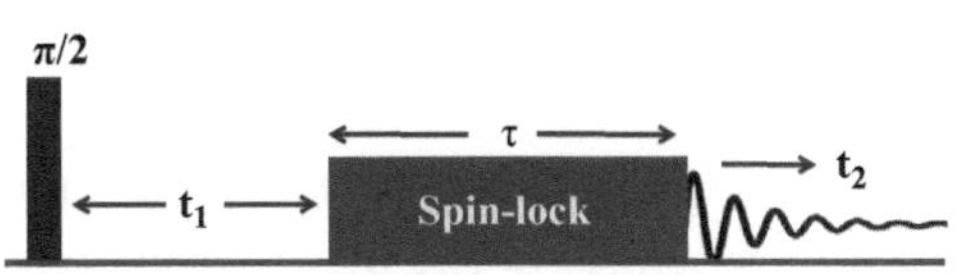

Figure 2.32 Total correlation spectroscopy and rotating frame nuclear Overhauser effect spectroscopy pulse sequence diagram.

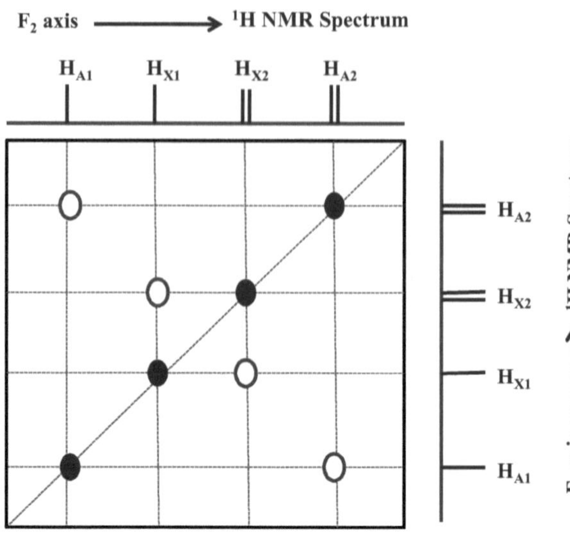

Figure 2.33 Schematic representation of a total correlation spectroscopy spectrum of two isolated spin systems.

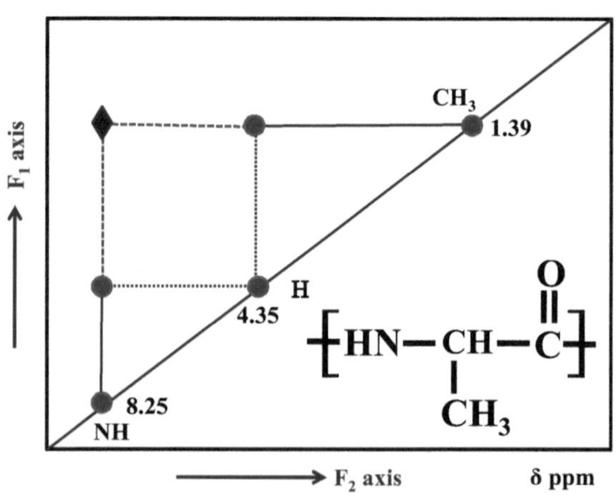

Figure 2.34 Total correlation spectroscopy (TOCSY) spectrum of alanine including correlations spectroscopy (·) and TOCSY peaks (◆).

2.18.2.1 Characteristics of Total Correlation Spectroscopy

1. By proper choice of the length of the mixing time (e.g., 10 ms), coherence transfer can be limited to just vicinal correlations, thus resulting in a COSY-like spectrum. For 80 ms mixing time, the correlations between all protons of the spin system can be studied.

2. Because the multiplet components are in-phase here, signal cancellation does not occur even if line widths become larger.
3. A short mixing time TOCSY is preferable to a COSY for larger molecules. For example, the schematic spectrum showing COSY and TOCSY peaks for alanine is shown in Fig. 2.34.

The rotating frame nuclear Overhauser effect spectroscopy (ROESY) pulse sequence is basically the same as the TOCSY experiment. Therefore, the effects from TOCSY will appear in the ROESY spectrum and vice versa.

2.18.3 [^{1}H, ^{1}H]—Exchange Spectroscopy

The exchange spectroscopy (EXSY) experiment is a homonuclear, shift correlated spectroscopy where coherence transfer takes place through chemical or conformational exchange. For example, the transfer of magnetization, however, is due to chemical exchange between methyl protons 1 and 2 in *N*, *N*-dimethylacetamide as a result of free rotation about N—C bond (Fig. 2.35). EXSY uses an identical pulse sequence to the NOESY experiment (Fig. 2.36). Because exchange is usually faster than the NOE buildup, shorter mixing times may be used for the EXSY. From recording a series of EXSY spectra with different mixing times, kinetics of exchange process may be obtained.

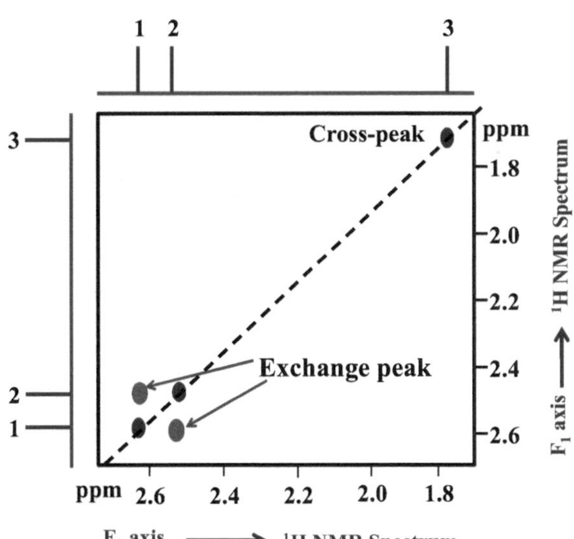

Figure 2.35 2D ^{1}H exchange spectroscopy spectrum of N,N-dimethylacetamide.

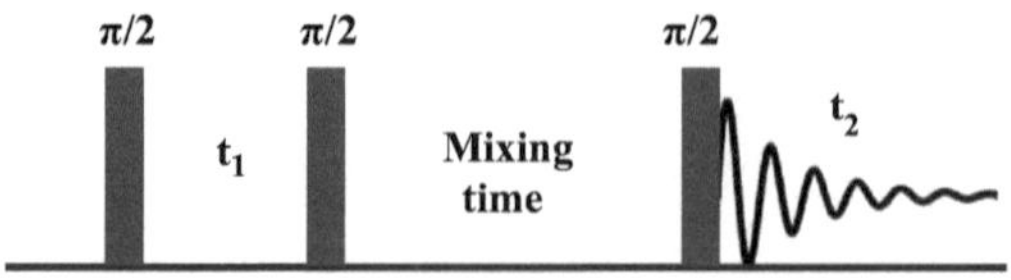

Figure 2.36 Nuclear Overhauser effect spectroscopy pulse sequence.

2.18.4 [^{1}H, ^{1}H]—Nuclear Overhauser Effect Spectroscopy

The underlying principle of the NOESY is the NOE. The NOESY is a homonuclear, shift correlated spectroscopy, in which cross-peaks result from *dipolar (spatial)* interactions between spins (not through scalar coupling). Dipolar couplings occur only through space interactions and thus depend on the distance between the spins but not on the number of intervening bonds. They are largely observable for nuclei separated by up to 4–5 Å. The efficiency of the NOE transfer additionally depends on the motional properties of the molecule (see Section 2.16). Therefore, the NOEs are generally stronger for larger molecules when there are restricted motions.

The NOESY spectrum has a diagonal set of peaks (filled circles) as well as peaks that are off the diagonal (open circles). The off-diagonal peaks occur at positions where a proton on the F_1 axis is close in space to a one on the F_2 axis. The schematic spectrum is shown in Fig. 2.37 where the off-diagonal signals indicate that H_A must be spatially located near H_B.

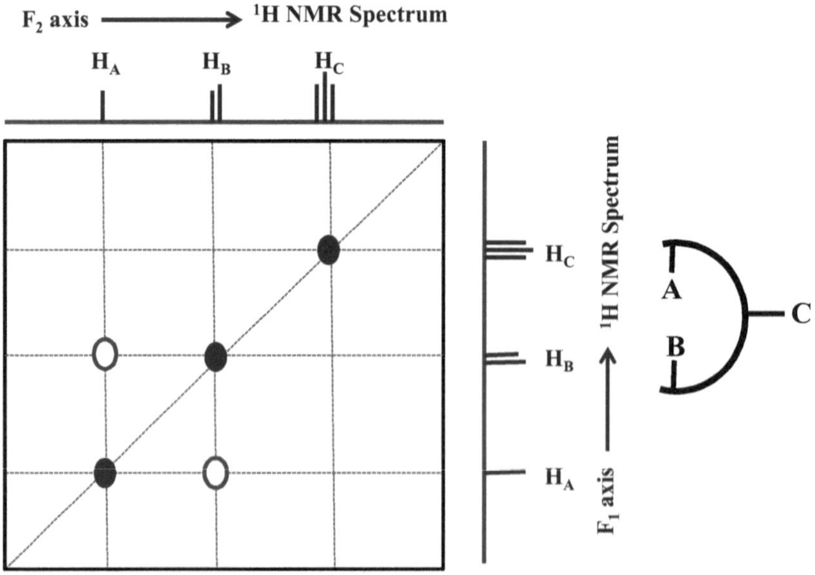

Figure 2.37 A schematic of nuclear Overhauser effect spectroscopy (NOESY) spectrum.

The NOESY spectrum is particularly useful for identifying the following:

1. The stereochemistry (e.g., the *cis/trans* configuration of a double bond or a ring junction) of a molecule where more than one possible stereoisomer exists.
2. It also gives the details of the residues that are adjacent to each other. It is hence used for sequence assignment in proteins.

2.18.5 [^1H, ^1H]—Rotating Frame Nuclear Overhauser Effect Spectroscopy

The ROESY experiment also called CAMELSPIN experiment is like the NOESY, a 2D homonuclear shift-correlated spectroscopy, in which coherence transfer is achieved through dipolar couplings for protons separated by less than 5 Å. Rotating frame Overhauser effect spectroscopy is an experiment in which homonuclear NOE effects are measured under spin-locked conditions. ROESY is especially suited for molecules with motional correlation times (τ_c) such that $\omega_{\tau c} \sim 1$, where ω is the angular frequency $\omega = \gamma B$. In such cases, the laboratory-frame NOE is nearly zero, but the rotating-frame NOE (or ROE) is always positive and increases monotonically for increasing values of τ_c. In ROESY, the mixing time is the spin-lock period (Fig. 2.32). During this time, spin exchange occurs in transverse plane among spin-locked magnetization components of different nuclei (whereas spin exchange in NOESY occurs while magnetization is aligned along the z-axis). Different spectral density functions are relevant for ROESY than for NOESY, and these cause the ROEs to be positive for all values of τ_c. The ROE buildup is twice as fast as the NOE buildup. The mixing time (t_m) should never be longer than 300 ms so as to not overheat the sample or spectrometer. For quantitative ROESY, a shorter mixing time should be preferably used.

2.18.5.1 Characteristics of Rotating Frame Nuclear Overhauser Effect Spectroscopy

The dipolar longitudinal cross-relaxation, also the NOE efficiency (ε), plotted against the correlation time $\left(\omega^2\tau_c^2\right)$ undergoes a zero crossing (Fig. 2.38). Thus, cross-peaks in a NOESY spectrum will have a different sign for small and large molecules and may disappear altogether for medium-sized molecules. This can be circumvented by using ROE. If a spin-locking field is applied to the spin system, the spins are static in the rotating frame and transversal cross-relaxation can happen. ROESY is also called as CAMELSPIN experiment (the latter acronym stands for cross-relaxation appropriate for minimolecules emulated by spin-locking).

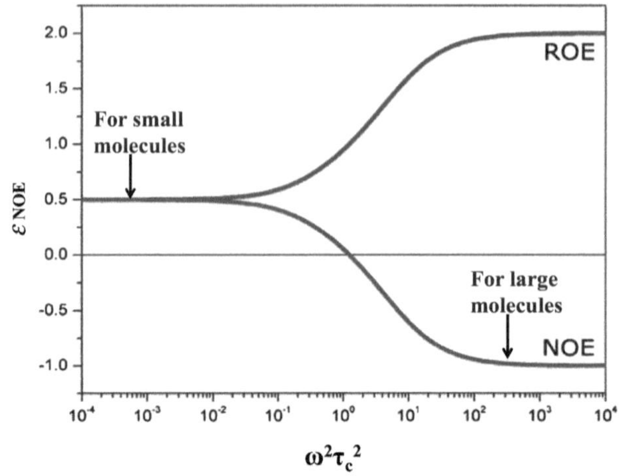

Figure 2.38 Effect of nuclear Overhauser effect (NOE) and rotating-frame NOE (ROE) on correlation time (τ_c).

The ROESY pulse sequence is basically the same as the TOCSY experiment. Therefore, the effects from TOCSY will appear in the ROESY spectrum and vice versa. To minimize the coherence transfer through J-couplings, the ROESY uses spin-locking pulses with less power. These are slightly off-resonance and reduce the magnetic equivalence and strong coupling effects of the TOCSY.

2.18.5.2 Advantages of Rotating Frame Nuclear Overhauser Effect Spectroscopy

1. It offers a simple way to obtain NOE information in a molecule by a single experiment and without prior knowledge of the spectral assignment or molecular structure.
2. In ROESY experiment, cross-relaxations are carried out in the rotating frame with spin-locked magnetization and this means that NOE in the transverse plane (ROE) is always positive (no nulling condition as in NOESY-type experiments) and,
3. In addition, chemical exchange can always be distinguished.

2.18.6 [^{13}C, ^{13}C]—Incredible Natural-Abundance Double-Quantum Transfer Experiment

Similar to the COSY, the incredible natural-abundance double-quantum transfer experiment (INADEQUATE) provides 2D spectra of the homonuclear, shift correlated type. INADEQUATE experiments are mainly used for correlation of ^{13}C, ^{13}C natural abundance ^{13}C molecules. It yields one-bond (J_{C-C}) correlations, i.e., carbon—carbon connectivity via spin—spin coupling. The INADEQUATE contains a very efficient filter to suppress ^{12}C signals of ^{13}C, ^{12}C isotopomers. This filter selects ^{13}C, ^{13}C double-quantum coherences, which can only be formed by a pair of directly coupled ^{13}C nuclei. The spectrum does not appear in the same form as COSY spectrum. There are no diagonal peaks because of double-quantum frequency, not single-quantum frequency. In F_1 axis, the double-quantum frequencies are recorded, i.e., the sum of the frequencies of the two carbons connected and F_2 axis is the ^{13}C chemical shift. Because ^{13}C, ^{13}C isotopomers are very rare ($0.01 \times 0.01 = 0.0001$) in ^{13}C natural abundance molecules, extremely concentrated samples are needed. This experiment is very useful for highly substituted compounds, in which proton density is low.

The 2D-INADEQUATE spectrum of *n*-butanol is shown in Fig. 2.39. The dashed line is pseudodiagonal. Here, the cross-peaks are split into doublets by the one-bond C—C coupling. The two coupled resonances can be recognized as two separate peaks at a common frequency in F_2 (on a horizontal line). In certain cases, one of the two peaks is missing due to low signal-to-noise. From the spectrum, it is observed that C_1 is attached to C_2. On jumping on the vertical axis from the C_2 carbon, the coupled partner

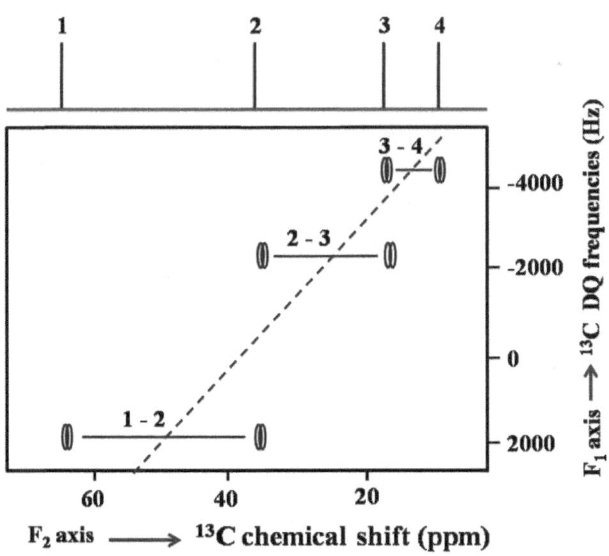

Figure 2.39 The 2D incredible natural-abundance double-quantum transfer experiment spectrum of *n*-butanol.

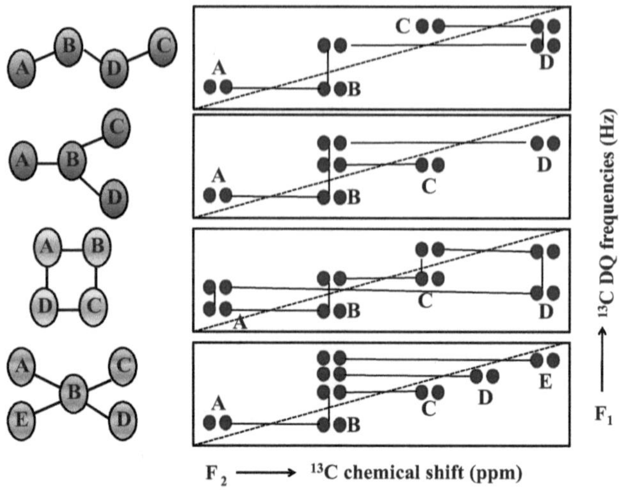

Figure 2.40 Schematic [^{13}C, ^{13}C] incredible natural-abundance double-quantum transfer experiment spectra for different carbon skeletons.

for C_2 is C_3 in the horizontal axis. Again going up vertically for C_3, its coupled partner is C_4 identified in the horizontal axis. The following Fig. 2.40 displays the schematic diagram of INADEQUATE for linear, branched, and cyclic carbon topologies. From the spectrum, each topology gives different cross-peaks useful for its identification.

2.18.7 Heteronuclear Correlation Spectroscopy

It is a two-dimensional heteronuclear correlation spectroscopy (HETCOR) in which one dimension is the chemical shift of one type of nucleus (e.g., proton) and the other dimension is the chemical shift of another nucleus (e.g., carbon-13) and are coupled to one another. Such spectra are often called shift correlation maps or shift correlation spectra. The one-bond coupling between the carbon-13 and proton directly attached to it is relatively constant (around 150 Hz). By utilizing this large difference, experiments can be devised, which give maps of carbon-13 shifts versus the shifts of directly attached protons. Hence, these spectra are very useful as aid to assignment; for instance, if the proton spectrum has already recorded and assigned, then simply recording a carbon-13 proton correlation experiment will give the assignment of all the protonated carbons. The schematic diagram of HETCOR is shown in Fig. 2.41. In this structure, H_A and H_B are coupled to C_Y, and H_C is coupled to C_X.

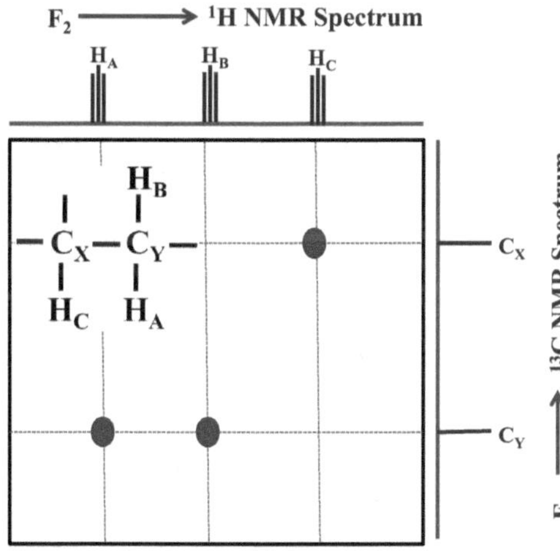

Figure 2.41 Heteronuclear correlation spectroscopy [^{1}H, ^{13}C] spectrum.

Only one kind of nuclear species can be observed at a time, so there is a choice as to whether to observe carbon-13 or proton when recording a shift correlation spectrum. Because HETCOR involves direct detection of ^{13}C nuclei, it is less sensitive than heteronuclear single quantum coherence (HSQC), heteronuclear multibond connectivity (HMBC), *etc.* For two reasons, it is very advantageous from the sensitivity point of view to record protons in the case of HSQC, HMBC, heteronuclear multiple quantum correlation (HMQC). (1) The proton magnetization is larger than that of carbon-13 because there is a larger separation between the spin energy levels giving, by the Boltzmann distribution, a greater population difference. (2) A given magnetization induces a larger voltage in the coil, as the NMR frequency becomes higher.

2.18.7.1 [^{1}H, ^{13}C]—Heteronuclear Single Quantum Coherence

The HSQC experiment is the most widely used heteronuclear shift correlation experiment. Compared with HETCOR, this method, including HMBC and HMQC, involves indirect (inverse) detection of H-nuclei. So, these techniques are more sensitive. This is also a CH correlation experiment that uses proton detection of the ^{13}C signals using an INEPT sequence (see Section 2.14). The nuclei usually separated by one bond are correlated via their scalar couplings (^{1}J). In [^{13}C, ^{1}H]-HSQC spectra, however, no correlations to quaternary carbons are observable. Because the one-bond proton—carbon or proton—nitrogen couplings are large and

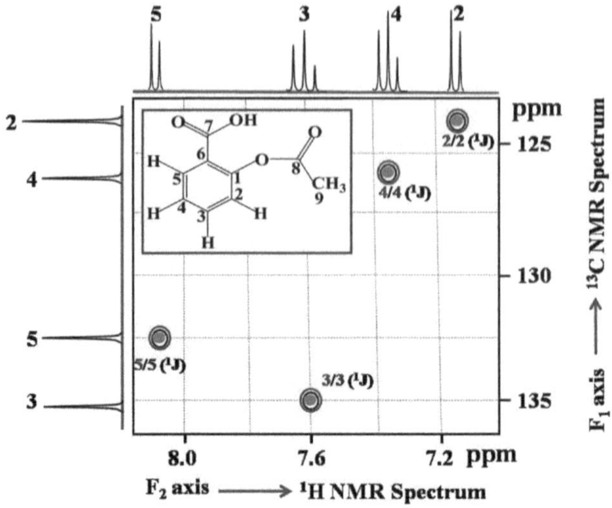

Figure 2.42 [^{1}H, ^{13}C] Heteronuclear single quantum coherence spectrum of aspirin and its structure.

rather uniform, these spectra are quite sensitive. It shows higher resolution in the C-dimension than does the related HMQC experiment (vide infra). For example, the HSQC spectrum of aspirin and its structure are shown in Fig. 2.42.

The type of spectrum is essentially observed the same as a COSY but herewith the correlation between two different nuclei. This means that there is no diagonal, but the correlation is identified by intensity at the cross-point.

2.18.7.2 [^{1}H, ^{13}C]—Heteronuclear Multibond Connectivity

The HMBC experiment like the HSQC gives heteronuclear shift-correlation spectra, but it is optimized to detect proton-carbon correlation for two (^{2}J) and/or three (^{3}J) bonds. In contrast to the HSQC, coherences are transferred through the much smaller long-range couplings. The long-range couplings span rather a wide range. The [^{1}H, ^{13}C] ^{3}J coupling, for example, displays a Karplus-type dependence on the dihedral angle (see Section 2.4.2). Depending on the system under study, the ^{2}J or the ^{3}J coupling may be larger, so that these spectra contain much ambiguity. Furthermore, this method uses relatively long delay times and thus much opportunity for loss of signal intensity by T_1 and T_2 relaxation. However, the HMBC is a very useful experiment because it contains correlations to quaternary carbons. If correlations over several bonds must be observed, a filter is used to remove all ^{1}J coupling of $J > 10$ Hz. This is known as HMBC or long-range HETCOR.

The HMBC is thus used to help establish the carbon skeleton through the multiple bond carbon to hydrogen connectivity. This experiment is relatively insensitive as compared with HMQC because multiple bond correlations are less efficient than one-bond correlations. Typical one-bond coupling constants are around 150 Hz, whereas multiple-bond coupling constants fall in the range of 5−15 Hz, which is similar to the range for H, H-homocoupling.

2.18.7.3 Heteronuclear Multiple Quantum Correlation

The **2D HMQC** experiment permits to obtain a 2D heteronuclear chemical shift correlation map between directly bonded ^{1}H and X-heteronuclei (commonly, ^{13}C and ^{15}N). It is widely used because it is based on proton detection, offering high sensitivity when compared with the conventional carbon-detected 2D HETCOR experiment. Similar results are obtained using the 2D HSQC experiment.

The advantage of HMQC over HETCOR is that in HMQC the nucleus with the highest γ (^{1}H) is detected, and so it is possible to obtain the highest sensitivity. The challenge of an inverse chemical shift correlation experiment, however, is that the large signals from ^{1}H's not coupled directly to a ^{13}C nucleus must be suppressed in a difference experiment. This poses a dynamic range problem: the signal of interest is that of ^{1}H's coupled directly to ^{13}C nuclei. However, the signal detected is dominated by the contribution of ^{1}H's bonded directly to ^{12}C nuclei. HMQC, which is based on multiple-quantum NMR, minimizes this dynamic range problem while optimizing the sensitivity of the experiment. The resonance frequency of low γ spins can be detected with enhanced sensitivity by the creation and ^{1}H detection of $^{1}H-^{13}C$ (or other X nucleus) multiple-quantum coherence.

For inverse detection methods, we can see that carbons are directly or remotely attached to protons. This means that quaternary carbons that do not have any attached protons do not appear in inverse detection experiments.

2.19 3D NUCLEAR MAGNETIC RESONANCE

A three dimensional NMR experiment can be readily constructed from a two-dimensional experiment. Here, an additional indirect evolution

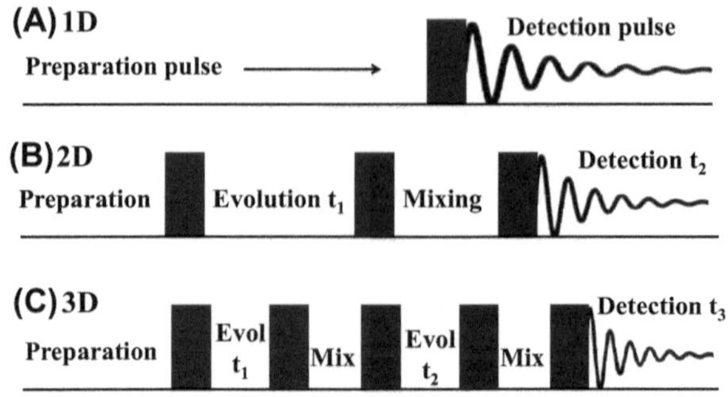

Figure 2.43 Pulse sequence diagram of (A) 1D, (B) 2D, and (C) 3D NMR.

time and a second mixing period are inserted between the first mixing period and the direct data acquisition (see Fig. 2.43). Each of the different indirect time periods (t_1, t_2) is incremented separately.

There are two principal classes of 3D experiments:

1. Experiments consisting of two 2D experiments one after another
2. The triple resonance experiments

2.19.1 3D NOESY–HSQC and TOCSY–HSQC

2D spectra (NOESY or TOCSY) of proteins are largely crowded with signals. Therefore, these spectra are spread out in a third dimension (usually ^{15}N or ^{13}C). Here, the signals are distributed in a cube instead of a square plane. This spreading out of signals is achieved by combining HSQC and NOESY in a single 3D experiment: The NOESY experiment is extended by an HSQC step. Instead of at the end of the NOESY mixing time, after the HSQC sequence, acquisition starts. The resulting experiment is known as 3D NOESY–HSQC. Similarly, a TOCSY–HSQC could be constructed by combining the TOCSY and the HSQC experiment into one 3D experiment.

2.19.1.1 HCCH–TOCSY

The HCCH–TOCSY and HCCH–COSY experiments are often used as alternatives to the ^{13}C-TOCSY–HSQC experiment (Fig. 2.44), which shows a markedly reduced sensitivity for larger proteins. In both the experiments, the magnetization is transferred via direct ^{1}J couplings between the atoms, allowing a much faster magnetization transfer than in the TOCSY–HSQC. The magnetization is transferred from a side-chain (or backbone) proton (red) to the directly attached carbon atom by ^{1}J coupling to the neighboring carbon atoms and finally to their attached protons. Thus, the HCCH–TOCSY looks in principle like a ^{13}C-TOCSY–HSQC experiment. The type of an amino acid can be recognized from the peak pattern in the usual way, as can be seen from the picture below.

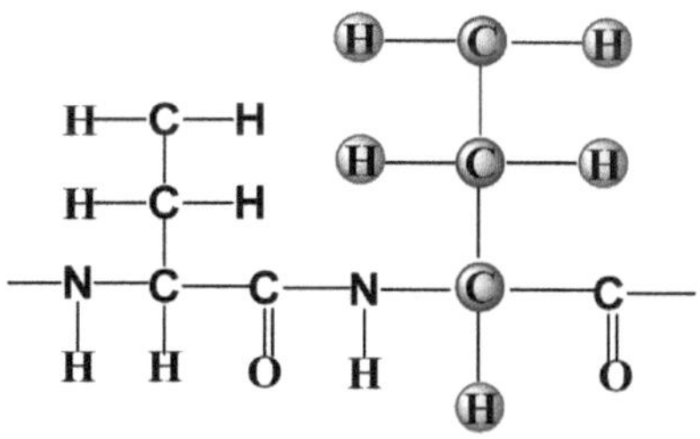

Figure 2.44 HCCH–total correlation spectroscopy experiment.

2.19.2 Triple Resonance Experiments

Triple resonance experiments are widely used as the method of choice for the sequential assignment of amino acids in large proteins (>150 AA). These experiments are known as "triple resonance" because three different nuclei (viz., ^1H, ^{13}C, ^{15}N) are correlated. The experiments are carried out on doubly labeled (^{13}C, ^{15}N) proteins. The most significant advantage of the triple resonance spectra is their simplicity. They contain only a few signals on each frequency—largely only one. Hence, the spectral overlapping is produced and therefore greatly reduced (this is the main reason why proteins of more than 20 kDa can be readily assigned with triple resonance experiments). However, the coordinates of clearly separated signals from different amino acids can accidentally be identical due to "degeneration" of signals. Therefore, the correct choice of connectivities between amino acids is a major drawback in the assignment of triple resonance spectra. Another significant advantage of triple resonance spectra is their high sensitivity, which is caused by an efficient transfer of magnetization especially via ^1J or ^2J couplings (i.e., directly via the covalent chemical bonds). Hence, the transfer times are shorter and the losses due to relaxation process are lesser than homonuclear experiments. The disadvantage of all triple resonance experiments is the necessity of doubly labeled proteins, the preparation of which is often expensive [20].

2.19.2.1 Nomenclature of Triple Resonance Experiments

The names of triple resonance experiments sound very cryptic at first glance, but they are very descriptive: The names of all nuclei, which are used for

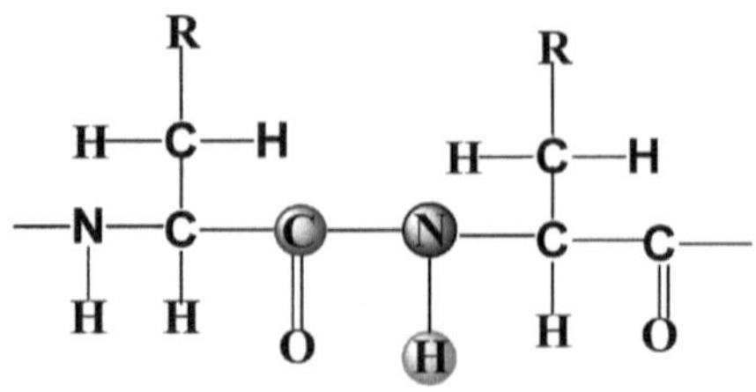

Figure 2.45 HNCO experiment.

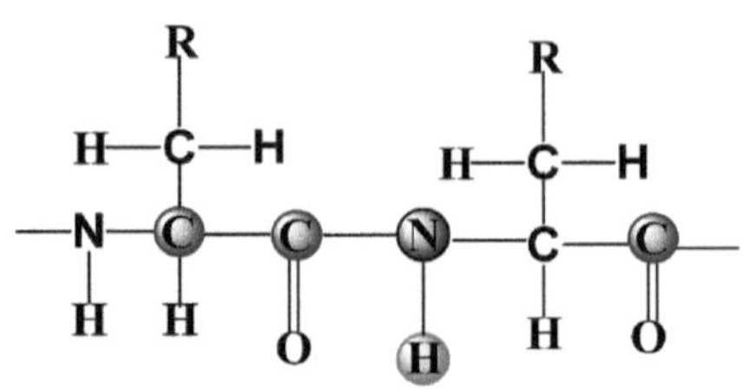

Figure 2.46 HN(CA)CO experiment.

magnetization transfer during the experiment, are listed in the order of their use, bracketing the names of nuclei, which are used only for transfer and whose frequencies are not detected. In the HNCO experiment (Fig. 2.45), the magnetization is transferred from the H^N(i) proton via the N(i) atom to the directly attached CO(i-1) carbon atom and returns the same way to the H^N(i) nucleus, which is directly detected. The frequencies of all three nuclei (red) are detected.

In the HN(CA)CO experiment (Fig. 2.46), the magnetization is transferred from the H^N(i) proton via the N(i) atom and the CA nucleus (C^{alpha}(i)) to the CO(i) carbon atom and back the same way. The CA atom (yellow) acts only as relay nucleus, and its frequency is not detected. It is only the frequencies of H^N, N, and CO (red) that are detected. This nomenclature has the advantage that the spectra can be easily imagined by their names: In HNCO, an amide proton is correlated with the CO atom of the proceeding amino acid, whereas in HN(CA)CO the correlation to the intraresidual CO atom is also visible.

2.19.2.2 The HNCA Experiment

The HNCA experiment is the prototype of all triple resonance experiments. Starting at an amide proton (H) the magnetization is transferred to the directly attached nitrogen atom (N), which is measured as the first spectral dimension (Fig. 2.47). Then the magnetization is transferred to the CA

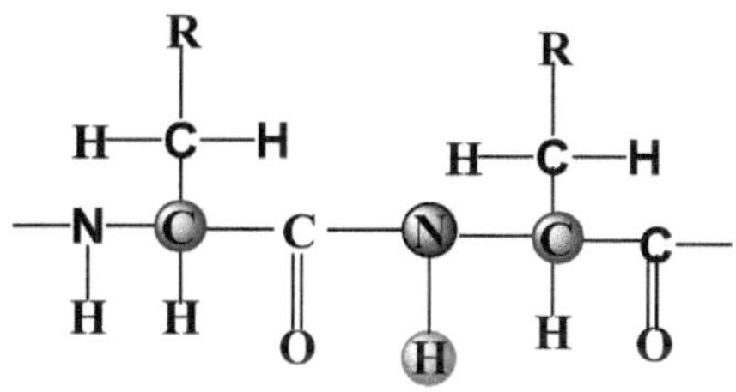

Figure 2.47 HNCA experiment.

nucleus, which is measured as second dimension. Afterward, the magnetization is transferred back the same way to the amide proton, which is measured as the third (direct) dimension.

In each step, the magnetization is transferred *via* strong ^{1}J couplings between the nuclei. The coupling that connects the nitrogen atom with the CA carbon of the proceeding amino acid ($^{2}J = 7$ Hz) is only marginally smaller than the coupling to the directly attached CA atom ($^{1}J = 11$ Hz). Thus, the nitrogen atom of a given amino acid is correlated with both CA, its own, and the one of the proceeding amino acid. Therefore, it is possible to assign the protein backbone exclusively with an HNCA spectrum. But usually more triple resonance experiments are needed because the cross-signals of the proceeding amino acid has to be identified and degenerate resonance frequencies have to be resolved.

REFERENCES

[1] L.D. Field, S. Sternhell, J.R. Kalman, Organic Structures from Spectra, fourth ed., John Wiley and Sons, Ltd., 2008.
[2] R.M. Silverstein, F.X. Webster, D.J. Kiemle, D.L. Bryce, Spectrometric Identification of Organic Compounds, eighth ed., Wiley, New York, USA, 2014.
[3] R.S. Macomber. A Complete Introduction to Modern NMR Spectroscopy, Wiley Interscience, New York, 1998.
[4] E. Breitmaier, G. Bauer, ^{13}C-NMR-Spektroskopie, Thieme Verlag, Stuttgart, 1977.
[5] L.M. Jackman, S. Sternhell, Application of NMR Spectroscopy in Organic Chemistry, second ed., 1969, pp. 281–283. Pergammon.
[6] H.J. Reich, Symmetry in NMR, 2017. https://www.chem.wisc.edu/areas/reich/nmr/05-hmr-08-symmetry.htm.
[7] H.J. Reich, Measuring ^{13}C NMR Spectra, University of Wisconsin, 2017. https://www.chem.wisc.edu/areas/reich/nmr/06-cmr-01-spectra.htm.
[8] T.D.W. Claridge, High Resolution NMR Techniques in Organic Chemistry, Pergamon Press, 1999.
[9] J. Cavanagh, W.J. Fairbrother, A.G. Palmer III, N.J. Skelton, Protein NMR Spectroscopy, Principles and Practice, Academic Press, 1996.
[10] J.D. Baldeschwieler, E.W. Randall, Chemical applications of nuclear magnetic double resonance, Chem. Rev. 63 (2) (1963) 81–110.
[11] G. Facey, INEPT. http://u-of-o-nmr-facility.blogspot.in/2016/04/inept.html.

[12] R. Freeman, A Handbook of Nuclear Magnetic Resonance, second ed., Longman Scientific & Technical (Wiley), 1997.
[13] G.A. Morris, R. Freeman, Enhancement of nuclear magnetic resonance signals by polarization transfer, J. Am. Chem. Soc. 101 (3) (1979) 760–762.
[14] I.V. Khudyakov, N. Arsu, S. Jockusch, N.J. Turro, Magnetic and spin effects in the photo initiation of polymerization, Des. Monomers Polym. 6 (1) (2003) 91–101.
[15] H.J. Reich, NOE, 2016. https://www.chem.wisc.edu/areas/reich/nmr/08-tech-02-noe.htm.
[16] J.K.M. Sanders, B.K. Hunter, Modern NMR Spectroscopy: A Guide for Chemists, Oxford University Press, Oxford, Oxfordshire, New York, 1987.
[17] R.R. Ernst, G. Bodenhausen, A. Wokaun, Principles of Nuclear Magnetic Resonance in One and Two Dimensions, Oxford Science Publications, 1987.
[18] G.D. Mateescu, A. Valeriu, 2D NMR, Density Matrix and Product Operator Treatment, Prentice Hall, 1993.
[19] Database. https://orgspectroscopyint.blogspot.in/2014/11/cosy-ethyl-acetate.html.
[20] H.J. Schirra, 3D-NMR Spectroscopy. http://www.cryst.bbk.ac.uk/PPS2/projects/schirra/html/3dnmr.htm.

CHAPTER THREE

Applications of Nuclear Magnetic Resonance

Chandran Karunakaran, Murugesan Balamurugan, Dhinesh Kumar Marimuthu
VHNSN College (Autonomous), Virudhunagar, India

Contents

Spin Resonance Spectroscopy
ISBN: 978-0-12-813608-9
https://doi.org/10.1016/B978-0-12-813608-9.00003-4

Abstract

This chapter is focused on applications of especially ^{1}H, ^{31}P, ^{19}F nuclear magnetic resonance (NMR) to inorganic structural elucidation using spin–spin coupling and the magnitude of coupling constants in distinction of stereoisomers. It also includes dynamic NMR spectroscopy also called temperature-dependent NMR to study the dynamics and kinetics, the rate of exchange/conformational processes/fluxionality in organometallic compounds. The mechanism of broadening of lines in solid-state NMR and its elimination using magic angle spinning are demonstrated. NMR of paramagnetic molecules, shift reagents including contact shift and dipolar shift, and the determination of their magnetic susceptibility are discussed. The application of NMR principles to magnetic resonance imaging (MRI), its theory and instrumentation are discussed. The different types of images based on contrast parameters viz., T_1, T_2, and proton-weighted images and pulse sequences viz., spin echo, gradient echo, turbo spin echo, etc., are highlighted. Recent advances viz., functional MRI, magnetic resonance spectroscopy, magnetic resonance angiography and contrast agents, and their mechanism are described.

Keywords: Contrast agents; DNMR; Exchange process; fMRI; Inorganic NMR; Magic angle spinning; MRS; Shift reagents.

3.1 APPLICATIONS OF (^{1}H, ^{31}P, ^{19}F, AND ^{15}N) NUCLEAR MAGNETIC RESONANCE TO INORGANIC STRUCTURAL ELUCIDATION

3.1.1 Spin–Spin Coupling to Structure Determination

The applications of spin–spin coupling to structure determination of especially inorganic compounds are illustrated here [1].

3.1.1.1 Metal-Hydrogen Bonding in $[H{-}Rh(CN)_5]^{3-}$

In general, protons attached to metal ions are highly shielded. In such cases, the resonance occurs at 10–20 ppm on the high-field side of the water peak at 5 ppm to the low-field side of tetramethylsilane (TMS). The metal atom thus provides to local environments of high diamagnetic shielding to the proton. This shielding can be attributed to the high electron density on H from the nonpolar nature of the M–H bond and from shielding resulting from electron density in the large and diffuse 4s and 4p orbitals of the metal. The proton nuclear magnetic resonance (NMR) of $[H\text{-}Rh(CN)_5]^{3-}$ exhibit a doublet (rhodium, I = ½), because of Rh–H coupling. The center of which occurs at −10 ppm on the high-field side of TMS. This spin–spin splitting and shift clearly establishes the existence of a bond between hydrogen and rhodium.

3.1.1.2 Determination of Structure of Phosphorous Acids and Its Fluorine Derivative

The characteristic chemical shift of hydrogen bonded to phosphorous occurs in a limited range. However, the proton NMR spectra give rise to fine structure corresponding to $J_{P\text{-}H}$ (^{31}P, I = ½). The phosphorous resonance of $HPO(OH)_2$ is reported to be a doublet and for $H_2PO(OH)$ a triplet supporting the structures as in Fig. 3.1. In hypophosphorous acid, the P atom is attached to two protons, which cause its resonance to be split into a triplet in the intensity ratio 1:2:1. The coupling of the hydroxyl protons with phosphorous is either too small to be resolved, or they are far removed or else it is not observed because of a fast proton exchange reaction.

Similar results obtained for fluorine NMR thus establish the structures of $FPO(OH)_2$ and $F_2PO(OH)$ containing one and two fluorine atoms attached to phosphorus, respectively.

3.1.1.3 ^{31}P Nuclear Magnetic Resonance of P_4S_3

The ^{31}P NMR spectrum of P_4S_3 consists of two peaks with intensity ratios of three to one. The more intense peak is a doublet and the less intense a quadruplet (four lines). Since I = 0 for S, both the spin—spin splitting and the relative intensities of the peaks indicate three equivalent and one unique phosphorus atoms. It is thus concluded that P_4S_3 has the structure as given in Fig. 3.2A.

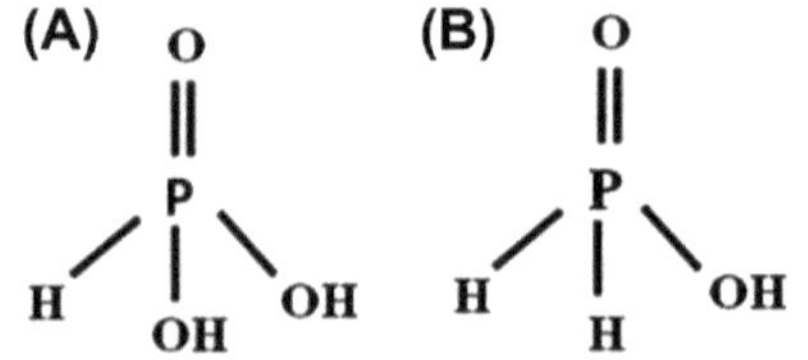

Figure 3.1 Structures of (A) phosphorous and (B) hypophosphorous acids.

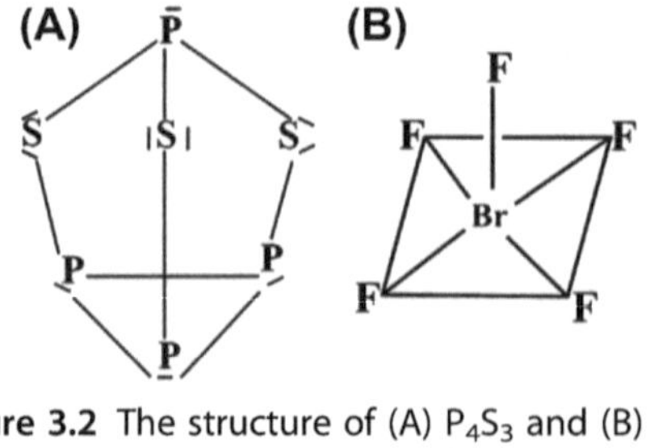

Figure 3.2 The structure of (A) P_4S_3 and (B) BrF_5.

3.1.1.4 ^{31}P Nuclear Magnetic Resonance of $[PF_6]^-$

The ^{31}P NMR spectrum of $[PF_6]^-$ consists a septet signal seven lines with coupling constant between ^{31}P and ^{19}F nuclei, $J_{P\text{-}F} = 708$ Hz. It reveals that all the fluorine atoms are equivalent. The observed large value of the coupling constants are characteristics of the directly attached nuclei. Therefore, the geometry of the complex is regular octahedron.

3.1.1.5 ^{1}H and ^{31}P Nuclear Magnetic Resonance Spectra of Phosphonium Ion PH_4^+

^{1}H and ^{31}P NMR spectra reveal that the P atom splits the resonance of the four equivalent protons in PH_4^+ ion into doublet, because both the ^{1}H and ^{31}P have I = ½, and the protons split the phosphorus resonance into a quintet in the intensity ratio 1:4:6:4:1. In other words, a doublet in ^{1}H NMR and quintet (five lines) in ^{31}P NMR were observed.

3.1.1.6 ^{31}P Nuclear Magnetic Resonance of Diphosphate Anion

Based on the proposed structure (Fig. 3.3), the observed ^{31}P NMR spectrum of diphosphate anion is interpreted using ABX system as follows [1]:

1. The two phosphorus atoms in different chemical environment give rise to two different chemical shifts represented by $\delta_{P(a)}$ and $\delta_{P(b)}$ (I).
2. Because the hydrogen is directly attached to $P_{(a)}$, $J_{P(a)\text{-}H} > J_{P(b)\text{-}H}$, the four lines obtained (II) are shown in Fig. 3.3.
3. The third step involves $P_{(a)}$–$P_{(b)}$ splitting and accounts for the final ^{31}P NMR spectrum for this ABX case (III).
4. However, two of the expected lines are not detected in the final spectrum because they are too weak (relative intensities 0.009) to be observed.
5. Furthermore, another pair falls so close together giving a single peak (with high intensity −2.3).

As the spectrum obtained is complicated, it is simplified by using double irradiation/resonance technique. The experiment is called double resonance because an additional rf field is used. For example, in an ABX type spectrum, it is difficult to extract coupling constants from the complicated NMR spectra. However, irradiation of the X nucleus yields a simple AB system from which spin–spin coupling constants can be readily obtained.

3.1.1.7 ^{31}P Nuclear Magnetic Resonance of $P_4N_4Cl_6(NHC_6H_5)_2$

The phosphorus NMR spectrum of $P_4N_4Cl_6(NHC_6H_5)_2$ shows two triplets of approximately equal intensity. This indicates that the compound possess the structure, as shown in the Fig. 3.4A, with the two aniline groups on

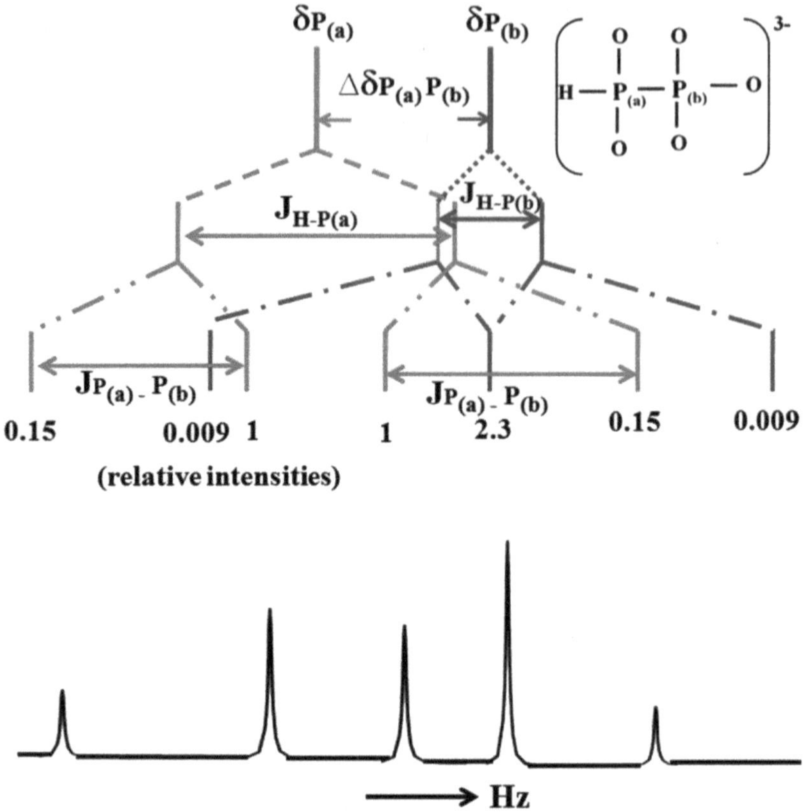

Figure 3.3 The stick diagram and phosphorus nuclear magnetic resonance spectrum of diphosphate anion $HP_2O_5^{3-}$.

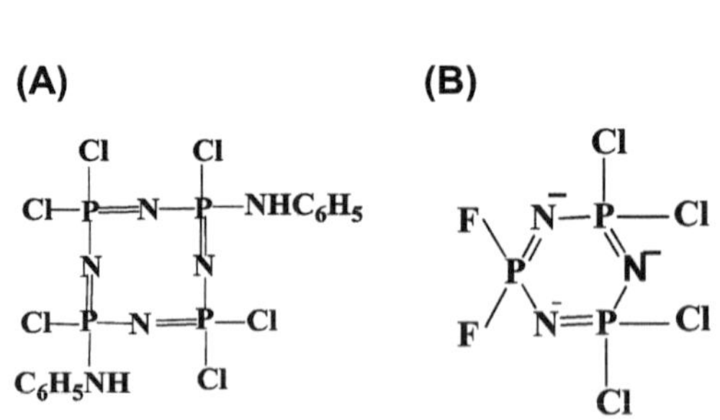

Figure 3.4 Structure of (A) $P_4N_4Cl_6(NHC_6H_5)_2$ and (B) $P_3N_3Cl_4F_2$.

the *trans*-phosphorus atoms. The aniline groups may also be considered on the same or opposite sides.

3.1.1.8 ^{31}P Nuclear Magnetic Resonance of $P_3N_3Cl_4F_2$

The phosphorus and fluorine NMR spectra of $P_3N_3Cl_4F_2$ have been investigated. The complex spectra obtained were analyzed as an AB_2X_2 spin system. Here, the two fluorine atoms are considered as X_2, the phosphorus to which the fluorines are attached is A, and the other two phosphorus atoms are B_2. From the NMR spectra, the structure of the compound was established as shown in Fig. 3.4B (1,1-difluoride 3,3,5,5-tetrachloride).

3.1.1.9 ^{19}F Nuclear Magnetic Resonance of BrF_5

The fluorine resonance in BrF_5 consists of two peaks with an intensity ratio of four to one. It indicates the presence of two kinds of F atoms in which four are one type and the fifth fluorine is another type. Furthermore, the intense line is a doublet and the weak line is a quintet (1:4:6:4:1). This supports that the molecule is a symmetrical tetragonal pyramidal geometry (Fig. 3.2B).

3.1.1.10 ^{19}F Nuclear Magnetic Resonance of TiF_6^{2-} and TiF_4

The fluorine NMR spectra of solutions of equimolar quantities of TiF_6^{2-} and TiF_4 in ethanol consist of two peaks with the intensity ratio 4:1 ($I = 0$ for Ti). The low-intensity peak is a quintet and the more intense peak a doublet. Hence, the structure $[TiF_5(HOC_2H_5)]^-$ containing octahedrally coordinated titanium was proposed [1].

The ^{19}F NMR spectra of solutions of TiF_4 in donor solvents recorded at −30°C exhibit two triplets of equal intensity. This is the spectrum expected for the *cis* isomer shown in Fig. 3.5. It consists of two sets of nonequivalent fluorine atoms with two equivalent fluorines in each. However, at 0°C

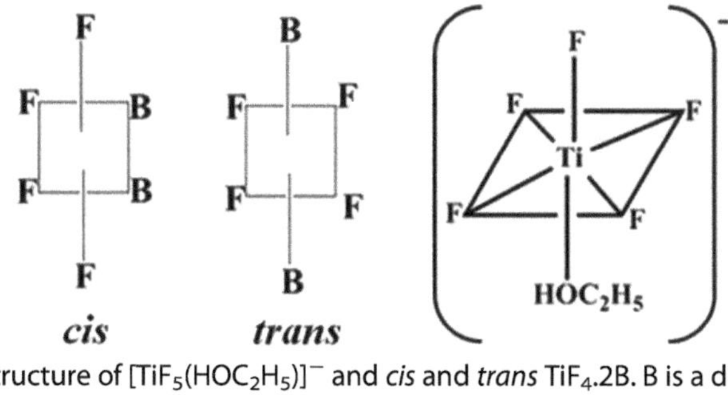

Figure 3.5 Structure of $[TiF_5(HOC_2H_5)]^-$ and *cis* and *trans* $TiF_4.2B$. B is a donor molecule.

only a single fluorine peak is observed. It is proposed that a rapid dissociation/association reaction occurs making all fluorines equivalent. However, at low temperature (−30°C), this reaction is slowed down so that the nonequivalence can be detected by NMR. Internal rearrangements and ionic exchange mechanisms are possible mechanism. This example, thus, clearly illustrate one of the possible pitfalls in structure determination using NMR spectroscopy. If only the high temperature (0°C) spectrum had been monitored, it could have been incorrectly assumed that the adduct had the *trans* structure on the basis of the observed single NMR peak. If the actual structure were *trans*, only a singlet fluorine resonance would be detected at both the low and high temperatures. So, it would be difficult to draw any structural conclusion because of the possibility that rapid exchange dissociation might be occurring at both temperatures.

3.1.1.11 Complex Nuclear Magnetic Resonance of ClF_3

Complex NMR spectra are usually obtained when the spin−spin coupling constant between two nonequivalent nuclei is of the same order of magnitude as the chemical shift difference (J ~ Δ) between them. In such cases, the spectra of ClF_3 were recorded at low- and high-frequency NMR to eliminate these complications [1].

1. Using ^{19}F NMR high-frequency probe (40 MHz) where $J \ll \Delta$, the first order spectrum in Fig. 3.6A is obtained. The ClF_3 molecule consists of two long Cl−F bonds and one short one, thus giving rise to nonequivalent fluorine atoms. The spectrum obtained using high frequencies exhibits two nonequivalent fluorine atoms splitting each other. The triplet observed is half the intensity of the doublet as expected.

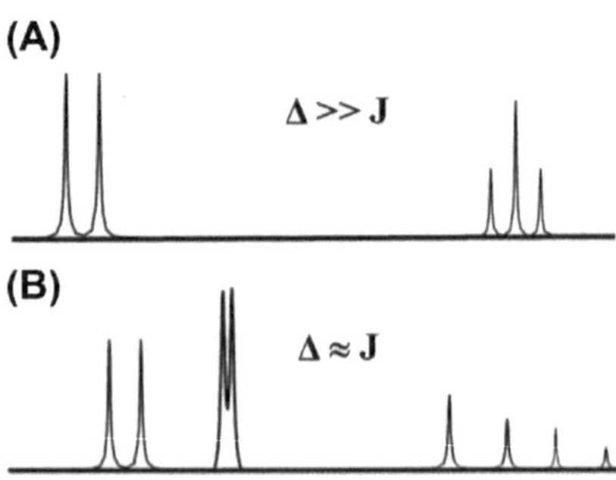

Figure 3.6 Fluorine nuclear magnetic resonance spectra of ClF_3 (A) at high (40 MHz) and (B) low frequency (10 MHz).

2. However, using a lower-frequency probe (10 MHz), the difference in Δ for the two nonequivalent fluorine is of the order of magnitude of J_{F-F} (because Δ is field dependent) and J is field independent. Hence, the complex second order spectrum is observed in Fig. 3.6B. Complex spectra result whenever the coupling constants between the nonequivalent nuclei are of the same order of magnitude of the chemical shift. So, the accurate meaningful coupling constants or chemical shifts cannot be extracted directly from the complex patterns. However, in such cases, a computer simulation of NMR spectrum is often used to obtain the NMR parameters.

3.1.1.12 Nuclear Quadrupole Relaxation Effect on Nuclear Magnetic Resonance

Another effect gives rise to spectra other than those predicted due to nuclear quadrupole relaxation of nuclei ($I \geq 1$). The spectra obtained do not exhibit expected splitting. Because, the nucleus ($I \geq 1$) to which the element being investigated is attached undergoes rapid relaxation. It results in a rapid change in the spin state of the nucleus, which is equivalent to rapid exchange. Hence, the average spin state is only detected. This rapid exchange often leads to a broadening of the resonance line. In certain instances, the proton resonance absorption is severely broadened due to this effect to such an extent that the signal cannot be distinguished from the background. The relaxation effects are largely encountered for nuclei ($I \geq 1$) which have a considerable quadrupole moment. Hence, these nuclei are very efficiently relaxed by the fluctuating electric field gradients resulting from the thermal motion of the polar solvent and solute molecules. For example, because of this effect, the proton spectrum of $^{14}NH_3$ ($^{14}N, I = 1$) consists of very three broad signals. However, the ^{1}H spectrum of $^{15}NH_3$ (^{15}N, $I = ½$) shows only a sharp doublet, where there is no quadrupole effect.

3.1.2 Applications Involving the Magnitude of Coupling Constants

Based on the magnitude of coupling constants, the stereoisomerism in inorganic complexes will be determined as follows.

3.1.2.1 Cis–trans *Isomers in $(R_3P)_2PtX_2$*

The Pt–P spin–spin coupling constants have been evaluated for a series of platinum complexes $[(R_3P)_2PtX_2]$. The platinum whose most abundant

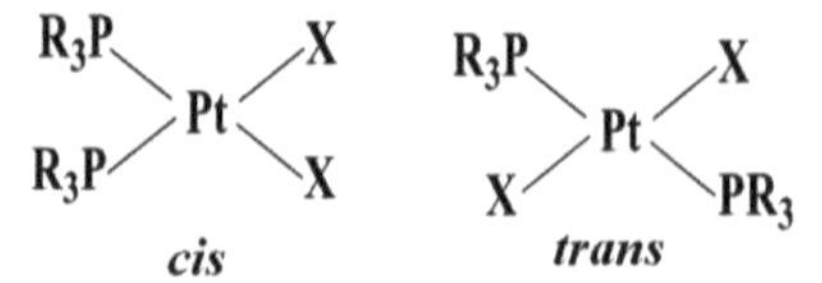

Figure 3.7 Structures of *cis* and *trans* isomeric complexes of $[(R_3P)_2PtX_2]$.

nucleus has zero spin does not show splitting in NMR spectrum. But two low-intensity peaks appear one on each side of the main phosphorus resonance due to the splitting of phosphorus caused by the ^{195}Pt isotope with spin 1/2. When *cis* and *trans* isomeric complexes exist as follows (Fig. 3.7), the coupling constant could well be used to distinguish between the two. The $J_{Pt\text{-}P}$ value is greater for the *cis* isomer than the *trans* complex. For example, in $(n\text{-}Bu_3P)_2PtCl_2$, $J_{Pt\text{-}P}$ is 3.62 Hz for the *cis* complex and 2.46 Hz for the *trans*. It is explained because of the π bonding effect on $J_{Pt\text{-}P}$. The d-orbitals of two *trans*-phosphorus atoms π overlap only with d_{xz} and d_{xy} orbitals on platinum (P located on the x-axis). On the other hand, the d orbitals of two *cis*-phosphorus atoms can overlap with d_{xy}, d_{yz}, and d_{xz}, metal-ligand π bonding is thus greater in the *cis* complex than in the *trans*. Consequently, $J_{Pt\text{-}P}$ is greater in the *cis* complex.

In the molecule $H_2C{=}CF_2$, where neither the two hydrogen nor the two fluorine can be considered equivalent (Fig. 3.8A). Each proton sees two nonequivalent fluorine (one *cis* and one *trans*), which have identical chemical shifts but different $J_{H\text{-}F}$ coupling constants. The simple 1:2:1 triplet expected (when both protons and both fluorine atoms are equivalent) is not observed in either the fluorine or proton NMR because of the magnetic nonequivalence that exists in the J values.

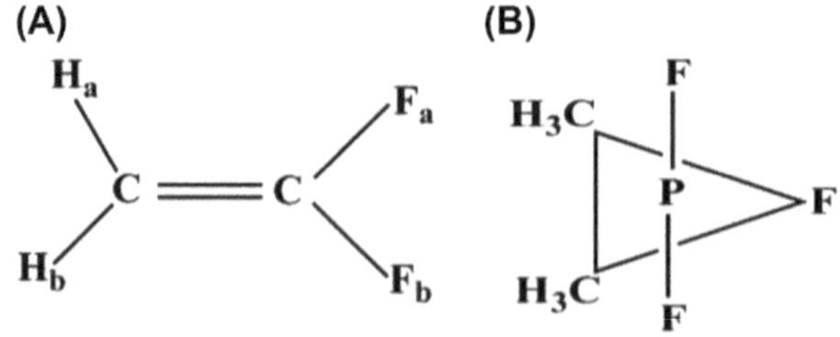

Figure 3.8 Structures of (A) $CF_2{=}CH_2$ and (B) $(CH_3)_2PF_3$.

3.1.2.2 Fluorine Nuclear Magnetic Resonance Spectra of $R_{5-n}PF_n$

The fluorine NMR spectra of a series of compounds of general formula $R_{5-n}PF_n$ (where R is an alkyl hydrocarbon/fluorocarbon, or aryl group) have been investigated. The number of peaks and the magnitude of the coupling constants in the spectrum are widely used to deduce their structures. For compounds of the type R_2PF_3, a trigonal bipyramidal structure is proposed. It is observed that the J_{P-F} values for equatorial fluorines are ~170 Hz greater than those for axial ones. The most electronegative groups are often found in the apical positions (Fig. 3.8B). The spectra and coupling constants, obtained on these compounds at low temperature, clearly establish that the two methyl groups and one F atom in $(CH_3)_2PF_3$ are equatorial and the other two F atoms (electronegative element) are axial. However, at room temperature (RT), rapid intramolecular exchange occurs and the effect of this exchange averages the coupling constants (see Section 3.2).

3.2 DYNAMIC NUCLEAR MAGNETIC RESONANCE SPECTROSCOPY

Dynamic nuclear magnetic resonance (DNMR) deals with the systems that undergo physical or chemical exchange with time. Consider, for example, the following two-site proton exchange reaction between A and B

$$A - H + B \underset{k_{-1}}{\overset{k_{+1}}{\leftrightarrow}} A + B - H$$

$$A - H_a + B - H_b \underset{k_{-1}}{\overset{k_{+1}}{\leftrightarrow}} A - H_b + B - H_a$$

The lifetimes of protons on states A and B, respectively, are given by

$$\tau_A = 1/k_{+1} \quad \text{and} \quad \tau_B = 1/k_{-1}$$

The NMR parameters that characterize both the states are (1) the chemical shift differences between the protons, (2) scalar spin–spin coupling of protons with its neighbors, (Fig. 3.10) and (3) the line width of the NMR peak (transverse relaxation time, T_2). The NMR timescale is long (milliseconds), so it can be used to study the slow exchange process. If the exchange process is faster than the milliseconds, the two protons are indistinguishable by NMR and give exchange averaged signal. On the other hand, if the exchange process is slower than the NMR timescale, the two protons are distinguishable to give separate signals. The effect of exchange rate (k) in relation to the NMR timescale on the magnitude of the NMR parameter is as follows [2].

NMR parameters	Exchange rates		
	Slow	Intermediate	Fast
Chemical shift difference (Δ)	$k \ll \delta_A - \delta_B$	$k = \delta_A - \delta_B$	$k \gg \delta_A - \delta_B$
Scalar coupling (J)	$k \ll J_{A\text{-}B}$	$k = J_{A\text{-}B}$	$k \gg J_{A\text{-}B}$
Transverse relaxation time (T_2)	$k \ll 1/T_{2A} - 1/T_{2B}$	$k = 1/T_{2A} - 1/T_{2B}$	$k \gg 1/T_{2A} - 1/T_{2B}$

The separation of the two signals *relative to the exchange rate* is influenced by two factors (1) temperature and (2) applied static field.

1. *Temperature-dependent NMR*
   Increasing temperature accelerates the exchange reaction and, therefore, pushes the system into the fast exchange regime.
2. *Applied static magnetic field*
   Increasing static fields increases the line separation (Δδ) and, therefore, drives the system into the slow exchange regime.

   So, DNMR involves recording spectra at various temperatures. The effect of chemical or conformational exchange on the chemical shift is shown below (Fig. 3.9). At high temperature a signal in the fast exchange regime is pushed into the slow exchange regime by decreasing the

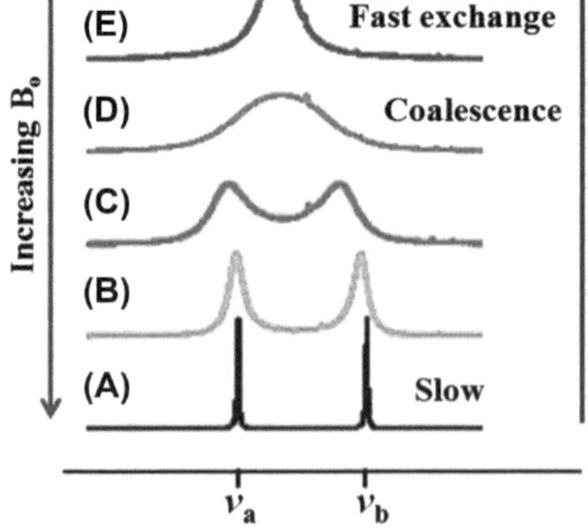

Figure 3.9 Dynamic nuclear magnetic resonance spectra of two-site exchange process.

temperature or increasing the field strength. At the **coalescence temperature**, a very broad line with a plateau is observed. There are four cases of exchange with respect to NMR timescale as follows.

a. ***Fast exchange***

i. At fast exchange, for $k \gg \delta_A - \delta_B$, a single resonance line is observed at the frequency,

$$\delta_{ave} = p_A\delta_A + (1 - p_A)\delta_B$$

It is obvious from the equation above that the resonance position of the averaged signal may be simply the arithmetic mean of the resonance positions of the two slow exchanging lines if the population of the two states (p_A and p_B) are equal. The exchange reaction causes an additional contribution to the line width of

$$\Delta\nu_{1/2} = (1/2)\pi(\nu_A - \nu_B)^2/k_{+1}$$

Thereby, the faster the reaction or the smaller the frequency difference (provided that a single line is observed, fast exchange regime) the narrower the line. Such an increase in reaction rate can be accomplished via a temperature increase and a decrease in frequency separation by lowering the magnitude of the static field.

ii. For $k \gg J_{A\text{-}B}$, a similar situation like above of *averaging of spin–spin couplings* occurs. For example, averaging of couplings is found below for hydroxyl protons of alcohols that exchange with solvent water protons. When the exchange is slow, both the lines of doublet are visible but for the fast exchange only the average singlet will be observed (Fig. 3.10). This mechanism has been termed **self-decoupling** and is the reason why OH protons in alcohols usually appear as (broad) singlets:

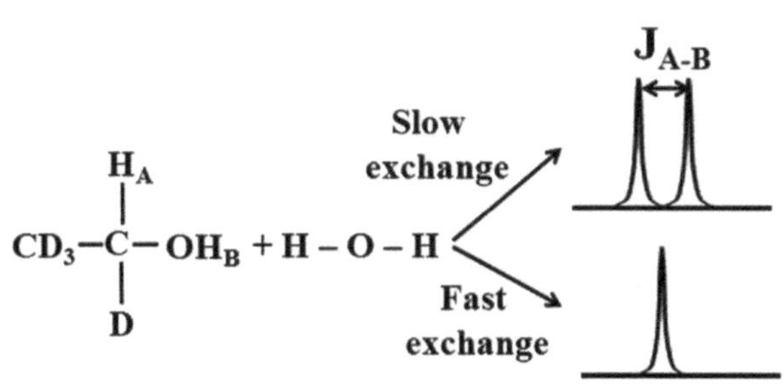

Figure 3.10 Exchange of hydroxyl proton with a proton from water causes averaging of spin–spin couplings.

b. ***Slow exchange limit*** ($k \ll \delta_A - \delta_B$)
Two separate signals are observed in the slow exchange. The additional contribution to the line width due to this slow exchange is

$$\Delta\nu = k_{+1}/\pi$$

Hence, temperature decrease or field increase will lead to a sharpening of the two lines.

A typical example where both situations (slow and fast exchange) are simultaneously encountered is exchange of amide and hydroxyl protons of peptides with water in aqueous solution. At slightly acidic pH, the amide protons give separate, sharp lines and the exchange rate is of the order of $1-10\ s^{-1}$. In contrast, exchange between hydroxyl and water protons is so fast that the hydroxyl protons coincide with the water chemical shift and cannot be observed separately.

c. ***Intermediate region*** ($k \approx \delta_A - \delta_B$)
In the region of partial collapse [spectra (b) to (d)], the lifetime of the proton on A, τ_A, is equal to the lifetime on B (provided the two sites equally populated with equal lifetimes), τ_B. The following expression can be obtained.

$$(\nu_A - \nu_B)_{obs}/\nu_A^o - \nu_B^o = \left[1 - (1/2)\pi^2\tau^2\left(\nu_A^o - \nu_B^o\right)\right]^{1/2}$$

$\nu_A^o - \nu_B^o$ is the separation of peaks for long lifetimes as in (a) in Hz; $(\nu_A - \nu_B)_{obs}$ is the separation of peaks which have begun to coalesce as in (c) and (d), and τ equals $\tau_A/2$.

d. ***Coalescence point*** ($k = \delta_A - \delta_B$)
At coalescence point the line shape is characterized by a flat top. In the spectrum (d), where the two peaks have just merged, $(\nu_A - \nu_B)_{obs}$ equals zero. The above expression is simplified to

$$\tau = \sqrt{2}/2\pi\left(\nu_A^o - \nu_B^o\right)$$

From these equations, it can be seen that for two separate peaks to be observed at frequencies ν_A and ν_B, the lifetimes of the two states (A and B) must be greater than $1/(\nu_A - \nu_B)$, i.e., $1/(\Delta_A - \Delta_B)$. For smaller values of τ (i.e., shorter lifetimes of a given state, as in (d)), the breath of the merged peak can be used to calculate the lifetime. If the exchange rate falls outside the limit that permits measurement by this technique, the rate may possibly be measured if conditions (temperature, concentration, and applied magnetic field) are changed to alter the rate.

For a *first-order process* in substances A, as, for example, the rotation about the C—N bond in amides, the average residence time, τ_A, is related to the specific rate of exchange by:

$$1/\tau_A = (1/[A]) \times (d[A]/dt)$$

Combining with the first order rate equation, $d[A]/dt = k[A]$,

$$1/\tau_A = k$$

For a *second order* reaction such as A + B, the rate equation is

$$d[A]/dt = k[A][B]$$

On combining the equation, we get

$$1/\tau_A = k[B]$$

Usually, the determinations of exchange rates by NMR involve assuming an exchange mechanism. To determine the order of a reaction to obtain the mechanism, the rate must be investigated as a function of the concentration of the reagents.

In general, changes in the concentrations of reagents, increasing acidity or basicity for acid or base-catalyzed reactions, or increasing temperature can give rise to the series of spectra, Fig. 3.9 (a-d), representing increasing rates of exchange. For example, the protons in the system $H_2O—H_2O_2$ are found to undergo rapid exchange to give rise to a single proton peak at RT under all conditions. From the width of the single peak (which varies with exchange rate), it is found that acids and bases catalyze the exchange reaction. By calculating the lifetime from the width of the peak as described above, the exchange mechanism was proposed.

3.2.1 Determination of Activation Parameters for Dynamic Process

The observation of dynamic processes in NMR spectra allows one to calculate the activation parameters for the temperature-dependent exchange process. The relation between the free energy of activation ($\Delta G^{\#}$) and reaction rate (***k***) for such an exchange process is given by Eyring equation as follows:

$$\boldsymbol{k} = \kappa(kT/h)e^{-\Delta G^{\#}/RT} \tag{3.1}$$

According to thermodynamics,

$$\Delta G^{\#} = \Delta H^{\#} - T\Delta S^{\#} \tag{3.2}$$

Hence,

$$\boldsymbol{k} = \kappa(\mathrm{kT/h})\mathrm{e}^{-\Delta \mathrm{H}^{\#}/\mathrm{RT}}\mathrm{e}^{\Delta \mathrm{S}^{\#}/\mathrm{R}}$$

$$\ln(\boldsymbol{k}/\mathrm{T}) = \ln \kappa(\mathrm{kT/h}) - \Delta \mathrm{H}^{\#}/\mathrm{RT} + \Delta \mathrm{S}^{\#}/\mathrm{R}$$

κ is the transmission coefficient (largely equal to one), k is the Boltzmann constant, T is the temperature, and h is the Planck constant.

A plot of ln ($\boldsymbol{k}$/T) versus 1/T yields a straight line with slope ($-\Delta H^{\#}/R$) and intercept ($23.76 + \Delta S^{\#}/R$). Thus the $\Delta H^{\#}$ and $\Delta S^{\#}$ parameters are calculated. Using Eq. (3.1), the free energy of activation, $\Delta G^{\#}$ value is obtained. Furthermore, the activation energy can also be estimated using the below equation from NMR in close agreement with the above value.

$$\Delta \mathrm{G}^{\#} \approx \mathrm{RT_c}[22.96 + \ln\{\mathrm{T_c}/(\nu_\mathrm{A} - \nu_\mathrm{B})\}]\ \mathrm{J/mol}$$

R is the molar gas constant and T_c is the coalescence point of temperature. Typical energy barriers measured and studied by NMR are in the range 30−100 kJ mol^{-1}.

3.3 DYNAMIC EXCHANGE PROCESS IN ORGANIC COMPOUNDS

There are several cases of dynamic and conformational exchange processes in organic and inorganic molecules. Some of them are as follows. (1) Restricted bond rotation, (2) Ring flipping (3) Pyramidal inversion at nitrogen, (4) Tautomerism viz., keto-enol, (5) Fluxionality of ligands in inorganic and in organometallic compounds, (6) Dissociation and recoordination of ligands, (7) Opening/closing of bridges, etc. A few examples are further illustrated here.

3.3.1 Hindered Rotation in N,N′-Dimethylformamide

A classic example of a fluxional molecule with hindered rotation is dimethylformamide (Fig. 3.11).

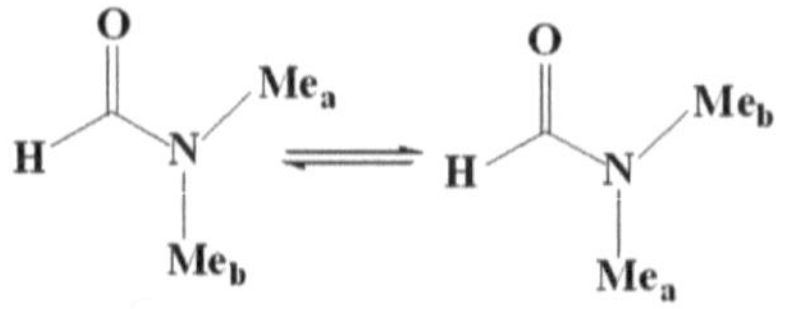

Figure 3.11 Hindered rotation in dimethylformamide.

This exchange process is used to demonstrate the calculation of rotational rate constant from temperature-dependent or DNMR spectra. It involves exchange between two sites that are equally populated and have equal lifetimes.

The ^{1}H NMR spectrum of N,N′-dimethylformamide, HC(O) $N(CH_3)_2$, exhibits three peaks at RT. Two of which corresponds to the two methyl groups (a and b) in the different environments on the nitrogen (one *cis* and one *trans* to oxygen). Because, the C—N bond has multiple i.e., double bond character and gives rise to an appreciable barrier to rotation about this bond. As a result of this barrier, the two nonequivalent proton signals due to $N-CH_3$ groups are detected. As the temperature is increased, the rate of rotation about the C—N bonds increases and the *N*-methyl resonances merge giving rise to a series of spectra similar to those in Fig. 3.9. The lifetime of a particular configuration can be determined as a function of temperature and thus the activation energy for the barrier to rotation can be evaluated. At temperatures near 100°C, the 500 MHz NMR spectrum of this compound shows only one signal for the two methyl groups. Near RT, however, separate signals are seen for the two nonequivalent methyl groups. The rate of exchange can be readily calculated at the temperature where the two signals are just merged. This "coalescence temperature" also depends on the measuring magnetic field. The relevant rate equation is as follows:

$$\tau = \sqrt{2}/2\pi\left(\nu_A^o - \nu_B^o\right)$$

$\Delta\nu_o = \nu_A^o - \nu_B^o$ the difference in Hz between the frequencies of the exchanging sites.

Similar studies have been carried out on other amides and on nitrosamines and the rates have been studied. Here, the assumption is made that δ do not change with temperature. Therefore, caution should be exercised in interpreting these results.

3.3.2 Effect of Chemical Exchange on Spin—Spin Splitting

Consider the ^{1}H NMR spectrum of CH_3OH in the absence of exchange. The doublet methyl group results from the OH splitting and the quartet hydroxyl from the splitting by the three equivalent protons on carbon. The exchange of hydroxyl protons with water as solvent may result in a change in the nuclear spin state of the proton attached on the oxygen. When rapid exchange occurs, i.e., the frequency of the exchange is greater than $\sqrt{2}\pi(\nu_A^o - \nu_B^o)$, only the average effect of the two spin states is observed. The spectrum obtained under such conditions of rapid exchange consists of only two sharp peaks one for each OH and CH_3 group protons and the methyl groups give a single line. Thus in the methanol−water system, the collapse of the methyl (doublet) peak can be used to calculate the total proton exchange (i.e., CH_3OH with CH_3OH and CH_3OH with H_2O). From the broadening of the water peak, the exchange rate of methanol with water can be calculated.

3.3.3 Conformational Exchange in Ethane and Cyclohexane

Conformational processes may lead to exchange broadening in NMR spectra when a molecule undergoes changes between two or more conformations. However, many of the conformational changes are so fast on the NMR timescale that normally only averaged spectra are observed. Especially, in molecules which are not sterically bulky, the rotation about C−C single bond is normally fast on the NMR timescale. For example, in ethane CH_3−CH_3, the hydrogen atoms of two methyl groups appear as singlet as a result of averaging of the various rotational conformers, i.e., fast exchange. On the other hand in molecules, if there are very bulky groups, steric hindrance may restrict the rotation about single (C−C) bonds and give rise to broadening in NMR spectra.

For example, in cyclohexane (Fig. 3.12), there are two types of hydrogen (1) axial and (2) equatorial. So, it is expected to give two signal separated by their chemical shift difference. However, it gives a single averaged resonance in the ^1H NMR at RT for both the axial and equatorial hydrogen because of rapid exchange i.e., ring flipping process. At very low temperature, two separate signals are observead in NMR (Fig. 3.9) [3].

From the exchange broadened spectra obtained at different temperature or magnetic field strengths, the rate constants for the exchange process (and hence the activation parameters $\Delta G^{\#}$, $\Delta H^{\#}$, $\Delta S^{\#}$) can be derived. When the signals coalesce (Fig. 3.9) from being two separate signals to a single averaged

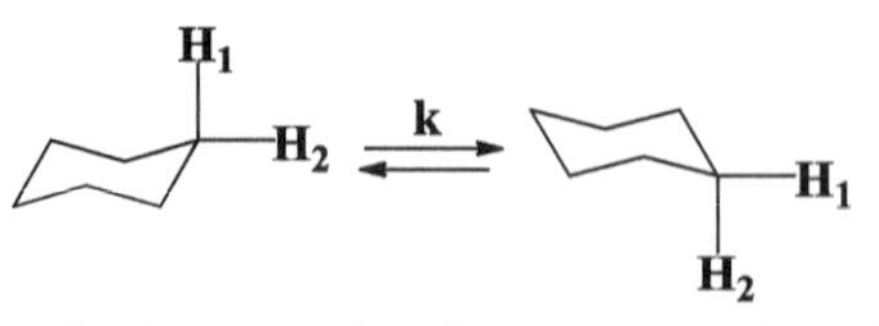

Figure 3.12 Conformational exchange process in cyclohexane.

signal, the rate constant for the exchange can be calculated using the equation as follows.

$$k = \pi \Delta \nu / \sqrt{2}$$

3.3.4 Intermolecular Interchange of Labile (Slightly Acidic) Protons

Functional groups such as −OH, −COOH, $-NH_2$, and −SH have labile protons which exchange with each other in solution. The −OH protons of a mixture of two different alcohols may give rise to either an averaged signal or separate signals depending on the rate of exchange. This depends on many factors including temperature, the polarity of the solvent, the concentrations of the solutes, and the presence of acidic or basic catalysts.

$$R - OH + R' - OH' \leftrightarrow R - OH' + R' - OH$$

3.4 NUCLEAR MAGNETIC RESONANCE OF FLUXIONAL INORGANIC MOLECULES

1. Fluxional molecules are molecules that undergo dynamics such that some or all of their atoms interchange between symmetry-equivalent positions.
2. Two or more molecules, which are fluxional, have comparable stability.
3. These fluxional molecules constantly change between or among these configurations at RT or at slightly elevated temperature. Hence, these molecules are also called stereochemically nonrigid.
4. Fluxional molecules have small HOMO−LUMO (highest occupied molecular orbital−lowest unoccupied molecular orbital) gap and hence one configuration can be transformed into another configuration with low-energy barrier to rotation.

Virtually, most of the molecules are fluxional in some respects, for example, exchange of positions between axial and equatorial in phosphorus pentafluoride (PF_5) and bond rotations in most organic compounds. Often, a molecule is called fluxional if its spectroscopic characteristics shows line-broadening (beyond that dictated by the Heisenberg Uncertainty Principle) due to chemical exchange. In some cases, where the rates are slow, fluxionality is not detected spectroscopically, but monitored by isotopic labeling method [4].

The pentacoordinate molecules of trigonal pyramidal geometry typically exhibit a kind of low-energy fluxional behavior called Berry pseudorotation (Fig. 3.13). Common examples of such molecules are iron pentacarbonyl ($Fe(CO)_5$) and PF_5 undergo pseudorotation. At high temperatures, only one signal is observed for the ligands (e.g., by ^{13}C or ^{19}F NMR), whereas at low temperatures, two signals in a 2:3 ratio can be resolved. In addition, the molecules that are not strictly pentacoordinate also undergo this process, viz., SF_4.

3.4.1 Berry Pseudorotation in PF_5, $Fe(CO)_5$, and SF_4

3.4.1.1 Phosphorus Pentafluoride and Pentacarbonyl

According to the valence shell electron pair repulsion theory theory, a pentacoordinate phosphorus atom is of a trigonal bipyramidal geometry, TBP (Fig. 3.13). There are two distinct sites on a trigonal bipyramid, the equatorial and the axial sites in ^{19}F NMR of PF_5. The bonding by equatorial sites utilizes sp^2-orbitals, and the axial sites involve the remaining p-orbital. Hence, the axial bonds have no s-character, and experience smaller $^1J_{P\text{-}F}$ coupling constants than the equatorial sites $[(^1J_{P\text{-}F})_{eq} > (^1J_{P\text{-}F})_{ax}]$.

Normally, no differences in the $^1J_{P\text{-}F}$ coupling constants between the axial and equatorial are observed in compounds with pentacoordinated phosphorus. Because, these compounds undergo a fast exchange of their equatorial and axial sites more rapidly than the NMR timescale known as pseudorotation, or Berry rotation through square pyramid (SP) intermediate in Fig. 3.13. The TBP and SP configurations differ only little in energy. These are readily interconverted by relatively small and simple angle deformations motions. However, at lower temperatures, two different $^1J_{P\text{-}F}$ coupling constants were observed, because of the absence of Berry rotation [4].

Analogously, $Fe(CO)_5$ also exhibits fluxionality in ^{13}C NMR spectrum and thus DNMR is used to monitor this pseudorotation and its rate constant.

3.4.1.2 SF_4

SF_4 has a C_{2v} structure with two sets of two nonequivalent fluorine atoms. The geometry is considered similar to TBP where one lone pair occupies

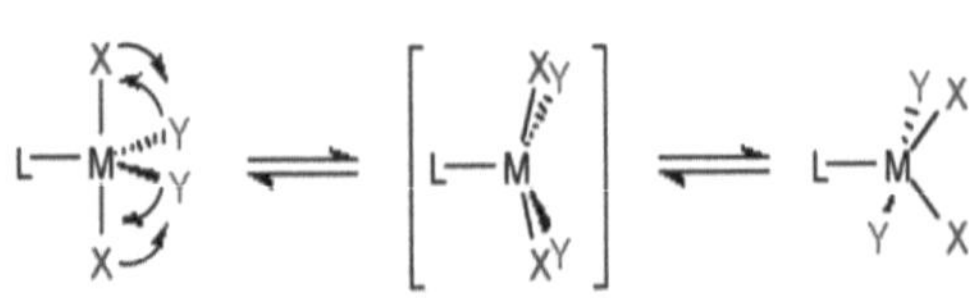

Figure 3.13 Berry pseudorotation leading to exchange of axial and equatorial ligands.

equatorial position and two fluorine each at axial and equatorial positions. A singlet peak is observed at RT due to dynamic effect. On lowering the temperature, the first broadens and split into two separate broad peaks. Further lowering the temperature to about −98°C, each broad peak is split into triplet. It was also observed that the rate of exchange depends on concentration of SF_4 supporting a bimolecular mechanism. Similarly, a series of group VI fluorides were investigated and it was observed that the exchange rate increases in the order $SF_4 < SeF_4 < TeF_4$.

3.4.2 Fluxional Behavior of η^3-Allyl Complexes

The η^3-allyl metal complexes also exhibit stereochemical nonrigidity called fluxionality at RT. So, the fluxionality of η^3-allyl complexes has been investigated extensively by ^{1}H NMR. In η^3-ally1 group (C_3H_5), there are three types of Ha, Hs, and Hc atoms as shown in Fig. 3.14.

1. **One Hc:** It is attached to the central carbon atom, C2.
2. **Two Hs:** These syn H atoms are present on the terminal carbon atoms (C1 and C3) and are toward the metal atom.
3. **Two Ha:** These anti-H atoms are again present on the terminal carbon atoms. However, these are away from the metal atom.

So, the metal coordinated η^3-ally1 group show three signals in ^{1}H NMR at very low temperature as follows: (1) **δHa** (doublet), (2) **δHs** (doublet), and (3) **δHc** (multiplet).

The NMR signals due to Ha and Hs atoms split up into doublet due to spin−spin interaction with Hc. It is called static NMR spectrum.

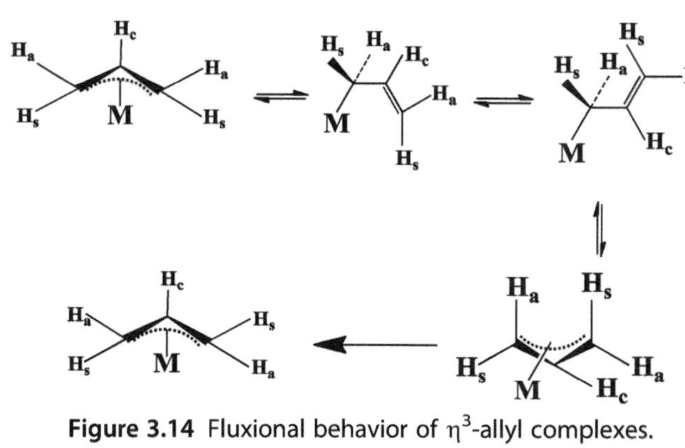

Figure 3.14 Fluxional behavior of η^3-allyl complexes.

3.4.2.1 *[Mn(CO)$_4$(η^3-C$_3$H$_5$)]*

For example, [Mn(CO)$_4$(η^3-C$_3$H$_5$)] shows three signals in the ratio 2:2:1 in ^{1}H NMR at low temperatures with.

$$\mathbf{\delta Hs} = 1.8\ \text{ppm (doublet)(2H)};\quad \mathbf{\delta Ha} = 2.8\ \text{(doublet)(2H)};$$

$$\mathbf{\delta Hc} = 4.7\ \text{(multiplet)(1H)}$$

On the other hand, at RT or elevated temperatures, it shows only two signals in the ratio 4:1 at **δHs**, **δHa** = 2.3 (doublet) (4H); **δHc** = 4.7 (multiplet) (1H).

The spectrum obtained is called DNMR spectrum. At higher temperatures, the *syn* and *anti*-H atoms are rapidly interchanging their positions probably via a short-lived η^1-allyl-metal intermediate as shown in Fig. 3.14.

3.4.3 Fluxionality (Ring Whizzing) in Cyclopentadienyl Complexes

In (η^1-C$_5$H$_5$)M complexes, there are three different environments for H atoms in the stereochemically rigid form as shown in Fig. 3.15. Therefore, it shows three signals in proton NMR in the intensity ratio 2:2:l.

2H$_A$: *quartet* in downfield; 2H$_B$: *triplet* in downfield; 1H$_C$: *triplet* in upfield.

But at RT, it exhibits one sharp singlet in both the ^{1}H and ^{13}C NMR spectra for the C$_5$H$_5$ group. That is, at high temperature (RT), the metal atom attached to the C$_5$H$_5$ ring carbon is shifting rapidly over the entire carbon atoms of C$_5$H$_5$ ring. So, all the H become time average equivalent on the NMR timescale and all the C atoms also become time average equivalent (due to 1,2-shift and 1,3-shift) and give singlet signal. Such behavior is called ring whizzing due to the motion of ring with respect to the metal atom.

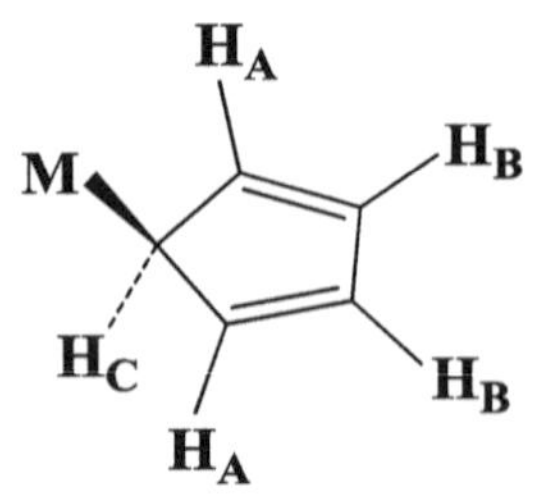

Figure 3.15 Ring whizzing in cyclopentadienyl complexes.

3.5 SOLID STATE NUCLEAR MAGNETIC RESONANCE AND MAGIC ANGLE SPINNING

Basically, NMR in the solid state gives very broad line due to increased line-widths. It is largely caused by two additional factors (1) dipolar interactions and (2) chemical shift anisotropy. These anisotropic effect are averaged out (to zero) by rapid random tumbling motion in solution NMR to give a series of very sharp isotropic transitions [5].

1. **Dipolar coupling** depends on the angle of orientation of the two magnetic spin vectors separated by distance (r) and on the magnetic moments (μ) of the two nuclei

$$H_{dd} = k\mu_1\mu_2\left(1 - 3\cos^2\theta\right)I_1I_2/r^3$$

2. **Chemical shift anisotropy**
   For nuclei with nonspherical electron distribution, the shielding effect depends on the orientation of the electron cloud with respect to the static field B_0. For example, for sp^2 carbons, such as carbonyl nuclei, the shielding depends on the orientation of the C=O bond relative to the B_0 (Fig. 3.16). In solution, the chemical shift observed is an average of the different orientation (isotropic chemical shift). However, when the rotational motion is hindered in the solid state, the line is very much broadened. Chemical shift anisotropy values are thus measured from solid state NMR (SS NMR).

3.5.1 Magic Angle Spinning Nuclear Magnetic Resonance

Let us consider two identical nuclei each with I = ½ (I = ½ + ½ = 1, i.e., $M_I = +1, 0, -1$) for simplicity. The energy level splitting diagram for this case is shown in Fig. 3.17.

$$E_{+1} = -\mu B_0 + \mu^2\left(1 - 3\cos^2\theta\right)/r^3$$

$$E_0 = -\left(2\mu^2/r^3\right)\left(1 - 3\cos^2\theta\right)$$

$$E_{-1} = \mu B_0 + \mu^2\left(1 - 3\cos^2\theta\right)/r^3$$

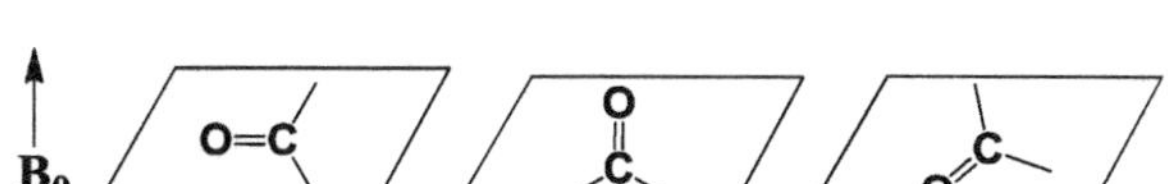

Figure 3.16 Chemical shift anisotropy due to different orientations.

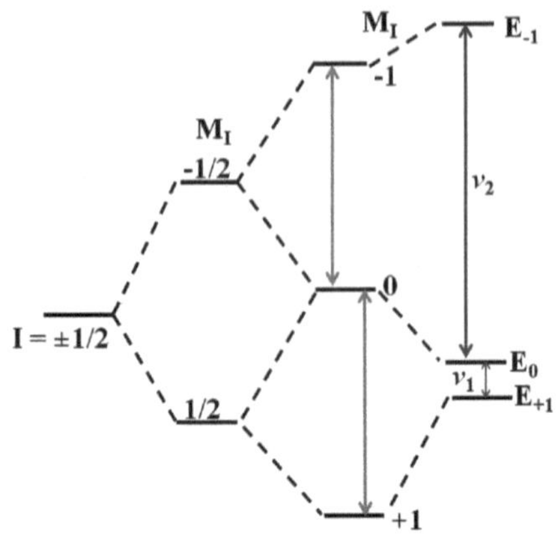

Figure 3.17 Energy level of two identical nuclei of spin I = ½ in the solid state in a magnetic field B_0 in the absence and in the presence of dipole–dipole interaction.

Using the selection rule for NMR $\Delta M_I = \pm 1$, the allowed two transitions are given as follows:

$$\nu_1 = (\mu B_0/h) - \left[3\mu^2\left(1 - 3\cos^2\theta\right)\right]/hr^3 \text{ and}$$
$$\nu_2 = (\mu B_0/h) + \left[3\mu^2\left(1 - 3\cos^2\theta\right)\right]/hr^3$$

An inspection of the above resonance equation shows that if the contribution from the factor $(1 - 3\cos^2\theta)$ is experimentally averaged to zero. It leads to the disappearance of the dipole–dipole factor and the spectrum will markedly resemble that of a liquid sample. So, a polycrystalline sample is rapidly rotated about an axis S, which makes an angle θ to the external magnetic field B_0 (Fig. 3.18).

If $\theta = 54.736°$, $(1 - 3\cos^2\theta) = 0$ which reduces the average value to zero. For this angle called magic angle, the term $(1 - 3\cos^2\theta)$ describing

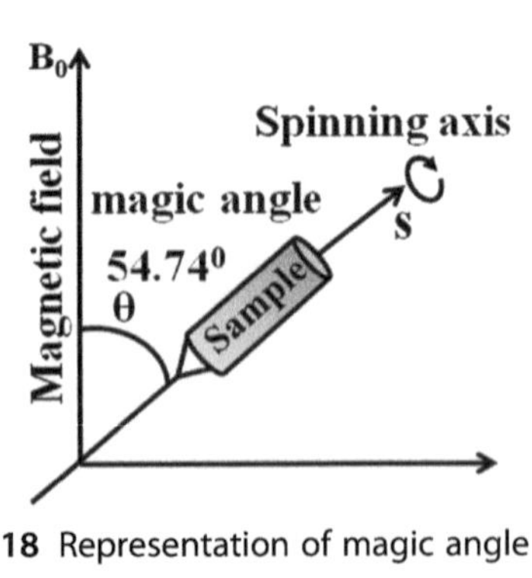

Figure 3.18 Representation of magic angle spinning.

the angular dependence in both the dipolar coupling and chemical shift anisotropy becomes zero. So, the powdered sample is placed in a cylindrical shaped rotor (910 mm length and 4 mm diameter) and rotated up to (30−50 KHz) about an axis which makes 54.74 degrees with the direction of the external field B_0. This method of rapidly rotating the sample about an axis 54.736 degrees to the field direction is referred to as magic angle spinning (MAS) NMR spectroscopy (Fig. 3.18). As a consequence, the spectrum thus exhibited line-narrowing effect, consisting of sharp peaks. This line broadening in SS NMR can be thus eliminated by spinning the sample about the magic angle 54.74 degrees. In other words, the contribution from dipole−dipole interaction averages to zero and the spectrum would be simplified to give isotropic.

In the Fig. 3.19 below are 81 MHz ^{31}P−CP MAS spectra of ammonium dihydrogen phosphate with high-power proton decoupling at various spinning rates [6]. When the spinning rate is less than the extent of the interaction (b−d), the spectrum is split into sidebands spaced at the spinning speed. When the spinning speed is zero, one can see the broad powder spectrum characterizing the chemical shift interaction in solids (a). On increasing the spinning speed, there is a center band at the isotropic chemical shift with sidebands whose intensities can be used to calculate the chemical shift

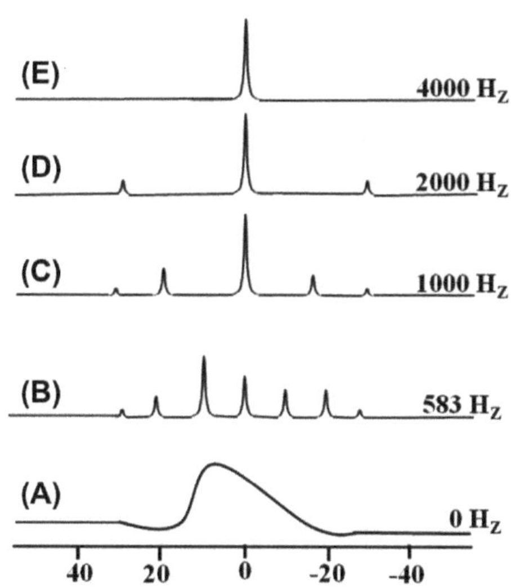

Figure 3.19 Magic angle spinning ^{31}P−MAS spectra of ammonium dihydrogen phosphate. *Courtesy of Glenn A. Facey, University of Ottawa.*

tensor (d). When the speed of spinning exceeds the span of the chemical shift interaction, the sidebands are very small and a liquid-like spectrum is obtained (e).

Hence, SS NMR using MAS is nowadays increasingly used to (1) probe properties in the material science (2) structure characterization of insoluble and also large compounds viz., synthetic polymers, lipid bilayer, etc.

3.6 NUCLEAR MAGNETIC RESONANCE OF PARAMAGNETIC MOLECULES

So far, NMR measurements on diamagnetic compounds were discussed. Paramagnetic compounds are also investigated, however, exhibiting a wide chemical shift range and broadened signals. The broadening effect in NMR signals is compensated in part by the wide range of chemical shift (often 200 ppm). There are two additional factors in the paramagnetic complexes causing this **isotropic shift** effect as follows. The isotropic shift is defined as the difference between the chemical shift of a given nucleus in a diamagnetic versus paramagnetic complexes or environments. The isotropic shifts arise due to two mechanisms, contact shifts and pseudocontact shifts.

1. *Contact shift* (also called scalar coupling) arises due to the unpaired electron wave function having a finite value at the nucleus and hence electron spin—nuclear spin coupling occurs. It results from spin polarization conveyed through the molecular orbitals of the molecule. This affects the energy difference between the nuclear spin states resulting in the change in chemical shift range for proton NMR much larger than the diamagnetic or free compound.
2. The second factor is the *dipolar mechanism or pseudocontact shift*, which largely broadens the spectrum and also causes a smaller shift. (1) This shift of nuclear Larmor frequency is caused by the dipolar coupling between the magnetic moments of the nucleus and of the unpaired electron. (2) This is like J-coupling in NMR where we would expect the peak to split into a doublet due to the coupling with the magnetic moment of the unpaired electron. (3) Because the rate of electron spin relaxation is much faster than the rate of nuclear spin relaxation, the expected doublet will not be observed. (4) Instead, one broad peak only at the position will be observed corresponding to the average of the available populations of spin states of the free electron, the chemical shifts of the nucleus, and also due to averaged overall possible molecular

orientations (due to the fast molecular tumbling). (5) The dipolar shift is given by $D_{Dip} = k\ (3\cos^2\theta - 1)/r^3$, where k is a proportionality constant which depends on the magnetic dipole moment of the paramagnetic metal ion. It shows an $1/r^3$ dependence and hence smaller, often within the normal 1–10 ppm range for ^1H NMR. (6) This effect is observable when the magnitude of the magnetic moment of the unpaired electron depends on the molecular orientation with respect to the magnetic field vector. Hence, the magnitude of the electron magnetic moment should depend on the orientation, i.e., anisotropic. (7) So, it is used for the determination of three dimensional (3D) structure of the molecules and also locate the unpaired electron center.

As the magnetic moment of an electron is $\sim 10^3$ times larger than the nuclear magnetic moment, tumbling of paramagnetic ions produces large magnetic fields which are very effective in causing dipolar spin–lattice relaxation, i.e., T_1 is decreased. This leads to a fluctuating magnetic field at the nucleus, which shortens T_1. In certain cases, the line broadening is very large that no well-resolved signals could be obtained [1]. Depending on the electron spin relaxation of paramagnetic compounds, there are four cases as follows.

1. If the electron relaxation is very slow, the lifetime of an ion in a given nuclear spin state will be long, and two resonances, corresponding to $S = \pm 1/2$, will be observed. This case is not very common.
2. When the paramagnetic state has a very short relaxation time, the magnetic nucleus sense only the time-averaged magnetic field of the two spin states of the electron and a single peak is observed.
3. The electron spin relaxation is often of such a rate as to be intermediate between these two extremes, in effect shortening T_2 and causing severe broadening, e.g., Cu(II) complexes and proteins.
4. **Contact scalar NMR shift**

When the electron relaxation is very rapid, in certain cases viz., Ni(II), Fe(II), Co(II) complexes, the broadening is minimized. The major effect on the spectrum from the presence of unpaired electrons is to change the magnetic field experienced by the magnetic nucleus. This causes a very large chemical shift sometimes 3000–5000 Hz of the resonance in the NMR spectrum. This shift is referred to as a **contact scalar NMR shift**.

The relationship between the contact shift, $\Delta\nu/\nu$, and the electron spin–nuclear spin coupling constant, A_N is given as follows:

$$\Delta B/B = \Delta\nu/\nu = -[2A_N\gamma_e g\beta I(I+1)S(S+1)]/9\gamma_N kT$$

In the specific case of proton magnetic resonance,

$$\Delta B/B = \Delta\nu/\nu = -[A_N\gamma_e g\beta S(S+1)]/6\gamma_N kT$$

γ_e is the magnetogyric ratio for the electron, γ_N is the magnetogyric ratio for the magnetic nucleus, I is the nuclear spin, S is the electron spin, ΔB is $(B_{complex} - B_{ligand})$ or $\Delta\nu = (\nu_{complex} - \nu_{ligand})$, β is the Bohr magneton, and g is the ratio of the magnetic moment to the total angular momentum of the electron.

The contact shift is to lower fields when A_N is positive and to higher fields when A_N is negative. The sign of the electron spin—nuclear spin coupling constant can be determined from NMR. Often all the terms in above equation are known, so A_N can be calculated from the measured contact shift.

The conditions requisite for observing a contact shift in NMR are that either of the following inequalities be satisfied:

$$\text{Rate of electronic relaxation,}\quad 1/\tau_S \gg A_N \quad \text{(or)}$$
$$\text{Rate of chemical exchange rate,}\quad 1/\tau_e \gg A_N$$

τ_S is the electronic relaxation time, τ_e is the chemical exchange time, and A_N is the constant (electron spin—nuclear spin coupling constant). Fortunately, most of the paramagnetic metal ions of the first transition series have sufficiently short electron spin lifetimes or sufficiently fast ligand exchange rates to allow NMR studies to be made on their complexes.

From the NMR contact shift, information such as magnetic susceptibility and bonding can be obtained from the electron spin—nuclear spin coupling constants (A_N), often called the *contact interaction constant*. The magnitude of this quantity is related to the unpaired electron spin densities in π orbital of sp^2 carbons by the equation below

$$A_N = Q\rho_N$$

ρ_N is the spin density at carbon, Q is the proportionality constant (−22.5 G for an aromatic C—H fragment and +27 G for C—CH_3 fragment). *Spin density* (ρ_N) is defined as that fraction of an unpaired electron which appears to be localized at site N. Spin densities may be either positive or negative. A *negative spin density* results when the electron spin at site N is antiparallel to the net spin of the paramagnetic species as a whole, whereas the opposite is true for *positive spin density*. Negative spin density is a concept foreign to Huckel molecular orbital theory. So, molecular orbital theory must

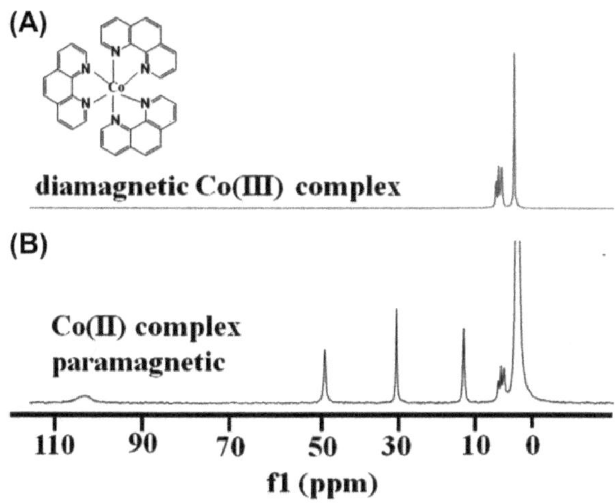

Figure 3.20 Nuclear magnetic resonance spectra of diamagnetic $[Co(phen)_3]^{3+}$ and paramagnetic $[Co(phen)_3]^{2+}$ dissolved in D_2O. *Courtesy of Paul S. Donelly, University of Melbourne.*

be modified by addition of π−π exchange interaction to account for negative spin densities.

For example, the ^1H NMR spectra of diamagnetic Co(III) low spin (S = 0) and paramagnetic Co(II) low spin complexes (S = 1/2), $[Co(phen)_3]^{3+/2+}$ (where phen = 1,10-phenanthroline) are shown in Fig. 3.20. The paramagnetic complex consists of four well-resolved and shifted signals corresponding to the four different aromatic protons in the complex. The chemical shifts of the signals lie in the range between 110 and 15 ppm. However, in the diamagnetic complex, the signals are not shifted compared to the pure ligand.

3.6.1 Determination of Magnetic Susceptibility Using Nuclear Magnetic Resonance (Evans Method)

As described earlier, the paramagnetic complex has significant influence on both the chemical shift and the relaxation of proton resonances causing broadening of peaks in the proton NMR spectrum of the complex. So, NMR could be applied to determine whether a complex is paramagnetic or diamagnetic. Furthermore, based on the shift in frequency, Evans published a method to estimate the magnetic susceptibility in turn the magnetic moment of the sample and thus the electronic structure of the complex [7,8].

He measured the difference in chemical shift of an inert reference compound (t-butanol) in the presence (υ_2) and absence (υ_1) of paramagnetic transition metal complexes and calculated the mass and molar magnetic susceptibility of the complex using the equations below.

$$\chi_g = \frac{3\Delta\upsilon}{2\pi\upsilon m}$$

χ_g is the mass susceptibility of the solute, $\Delta\upsilon$ is the observed shift in the frequency of the reference compound (t-butanol), υ is the frequency of the spectrometer, and m is the mass of the complex per cm^3 of solvent.

$$\chi_m = \frac{6}{1000}\frac{1}{c}\frac{\Delta\upsilon}{\upsilon}$$

χ_m is the molar susceptibility of the complex ($m^3\ mol^{-1}$) and c is the concentration of the complex (mol dm^{-3}).

In his experiment, NMR tube with a coaxial insert was used to determine the difference in chemical shift of the reference compound (t-butanol) caused by paramagnetic compound. The NMR tube contains a solution of the paramagnetic substance with 2% reference and the coaxial insert contains only the reference in solvent. The effective magnetic moment, μ_{eff} is calculated from χ_m using the following equation:

$$\mu_{eff} = \sqrt{\frac{3k}{N_0\mu_0\mu_B^2}}\sqrt{\chi_m T}$$

k is the Boltzmann constant, N is Avogadro's number, μ_0 is the permeability of a vacuum, and μ_B is the Bohr magneton. On substituting the values, the simplified equation is obtained as follows:

$$\mu_{eff} = 798\sqrt{\chi_m T}$$

T is the temperature in Kelvin and χ_m is the molar susceptibility determined by Evans method.

$$\mu_{eff} = \sqrt{n(n+2)}$$

From the μ_{eff}, the number of unpaired electrons on the metal atom of a complex (n) is determined.

For example, the difference in chemical shift of an inert reference compound (t-butanol) in the presence (υ_2) and absence (υ_1) of $[Fe(acac)_3]$ complex using $\upsilon = 42.5$ MHz and $c = 0.040$ mol dm^{-3} (Fig. 3.21).

$$\Delta\upsilon = 57.85 - 16.84 = 41.01 \text{ Hz}$$

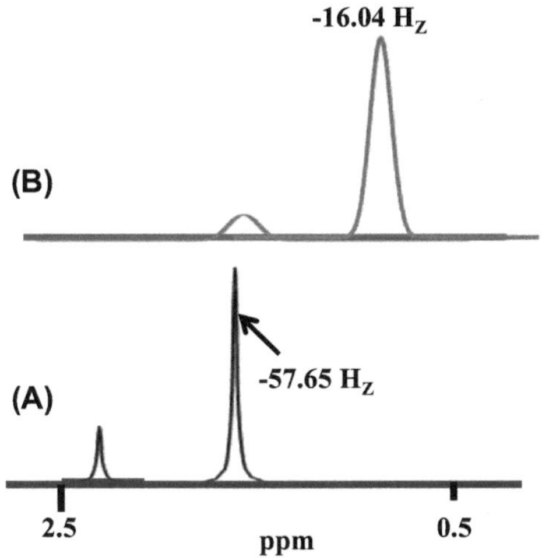

Figure 3.21 [1]H-nuclear magnetic resonance spectra of the reference sample (A) and the reference with $Fe(acac)_3$ (red dark gray in print versions) (B).

The molar susceptibility of $[Fe(acac)_3]$ is calculated, $\chi_m = 1.46 \times 10^{-7}\ m^3\ mol^{-1}$. The effective magnetic moment is determined, $\mu_{eff} = 5.23\ \mu B$. The high spin complex has five unpaired electrons and the low spin complex has one. So, for high spin, $n = 5$, $\mu_{eff} = 5.92\ \mu B$ and for low spin, $n = 1$, $\mu_{eff} = 1.73\ \mu B$. The observed magnetic moment of the high spin complex is much closer to the value for spin, therefore, the complex is in high spin state.

3.7 SHIFT REAGENTS

Paramagnetic reagents, also known as shift reagents, are used to disperse the overlapping proton NMR spectra. It causes a shift of the signals due to rapid electron spin relaxation than contrast or relaxation agents where the electron spin relaxation is slower. But, the shift is not equal for all resonances and hence useful in spectral analysis. Thereby, a 2 ppm shift range for aliphatic methylene protons can be dispersed over 6 ppm after addition of the shift reagent. Paramagnetic ions of Ni(II) and Co(II) were the first used shift reagents. However, they exhibit severe line broadening. In 1969, Hinckley discovered that paramagnetic lanthanide ions gave shifts without significant line broadening [9].

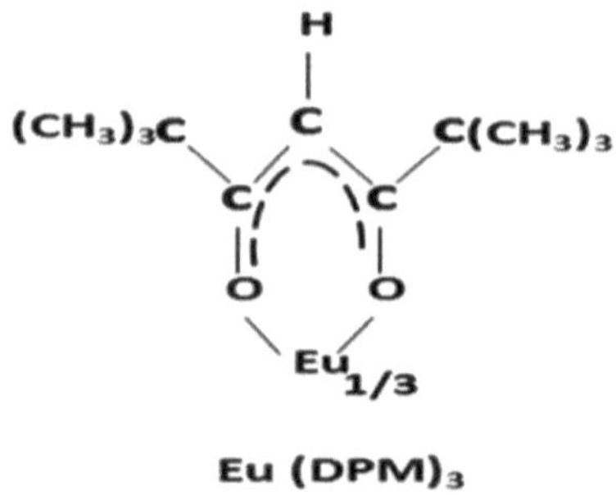

Figure 3.22 Structure of (dipivalomethanato)Eu(III).

For example, 90 MHz ^{1}H-NMR spectrum of 1-hexanol are shown in the absence and presence of chelate complex, Tris(2,2,6,6-tetramethyl-3,5-heptanedionato)europium(III) in Fig. 3.22.

It is observed that all the protons become less shielded; all the CH_2 groups become separated. The shifts increase with the proximity of the protons to the OH group of alcohol (Fig. 3.23). Shift effect or mechanism occurs due to the following two types of interactions between nuclear spin and the spin of the unpaired electrons of paramagnetic ions.

1. The contact interactions (direct contact) and
2. The pseudocontact or dipolar interactions (through space)

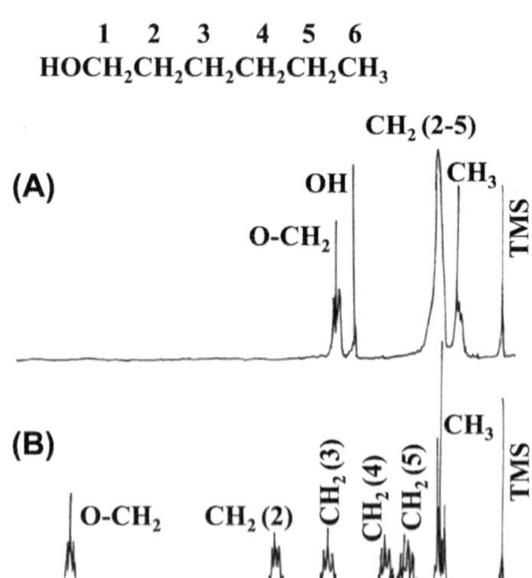

Figure 3.23 Nuclear magnetic resonance spectra of 1-Hexanol in the absence (A) and presence (B) of Eu(DPM)$_3$. *Reprinted with permission from the American Chemical Society.*

These two interactions depend on the dynamic equilibrium and stability of formation of a complex between the substrate or ligand, S, and the paramagnetic metal ion, M.

$$M + S \leftrightarrow MS$$

1. *Contact term*
   The characteristics of contact term also called contact shift is (see Section 3.6) as follows:
   a. It is based on the contact interaction due to the transfer of spin density of the unpaired electron to the substrate molecule.
   b. The electron spin density is not the same at all positions of the observed nuclei throughout the molecule.
   c. In saturated compounds, the most affected ^{13}C nuclei are those in α- and β-positions relative to the complexing center (e.g., O, N, or S).
   d. In conjugated systems, more distant positions could be affected as well.
   e. The contact term is also very important in ^{13}C NMR as it suppresses nuclear overhauser effect (NOE).
2. *Pseudocontact term*
   It is used to describe a dipolar interaction between the magnetic dipole field of the unpaired electron and that of the observed nucleus. In contrast to contact interaction, the interaction here is transmitted through space. This type of interaction is of greater importance in ^{1}H NMR than the contact term. The information about the geometry of the complex is obtained from the pseudocontact term. The shift in the resonance frequency of the observed nuclei depends on the geometry of the complex and given by dipolar shift,

$$D_{Dip} = k\left(3\cos^2\theta - 1\right)/r^3$$

   k is a proportionality constant which depends on the magnetic dipole moment of the paramagnetic metal ion. This equation is valid if the complex is symmetrical about the M–L axis. Furthermore, this shift decreases in inverse proportion to r^3. It can be positive or negative (depending on the sign of ($3\cos^2\theta - 1$) term.

3.7.1 Applications of Shift Reagents

Shift reagents are widely used in the following:

1. Simplifying complicated spectra.
   a. Separation of overlapping signals.
   b. Easier assignment.

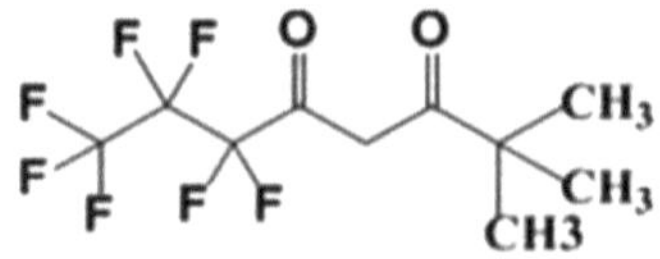

Figure 3.24 1,1,1,2,2,3,3-heptafluoro-7,7-dimethyl-4,6-octandione (FOD).

   c. Integration of signals which are otherwise overlapped.
   d. Decoupling experiments.
2. Useful in integration in ^{13}C NMR as it suppresses the NOE.
3. It causes a shift of the signals but not equally for all resonances and hence useful in spectral analysis.
4. Determining accurate geometrical data for the metal-substrate complex from the pseudocontact interactions.

3.7.1.1 Demerit

However, olefinic and aromatic compounds protons do not show lanthanide-induced shifts because they do not form complexes with lanthanide ions. Recently, a solution is found for this problem using silver (I) ions. Obviously, the silver ions are able to transmit the shift effect because,

- Ag(I) ions form complexes with π-electrons of olefinic and aromatic compounds.
- They are added in the form of AgFOD (Fig. 3.24) to the solution containing these substrates and the shift effects are observed for olefines and arenes.

3.8 NUCLEAR MAGNETIC RESONANCE IMAGING

Nuclear magnetic resonance imaging (NMRI) is a new noninvasive diagnostic technique for mapping the internal structure and certain aspects of function within the body. It has revolutionized medical imaging and biomedical research. Although it is used mainly for medicinal and clinical applications, its potential for nonmedical applications has now been recognized. Unlike its predecessor, X-ray computerized tomography (computerized tomography (CT) and positron emission tomography scanning), it does not use ionizing electromagnetic radiation and considered risk-free for all patients. NMRI is based on the principles of NMR. The biological samples viz., cells and tissues contain abundant water (63%) and other fluids with protons ($I = 1/2$). In addition, the other magnetically active nuclei ^{31}P, ^{19}F, and ^{14}N could be imaged [10–13].

3.8.1 Nuclear Magnetic Resonance Versus Magnetic Resonance Imaging

In NMR, the resonance absorption is recorded for very small bulk sample by applying a homogeneous magnetic field over it. In NMRI, one has to measure the net magnetization at each location in the object. This is done by "spatial encoding" the signal in such a way that it's spatial dependence (x, y, z) can somehow be decoded. This is achieved by using the magnetic gradients produced by suitably placed electrical coils (Fig. 3.25). These gradients give rise to a spatially-dependent distribution of NMR frequencies. They are generally varied in the 1 G cm^{-1} range, thereby producing magnetic fields considerably smaller than that of the primary field (2T). NMRI is thus accomplished by making the field inhomogeneous. However, the conventional high resolution NMR uses homogenous magnetic fields.

The contrast parameters normally used for NMRI are proton spin density, relaxation times T_1 and T_2, chemical shift, and the fluid flow velocity of the object. Relaxation times vary over a wide range of values (0.3−1.2 s at 100 MHz) and are dependent on the pathological sate of the tissues. Therefore, relaxation times are often preferred as contrast agents for the discrimination of anatomical features. As already discussed in Chapter 1, the spin−lattice relaxation time T_1 characterizes the time-dependent return of the z component of magnetization M_z to M_o, the equilibrium value. If a spin system is subjected to 90 degrees pulse, the magnetization is given by

$$M_z(t) = M_o[1 - \exp(-t/T_1)]$$

To differentiate tissue types, the magnetic resonance imaging (MRI) systems analyze the magnitude of the received signals. Excited nuclei continue

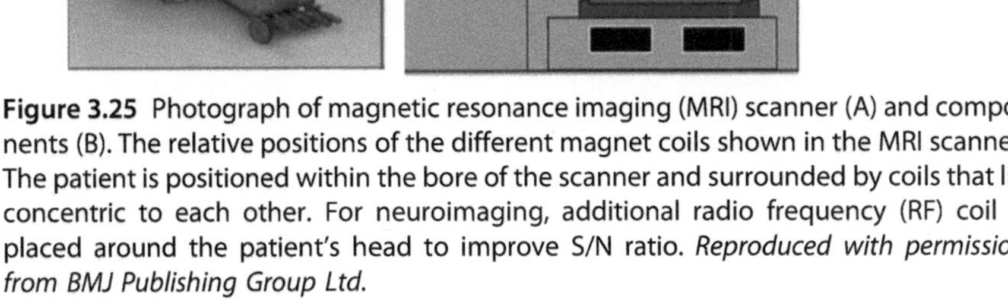

Figure 3.25 Photograph of magnetic resonance imaging (MRI) scanner (A) and components (B). The relative positions of the different magnet coils shown in the MRI scanner. The patient is positioned within the bore of the scanner and surrounded by coils that lie concentric to each other. For neuroimaging, additional radio frequency (RF) coil is placed around the patient's head to improve S/N ratio. *Reproduced with permission from BMJ Publishing Group Ltd.*

to radiate a signal until the energy absorbed during the excitation phase has been released. The time constant T_1 of these exponentially decaying signals ranges from tens of milliseconds to over a second. The recovery time is a function of field strength and the type of tissue. It is the variations in this time constant that allow different tissue types to be identified. Thus, different tissues have different values of T_1, and diseased tissues often have substantially altered T_1 relaxation time hence different contrast compared to the healthy tissue.

The spin-spin relaxation time T_2 governs the decay of any transverse magnetization to zero. In other words, it governs the return of the M_x and M_y components of magnetization to their thermal equilibrium values of zero. Again for a 90 degrees pulse,

$$M_{x,y}(t) = M_o \exp(-t/T_2)$$

As is the case for T_1 relaxation times, different tissues in the body have different values of T_2, and these can also be used to differentiate between healthy and diseased tissues in clinical images.

3.8.2 Magnetic Resonance Imaging Methods

The different available imaging methods are classified according to the spatial element that is detected. They are (1) sequential point, (2) sequential line, (3) sequential plane, and (4) simultaneous volume method.

1. Sequential point imaging involves obtaining the NMR response from only a single point in the sample at any one time. The image is then formed by sequential scanning of the selected point over the entire region of interest.
2. In the sequential line method, a line-by-line scan is performed through the region of interest. This method is more efficient and sophisticated than the previous one.
3. In a similar way, an image can be produced by selecting sequential planes or
4. By simultaneously imaging the whole volume. The selection of the technique depends on the S/N ratio, freedom from image distortion and minimum imaging time, etc.

3.8.3 Magnetic Resonance Imaging Scanner

MRI measures the NMR signal and properties of protons at different location in the patient's body in a high magnetic field. Therefore, the patient is placed into a long tubular magnet. The patient can be easily transported in and out of the magnet with a motor-driven table as shown in Fig. 3.25. The main magnetic field in the MRI scanner usually has field strength of about 2T. That is roughly 50,000 times stronger than the Earth's magnetic field of about 0.5×10^{-4} T. The magnet employed is typically a *super-conducting* magnet, with the coils cooled to about 4 K by liquid helium. In these conditions, the coils have no resistance to electrical current. The coils and the liquid helium are situated in a big thermostat to prevent heat transfer to and from the outside. Otherwise, the patient would freeze immediately. Nowadays, scan times of 0.5 s per slice with 512×512 pixels are common. The slice thickness usually ranges between 0.5 and 2 mm.

The major components and their functions in the MRI scanner are described as follows.

3.8.3.1 *Static Magnetic Field*

MRI imaging requires the patient to be placed in a strong magnetic field to align the hydrogen nuclei. Localization of MRI signal to specific part of the body (selecting a slice) and the ability to create a 3Dimage depends on the fact that the Larmor frequency of a nucleus is governed in part by the strength of the external magnetic field. There are typically three methods to generate this field: (1) fixed magnets, (2) resistive magnets (current passing through a traditional coil of wire), and (3) superconducting magnets. Fixed magnets and resistive magnets are generally restricted to field strengths below 0.4 T and cannot generate the higher field strengths typically necessary for high-resolution imaging. As a result, most high-resolution imaging systems use superconducting magnets. However, the superconducting magnets are large and complex. The magnetic coils are to be soaked in liquid helium to reduce their temperature to a value close to absolute zero.

The magnetic fields generated by these methods should not only be strong but also highly uniform in space and stable in time. A typical system must have less than 10 ppm variation over the imaging area. To achieve this accuracy, *shim coils* are used to generate weaker static magnetic fields and to "shim" or "tweak" the static field from the superconductor and thereby correct for field inaccuracies.

3.8.3.2 *Magnetic Gradients Coils for Position Encoding*

In 2003, Nobel Prize winners—Lauterbur and Mansfield thought of an idea and implemented in MRI to know how much signal comes from where. He made the left side of the field around the patient a little bit stronger than the right side. This was done with a *gradient* field. So, in a 1.5 T magnetic field, the left side starts at 1.51 T and the right side ends at 1.49 T. This means that different frequencies are measured from the relaxing protons, i.e., a little higher frequency on the left, and with decreasing frequency to the right. So the frequency codes for the position of one dimension (x) and the strength of the signal denotes how many protons participate. This process was repeated for the other two directions: up—down (y) and front—back (z). The different frequencies are thus measured from so-called *Fourier spectrum*.

For example, there are two locations (a and b) containing protons of same kind. When the uniform magnetic field is applied along the axis (shown in Fig. 3.26), the free induction decay (FID) obtained on Fourier transform (FT) gives only one frequency signal for two locations. Thus, the spatial information is not observed. However, when the varying magnetic field is applied, the FID obtained using the gradient field on FT display two different frequencies corresponding to two different locations, along x-axis. This demonstrates the spatial encoding in x-direction. In a similar way, the y- and z-gradients encode the other dimensions.

So an MRI system must have x, y, and z gradient coils to produce gradients in three dimensions and thereby create an image slice over any plane within the patient's body. The application of each gradient field and the excitation pulses must be properly sequenced, or timed, to allow the collection of an image data set (Fig. 3.29). Gradient coils generate spatially

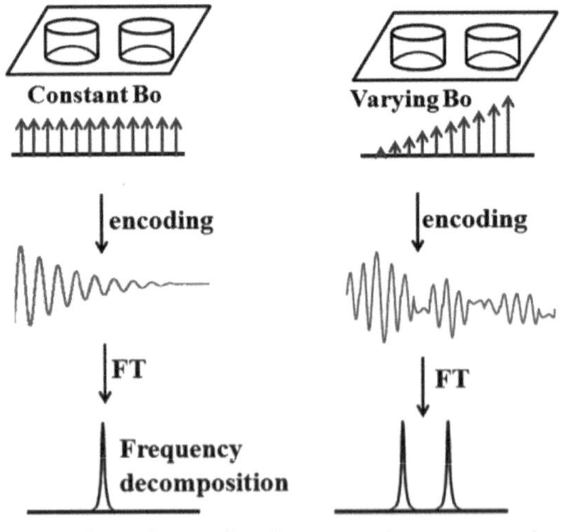

Figure 3.26 Spatial encoding in magnetic resonance imaging.

varying magnetic field so that spins at different location precess at frequencies unique to their location, allowing us to reconstruct two dimensional (2D) or 3D images. If a field gradient is applied across an object, the nuclear spins will no longer precess at the same Larmor frequency. Those in a stronger total field will precess more rapidly than those in a weaker total field.

These magnetic gradient coils vary the magnetic field within the subject for spatial encoding of protons by successive application of three field gradients along x (frequency encoding), y (phase-encoding), and z (slice encoding/selection) axes. These gradient coils are designed so that the spatial variation in magnetic field is linear (Fig. 3.27) with respect to spatial location, i.e.,

$$\delta B_z/\delta_z = G_z, \quad \delta B_z/\delta x = G_x, \quad \delta B_z/\delta_y = G_y \tag{3.3}$$

G represents the gradient measured in T/m. The three separate magnetic field gradients are produced by passing a direct current through separate coils of wire. The gradient coils are also designed such that there is no additional contribution to the magnetic field at the isocenter ($z = 0$, $y = 0$, $x = 0$) of the gradients, which means that the magnetic field at this position is simply B_0 (Fig. 3.27). By convention, the y-axis corresponds to the anterior/posterior direction, and the x-axis to the left/right direction of the patient lying in the magnet (Fig. 3.28B).

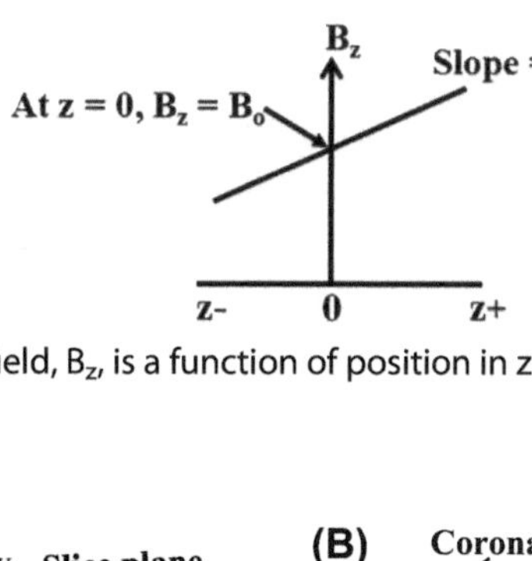

Figure 3.27 Magnetic field, B_z, is a function of position in z. The slope of the graph is G_z, the z-gradient.

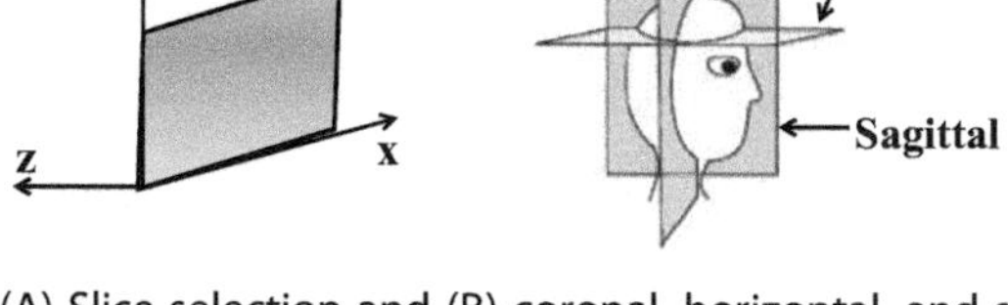

Figure 3.28 (A) Slice selection and (B) coronal, horizontal, and sagittal planes.

The magnetic field, B_z experienced by protons with a common z-coordinate is given by:

$$B_z = B_0 + zG_z \tag{3.4}$$

From the graph shown in Fig. 3.27 at position $z = 0$, $B_z = B_0$; for all positions $z > 0$, $B_z > B_0$, and for the position $z < 0$, $B_z < B_0$. The Larmor precession frequencies (ω_z) of the protons, as a function of their position in z, are given by:

$$\omega_z = \gamma B_z = \gamma(B_0 + zG_z) \tag{3.5}$$

Analogous expressions can be obtained for the spatial dependence of the resonant frequencies in the presence of the x- and y-gradients.

As mentioned earlier, the magnetic resonance (MR) signal is localized in three dimensions using three separate magnetic field gradients termed (1) slice-selection gradient (z-axis) (2) phase-encoding gradient (GP) (y-axis), and (3) frequency-encoding gradient (GF) (x-axis).

3.8.3.2.1 Slice-Selection Gradient

Slice localization is achieved by using gradient coils to generate a gradient field orientated along a chosen axis. By applying a gradient in the z direction slice selection (x, y plane), for example, one can change the resonant frequency required to excite a 2D slice in that plane (Fig. 3.28A). For example, coronal, axial, or sagittal images corresponding to slice selection in the y-, z-, or x-directions, respectively, can be chosen (Fig. 3.28B). This gradient field alters the strength of B_0 in the chosen direction, so that protons within the gradient field have different Larmor frequencies. Typically, the radio frequency (RF) pulse is applied as a small range of frequencies (bandwidth) rather than as a single frequency. This excites a slice of a certain thickness. Hence, slice thickness can be altered in two ways: (1) by changing the bandwidth of the RF pulse or (2) by changing the steepness of the gradient field. With a given slice selectively excited, the signals arising from each slice element (pixel) within that section need to be spatially encoded. This is achieved using GP and frequency-encoding gradients as follows.

3.8.3.2.2 Phase Encoding

After an axial slice selection, in which the slice-select gradient was applied in the z-direction, the y- and x-direction gradients are now applied to the other dimensions encode via the phase and frequency of the MR signal (Fig. 3.29). Applying a new phase gradient magnetic field along y-axis

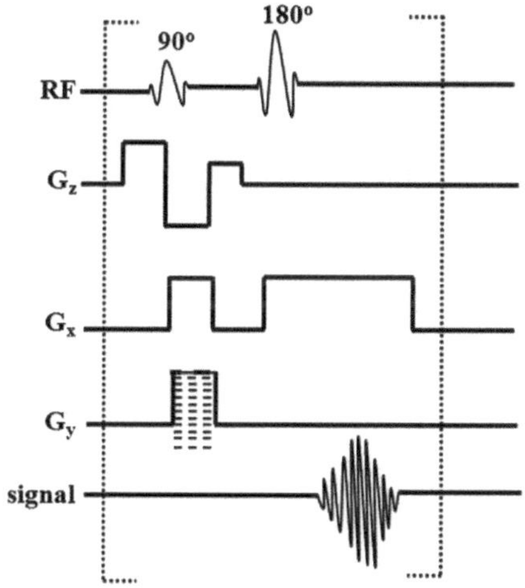

Figure 3.29 Spin echo pulse sequence for imaging. An RF pulse is applied while various gradients are turned on and off.

will make some of the protons precess faster than others depending on their position within the gradient. The gradient applied is called the G_y. Thus, spin phases will vary linearly over the phase-encode direction and thus encoded in y-dimension.

3.8.3.2.3 Frequency Encoding

The x-dimension is encoded by applying a frequency-encoding gradient (G_x). The G_x is applied following and perpendicular to the GP. It causes the protons to rotate at different frequencies according to their relative position along the gradient permitting differentiation of the signal in the third dimension (x).

3.8.3.2.4 Image Signal Processing

Both frequency and phase data are collected in what is commonly referred to as the k-space. A 2D Fourier transform of this k-space is computed to produce a gray-scale image.

3.8.3.3 Transmit/Receive/Shim Coils

Transmit and receive coils are used both to stimulate the hydrogen nuclei and to receive the signals generated as the nuclei recover. The MRI transmitter generates the RF pulses necessary to resonate the hydrogen nuclei. The range of frequencies in the transmit excitation pulse and the magnitude of the gradient field determine the width of the image slice. These coils must be optimized for the particular area of body to be imaged, so they are available in a wide variety of configurations. Depending on the area to be imaged, either separate transmit and receive coils or combined transmit/receive coils are used. In addition, to improve image acquisition times, MRI systems use multiple transmit/receive coils to recover more information in parallel, thus utilizing the spatial information associated with the location of the coils.

Shim coils: Localization of the MR signal requires good homogeneity within the local magnetic field. In other words, the more uniform the magnetic field the better the image. However, placement of an object (including a patient) within the main B_0 field creates local susceptibility effects and reduces homogeneity. Shimming coils are used to make adjustments to the magnet to improve its homogeneity.

3.8.4 Image Formation/Image Acquisition Time

The process of image formation can be thus broken down into three separate independent components, slice-selection, phase-encoding, and frequency-encoding. An overall imaging "pulse sequence" is shown in Fig. 3.29. The transmitter line indicates that an RF pulse is applied, and the length and power of the pulse are adjusted to give an indicated flip angle. For each gradient line, the height of the gradient pulse indicated its strength and the polarity (positive or negative) indicates which direction current is flowing through the particular gradient coil. The entire sequence of RF pulse and three gradients have to be repeated a number of times (N_P typically is between 128 and 512) to build-up a 2D data set.

As each phase-encoding step requires a new pulse sequence (Fig. 3.29) the total image acquisition time will depend on the product of TR (time interval between pulse sequences) and NP (number of phase-encoding steps). Conventional pulse sequences such as spin echo (SE) and gradient echo (GE) acquire only one phase-encoding step (one line of k-space) per TR, making the image acquisition time considerably long. This limitation is overcome by faster imaging techniques that acquire multiple lines of k-space per TR. The so-called Turbo (or fast) SE (TSE) and GE techniques are now used in MRI.

3.8.5 Types of Images Based on Pulse Sequences

3.8.5.1 Spin Echo Images

A SE method uses a 90 degrees RF pulse along with a slice-selective gradient (Fig. 3.29). This selective 90 degrees excitation pulse flips the magnetization within the slice to the transverse plane magnetization which is dephased by the first gradient. A 180 degrees pulse is applied at the middle of the sequence and the magnetization is replaced by the second gradient. The amplitude of the SE is affected by T_2 *weighted*. The degree of T_2 weighting is determined by the value of echo time (TE), which may vary from few milliseconds to hundreds of milliseconds.

SE sequence employs large flip angles, so it requires long recovery time TR to allow adequate recovery of longitudinal magnetization. Typically, TR values range from hundreds of milliseconds to seconds. As the total scan time is dependent on TR, SE sequences can be lengthy. SE is the least artifact-prone sequence and generates a high signal-to-noise ratio. However, the S/N is higher in SE due to both 90 and 180 degrees RF pulse [10,12].

3.8.5.2 Gradient Echo Image

SE sequences with their relatively long TR and TE are time consuming. This limits the number of patients who can be scanned in a session and also risks movement artifact by a restless patient. So, gradient echo image (GE) sequence is used to overcome the above.

A GE pulse sequence uses a variable RF excitation pulse (generally less than 90 degrees). Hence, the magnitude of transverse magnetization is less than SE method, where all the longitudinal magnetizations are converted to a transverse magnetization. After the RF pulse is applied, the magnetization in the transverse plane is dephased by the gradient and then rephased by the second gradient. The term TE is the interval between RF excitation and the center of the GE. The value of TE is important in determining the signal contrast of the image. As the transverse magnetization is subject to T_2 dephasing, regions of tissue whose T_2 value is short compared with TE will exhibit greatly attenuated signals. By contrast, regions with longer T_2 will have somewhat higher signals.

GE sequence often employs very short TR values, and images thus exhibit T_1 *weighting*. Tissue with short T_1 appears brighter because their longitudinal magnetization is less easily saturated. The degree of T_1 weighting also increases with flip angle, because higher flip angles cause greater saturation. The flip angle typically used in GE is within the range of 20—45 degrees. GE can provide faster imaging using shorter TR and shorter TEs than SE. In GE, less energy deposit occurs in the body due to use of low flip angle ($>$90 degrees). In GE, more slices per TR are generated than SE. GE is thus more compatible with 3D acquisitions. Chemical shifts are not refocused in GE, and this is the most important differences with an SE sequence.

3.8.5.2.1 Difference Between Spin Echo and Gradient Echo

There exist several differences between the two techniques.

1. TE can be shorter with GE as only one RF pulse is used.
2. GE sequences are typically used with low flip angle excitations (e.g., 5–40 degrees) as opposed to the 90 degrees flip angle used in SE. As a consequence, longitudinal magnetization is not completely abolished with GE and will provide reasonable signal even at very short TR. This combination of short TE and short TR permits rapid signal acquisition with GE imaging.
3. The price of short acquisition time in GE, however, is greater signal loss in the presence of magnetic susceptibility effects.
4. Gradient reversal refocuses only those spins that have been dephased by the action of the gradient itself and unlike SE, where phase shifts resulting from local magnetic field inhomogeneity are minimized through the 180 degrees RF pulse, this inhomogeneity persists in GE. Hence, image contrast in GE is influenced by T_2* effects and is not confined to true T_2 relaxation [10,12].

So, in GE imaging, T_1- and T_2*-weighted images will be obtained by manipulating the flip angle, TE, and TR.

3.8.5.3 Turbo Spin Echo Image

The TSE sequence applies multiple 180 degrees pulses after an initial 90 degrees pulse (as opposed to the single 180 degrees pulse used in a conventional SE sequence). Each 180 degrees pulse generates an echo, which (after phase encoding) is used to fill a new line of k-space. The number of 180 degrees refocusing pulses (and, thus, the number of echoes generated) during one TR of the TSE sequence is termed the echo train length (ETL), echo factor, or turbo factor. ETL specifies the factor by which the pulse sequence is accelerated (e.g., an ETL of 16 means that the TSE sequence is 16 times faster than a conventional SE sequence). The signal amplitude of each successive echo generated by the TSE sequence will be smaller than the last by way of T_2 decay effects. The echoes will also have different TEs. This means that in the traditional sense, a single TE does not exist for the TSE sequence. Instead, an effective TE or pseudo-TE is ascribed.

3.8.5.4 Echo-Planar Imaging

Echo-planar imaging (EPI) is a fast MRI technique in which the rapid oscillation of high-amplitude gradients is used to generate multiple GEs per TR. In single-shot EPI, all of the phase-encoding steps are acquired in a single

TR. Multishot EPI uses a few TRs to acquire all phase-encoding steps. It is a rapid MRI technique that is capable of producing tomographic images at video rates.

3.8.6 Types of Images Based on Contrast Parameters

There are three types of images obtained based on the measurement of the tissues relaxation times T_1 and T_2 and proton density (PD).

3.8.6.1 T_1-Weighted Image

A T_1-weighted image is produced by a short repetition time (TR) between RF pulses and a short signal recovery time (TR). Because T_1 is all exponential growth time constant, a tissue with short T_1 produces all intense MR signal and is displayed as bright white in a T_1-weighted image. A tissue with long T_1 produces a low intensity signal and appears dark in MR image. A short TR of 500 ms between pulses and a short echo of signal TE of 20 ms produces T_1-weighted image. T_1-weighted images are called ***fat images*** because the fat has the shortest T_1 relaxation time and thus appear bright in the image. High anatomic detail is possible in this type of image because of good image contrast. T_1-weighted images are thus useful for depicting small anatomic regions where high spatial resolution is required.

3.8.6.2 T_2-Weighted Image

The magnetic moments of adjacent hydrogen nuclei begin to interfere with one another; this causes the nuclei to dephase, with a resultant loss of transverse magnetization. The time constant that describes the rate of loss of transverse magnetization is called T_2 relaxation time/transverse (Spin—spin) relaxation time. The transverse magnetization rapidly decays (exponentially) to zero. A T_2-weighted image is usually acquired using a long repetition time between RF pulses and a long signal recovery time. A tissue with a long T_2 produces a high-intensity signal and is bright in the image. One with short T_2 produces a low-intensity signal and is dark in the image. This property is opposite of T_1.

A long TR (2000 ms) and a long TE (80 ms) produces T_2-weighted images for every diagnostic task. The operator must decide which imaging sequence will bring out optimal image contrast. T_2-weighted images are called ***water images*** because water has the longest T_2 relaxation time and thus appear bright in the image. In general, the T_2 time of abnormal tissues is longer than that of normal tissues. T_2-weighted images are most commonly used for inflammatory changes and tumors.

3.8.6.3 Proton Density Weighting Image

A PD image is one where the difference in the numbers of protons per unit volume in the patient is the main determining factor in forming image contrast. PD weighting is always present to some extent. To achieve PD weighting, the effects of T_1 and T_2 contrast must be diminished, so that PD weighting can dominate. A long TR allows tissues, e.g., fat and water to fully recover their longitudinal magnetization and therefore diminishes T_1 weighting. A short TE does not give fat or water time to decay and therefore diminishes T_2 weighting. Fig. 3.30 below shows a comparison of T_1, T_2, and PD weighting images.

The relationship between TR and TE and tissue contrast are shown in Fig. 3.30 and in Table 3.1.

3.8.7 Recent Advances in Magnetic Resonance Imaging

Every year seems to bring a new innovations and applications of MRI or a new pulse sequence that opens up new imaging opportunities and applications of MRI.

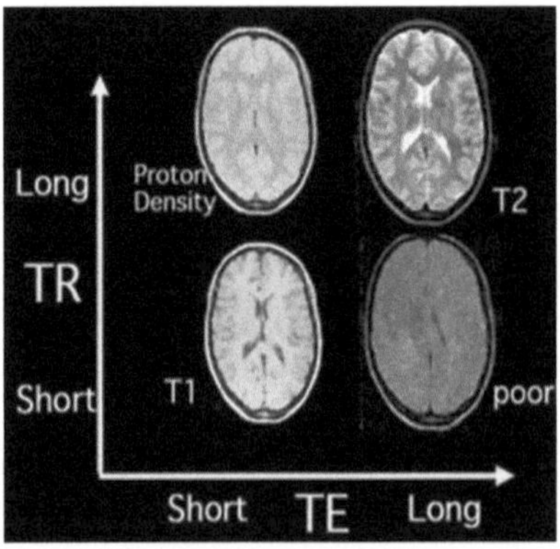

Figure 3.30 T_1, T_2, and PD images. *PD*, proton density.

Table 3.1 Comparison of TR and TE for T_1, T_2, and PD weighted sequences

Weighting	TR	TE
T_1	Short	Short
T_2	Long	Long
PD	Long	Short

3.8.7.1 Volume Imaging—Three Dimensional Imaging

Volume imaging is the acquisition of magnetic resonance data from a volume rather than a tomographic slice. It can be thought of as collecting several contiguous slices through a region of imaged object.

3.8.7.2 Echo-Planar Imaging (Functional Magnetic Resonance Imaging)

It is based on EPI method. (see Section 3.8.5.4) It is a rapid MRI technique which is capable of producing tomographic images at video rates. Its greatest application appears to be in the area of functional MRI (fMRI) of the brain. Functional imaging is the imaging that relates body function or any particular thought to activation of specific locations in the brain.

Here with an MRI scanner, one can even measure *where we think* in the brain for that thought. When brain cells are active, for example, when we solve a crossword puzzle, the brain cells need energy. It turns out that near the active cells the very small blood vessels (the capillaries) become a little wider to provide the extra oxygen required. So the amount of oxygen locally increases, and this is exactly what the MRI scanner can measure; oxygen is a paramagnetic substance, which means that differences in concentration can be seen by the scanner. The difference with the nontask situation is small, so normally one makes two scans: one with the person doing nothing and one with the person performing the thinking task. When the two images are subtracted, the difference shows up easily. This method is called fMRI. It plays an important role in brain research, and in finding the critical areas for language, speech disorders, etc., and in brain surgery patients. When the location of these critical areas is identified, they can be carefully avoided during surgery. So, fMRI is based on,

1. Hemoglobin has different magnetic properties when bound to oxygen that can be distinguished by fMRI.
2. Areas of brain activity have a surge of oxygenated blood.
3. fMRI can identify areas of the brain with high oxyhemoglobin content, which correlates to areas of heightened brain activity.

3.8.7.3 Phase-Contrast Imaging

It is used to obtain quantitative MRI flow information. Phase-contrast imaging (PC MRI) is a technique which images the magnetized moving spins thereby obtaining quantitative flow information. By using PC MRI, flow can be directly quantified on the property of protons moving in a magnetic field causing phase shift, and the magnitude of this phase shift is proportional to their velocity. Thus, the velocity can be measured and imaged. This is achieved by the introduction of a *bipolar flow encoding gradient*, after the RF excitation pulse and slice selection gradient, to encode the velocity of the moving spins. The velocity of the moving spins is translated into phase of the image by the flow encoding gradients and is directly proportional to the velocity of the moving spins. The flow gradients can be placed in any physical gradient axis to achieve flow sensitivity in the respective direction. In fact, flow gradients can even be applied simultaneously in multiple axes to obtain flow sensitivity in an arbitrary direction. Phase data obtained from the different MRI techniques is influenced by factors such as B_0 inhomogeneities and eddy currents. To reduce these effects in PC MRI, two images are acquired. The first image is acquired with the flow gradient and the second image either without or toggled flow gradient. PC MRI relies on two image acquisitions with and without velocity-encoding gradient.

3.8.7.4 Magnetic Resonance Angiography

Angiography is a flow imaging method. The magnetic resonance angiography (MRA) produces images of flowing blood in the arteries and veins of the body. The intensity in these images is proportional to the velocity of the flow. There are three general types of MRA—(1) time of flight, (2) phase-contrast angiography and (3) contrast-enhanced angiography.

The vascular system (the network of blood vessels) can be visualized with MRI in two ways. Injection of the contrast agent gadolinium (Gd) in the blood vessels is one option called contrast enhanced [14—16]. This contrast agent has a particular magnetic property by which it changes the duration of the relaxation times because it influences the surrounding magnetic field of the protons in the patient's body. This method is very sensitive and produces images with high contrast as shown in Fig. 3.31. The problem with this method is that it is very expensive: one injection costs around 100 Euro.

Contrast agent always brings some risks, because a patient may be allergic to it. The second phase-contrast MRA method has the advantage of not using any contrast agent. This method is based on exciting the protons in one thin slice only. If this slice contains an artery or vein, new unexcited blood will flow in from the neighboring slices. A short while after the excitation an MR image is made of the excited slice in which all tissues are still excited, except the locations of the blood vessel where the excited protons have moved out. This results in many small "zero intensity areas," where the blood flow is located. Applying this method to all slices, a 3D image can be reconstructed. In this way the whole vascular system can be mapped,

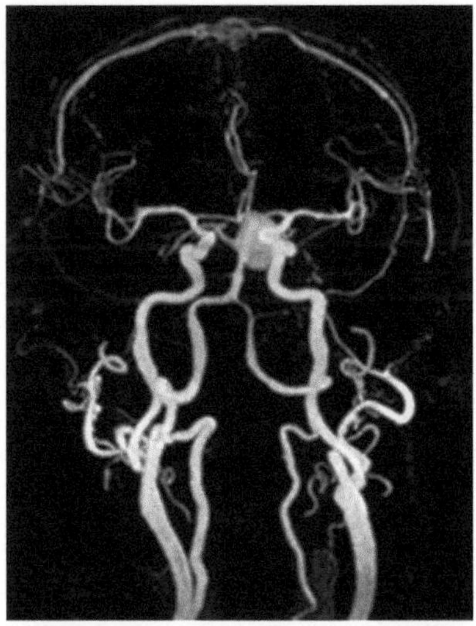

Figure 3.31 Contrast-enhanced magnetic resonance angiography showing the supra-aortic arteries. *Reproduced with permission from American Society of Neuroradiology.*

without the need for an expensive and potentially risky contrast medium. However, the main disadvantage of this method is that an artery or vein may be streaming in the direction of the slice. Using these methods, the occurrence of a *stenosis* (narrowing of a vessel) might be detected in an early stage.

3.8.7.5 Chemical Shift Imaging (Fat Suppression)

Chemical shift imaging (fat suppression) is the production of an image from just one chemical shift component in a sample.

3.8.7.6 Magnetization Transfer Contrast

It is a method of increasing the contrast between tissues by physical means rather than using chemical contrast agents.

3.8.7.7 Magnetic Resonance Elastography

It is the imaging of shear waves using MRI. Contrast in magnetic resonance elastography is related to the elastic modulus of the tissue. MRI is recorded, whereas ultrasound waves are being sent into the imaged volume. This technique is expected to find applications in locating pathology in soft tissue based on difference in the elastic modulus of tissues. Hence it is referred to as *magnetic resonance palpation*.

3.8.7.8 Magnetic Resonance Spectroscopy

It is a noninvasive tool to measure the chemical composition of tissues (in vivo) and characterize functional metabolic processes in different parts of the human organs. It provides vital biological information at the molecular level. Combined with MRI, an integrated MRI/MRS (magnetic resonance spectroscopy) examination provides anatomical structure, pathological function, and biochemical information about a living system. MRS provides a link between the biochemical alterations and the pathophysiology of disease. The same scanner is used for both MRI and MRS studies. Both techniques are based on the same physical principles. However, a special RF coil is placed around the patient's head or that particular organ and MRI/MRS experiments are performed [12,13].

MRS is generally less sensitive than MRI because the concentration of nuclei (^{1}H, ^{31}P, and so on from the neurometabolites) as measured by MRS are orders of magnitude less concentrated compared to the concentration of hydrogen (from water) generally involved in MRI. MRS provides chemical composition of tissues from the particular region of interest. An MRI image is first generated, the voxel is selected at the site of interest, and then MRS spectra recorded from that voxel (Fig. 3.32). The use of spatial localization is essential for in vivo MRS for selection of single voxel from a particular region of interest.

MRS requires a magnetic field and a RF transmit pulse at a particular resonant frequency to observe the signal of a specific nuclei (e.g., ^{1}H, ^{31}P, ^{13}C, etc.) in the region of interest. The product of MRS is a "spectrum" with a frequency axis in parts per million (ppm) and a signal amplitude

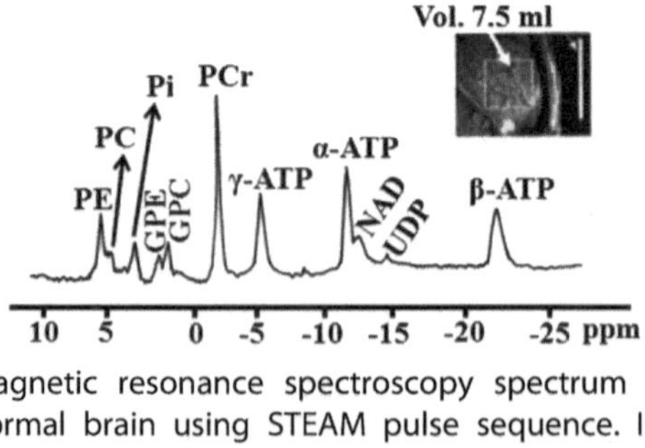

Figure 3.32 ^{31}P magnetic resonance spectroscopy spectrum from parietal white matter at 7 T in normal brain using STEAM pulse sequence. Inset indicates voxel location [12]. *GPC*, glycerophosphorylcholine, *GPE*, glycerophosphorylethanolamine, PC, phosphorylcholine, *PE*, phosphorylethanolamine, *STEAM*, STimulated Echo Acquisition Mode, *UDP*, Uridine diphosphate. *Reproduced with permission from John Wiley and Sons.*

axis similar to 1D NMR. The signal amplitude (area) is a measure of a particular metabolite concentration (Fig. 3.32).

The following metabolites are detected by ^{1}H MRS as follows:

1. *N*-acetyl aspartate is a neuronal marker seen only in nervous tissue.
2. Glutamate (Glu) and glutamine complex is a mixture of peaks that helps to monitor Glu metabolism in the brain for chronic epileptic activities, amyotrophic lateral sclerosis and in other neuronal disorders.
3. Lactate is a highly specific marker of cell death as well as tissue necrosis.
4. Creatine is a marker of energetic status of cells.
5. Choline, an indicator of membrane activity, is often elevated in the presence of malignant processes.
6. Myoinositol, a sugar alcohol, is a marker of astrocytic activity and is often higher in conditions such as Alzheimers disease and malignant tumors.

3.8.7.9 Electron Spin Resonance or Electron Paramagnetic Resonance Imaging

It is based on the electron spin of the paramagnetic species rather than the nuclear spin. Electron spin resonance (ESR) imaging is the study of the spatial distribution of ESR signal bearing substance. However, very few substances are paramagnetic and can be studied with ESR (see Section 5.11). For example, nitroxide spin probes and some transition metals exhibit an ESR signal. So, these substances have been studied directly by ESR and are commonly used to probe biologic processes, oximetry, and redox status with ESR imaging.

3.8.8 Magnetic Resonance Imaging Demerits and Safety

3.8.8.1 Demerits

1. Because of the strong magnetic field, patients having electrically, magnetically, or mechanically activated implants such as cardiac pacemakers, implantable defibrillators, and some artificial heart valves may not be investigated safely by MRI.
2. The MRI image becomes distorted by metal, so the image is distorted in patients with surgical clips or stents.
3. The very powerful magnets can pose problems using any equipment.
4. MRI scanners are noisy.
5. Patient could develop an allergic reaction to the contrasting agent, or cause a skin infection.
6. MRI equipment is expensive to purchase, maintain, and operate. Hardware and software are still being developed.

3.8.8.2 Safety

1. Several deadly MRI related accidents have occurred.
2. The most dangerous part of an MRI scanner is its powerful magnetic field. The magnetic field in an MRI scanner is stronger than those used in industry to move cars. This magnetic field will cause objects made of ferromagnetic materials to be pulled into the scanner. It will also pull on ferromagnetic objects implanted in the body.
3. The following metals are ferromagnetic and are not MRI compatible:
   a. Iron
   b. Nickel
   c. Cobalt

   The objects made of these materials should not be taken into the MRI scanner room.
4. Patients should be screened for the following:
   a. Pacemakers
   b. Surgical clips
   c. Implants made of a ferromagnetic metal.

3.8.9 Contrast Agents

One of the strengths of MRI is the significant amount of intrinsic contrast between tissues. This contrast is based on differences in signal intensity between adjacent pixels in the image. It is a result primarily of differences in the T_1 and/or T_2 relaxation times of the tissues under observation. Pathologic tissue may or may not have significant differences in T_1 or T_2 from the surrounding normal tissue. For this reason, there may be little signal difference between normal and pathologic tissue in spite of the inherent high contrast in the images. This signal difference can be increased through the administration of a contrast agent [11,15,17].

MRI contrast agents are also known as relaxation reagents (slower electron spin relaxation) administered to enhance natural contrast to improve MR images by adding functional data to the morphological information gathered from unenhanced images and obtain dynamic information (Fig. 3.31). MR images are created from proton relaxation times (T_1 and T_2) and from proton concentrations. The presence of a biological marker of interest can be imaged if it alters any of these two variables. Most contrast agents are chosen for their ability to alter T_1 or T_2 of the protons in their local environment. T_1 and T_2 can be affected by the various properties of the contrast agents but usually to the different extent. Thus, the tissue will

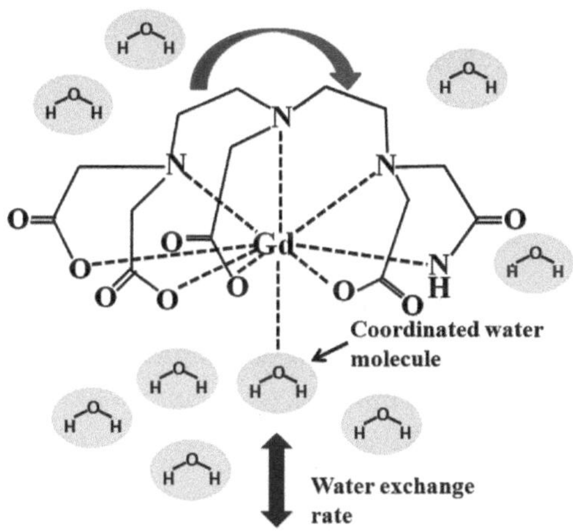

Figure 3.33 Coordinated water molecule for diethylenetriaminepentaacetate-Gd, water exchange, and rotational correlation time.

experience a varying magnetic field that depends on the magnetic moment of the contrast agent, its electron spin relaxation time, and the nature of the contrast agent's molecular interactions with the protons. These molecular interactions include inner-sphere and outer-sphere coordination and diffusional translation (Fig. 3.33).

3.8.9.1 Classification of Magnetic Resonance Imaging Contrast Agent

They are classified into two major classes based on (1) paramagnetic properties and (2) route of administration.

3.8.9.1.1 Based on Paramagnetic Properties

There are two major types of MRI contrast agents based on paramagnetic properties.

(1) Paramagnetic contrast agent and (2) Superparamagnetic contrast agent.

3.8.9.1.1.1 Paramagnetic Contrast Agent These are paramagnetic ions or compounds called *positive enhancers* because they increase the MRI signal intensity. They are largely extracellular nonspecific. These include manganese(II), iron(III), Gd(III) compounds, and also stable free radicals such as nitroxides. They are categorized as T_1 *contrast agents* based on their primary effect of shortening T_1 relaxation time. These contrast agents have several unpaired electrons highly paramagnetic and, therefore, cause large field fluctuations. The magnetic moments of these contrast agents are about 1800 times larger than the hydrogen proton. If close enough, the relaxation rate of neighboring protons significantly modified

$$1/T_1 = (1/T_1)_{\text{inner sphere}} + (1/T_1)_{\text{outer sphere}}$$

Gd is most well-known currently the only MRI contrast agent approved for human use and is therefore more thoroughly studied in the literature. Gd-based pharmaceuticals alter the relaxation time of those protons in water molecules that are transiently attached to the paramagnetic Gd resulting in contrast enhancement of the MRI image. This allows especially visualization of some brain tumors.

3.8.9.1.1.2 Superparamagnetic Contrast Agents It consists of contrast agents that decrease the MRI signal intensity by increasing the magnetic field inhomogeneity. They are also called *negative* MRI signal enhancers by shortening or decreasing $T_2/T_2^\star$. These include metal ions such as Dy(III) and superparamagnetic particles containing several iron atoms as clusters such as iron oxide nanoparticles. These are intravascular (so-called blood pool contrast agent) and highly tissue specific. These particles alter proton relaxation times by creating microenvironments of magnetic field inhomogeneity. Protons located in or near these microenvironments will experience this field inhomogeneity, called the *outer-sphere* effect or *susceptibility* effect, and their spins will rapidly dephase resulting in a decrease in T_2. These effects are best seen using T_2-weighted and gradient-echo pulse sequences.

The mechanistic effect of contrast agents (Fig. 3.33) include (1) water exchange rate, which is the rate of the coordinated water molecule exchanging with the bulk (also called inner-sphere effects), (2) rotational correlation time, which is how quickly the contrast agent is tumbling in solution, and (3) susceptibility effect ($T_2^\star$), which is the effect of the contrast agent on the bulk water (noncoordinated water molecules, also called outer-sphere effect). (4) These effects alter either T_1 or T_2 relaxation times. (5) It also leads to resonance frequency shifting. (6) However, the difference of PD cannot be modified by contrast agents. (7) Most MRI contrast agents are indirect agents in that they are never visualized directly in the image, but affect the relaxation times of the water protons in the nearby tissue [11,15,17].

3.8.9.1.2 Characteristics of Contrast Agents

The essential properties of contrast agents used in MRI are as follows:

1. They possess the ability to modify/alter T_1, T_2, or some aspect of the observables/tissue property.
2. They must be tissue specific, for example, iron complexes are accumulated in the cancer cells.
3. It should be fairly stable and also has reasonable clearing period in our biological system.
4. It should not be toxic and should be excreted through the renal system within 3—4 days.
5. It must possess long shelf life.
6. The concentration and dosage for MRI agents are also significantly lower than for a CT agent, which in part explains the lower occurrence of adverse reactions to MRI agents.

3.8.9.1.3 Based on Route of Administration

Contrast agents may also be grouped depending on the route of administration into two types (1) intravenous and (2) oral agents. A brief discussion of MR contrasts agents with emphasis on those in current clinical use is as follows.

3.8.9.1.3.1 Intravenous Agents These are administrated through veins by means of injection. Among the intravenous contrast agents, they are further classified into (1) T_1 relaxation agents and (2) T_2 relaxation agents.

3.8.9.1.3.1.1 T_1 Relaxation Agents Most intravenous contrast agents currently in clinical use are T_1 relaxation agents. A variety of formulations are available, all of which consist of one or more paramagnetic metal ions that contain one or more unpaired electrons. The metal ions are either bound to a chelate complex or contained within a macromolecule.

1. Intravenous T_1 contrast agents are available in both ionic and nonionic forms. The most common agents use Gd as the metal ion with,
   a. *Ionic*—diethylenetriaminepentaacetate (DTPA) *Magnevist*,
   b. *Nonionic*—(1) DTPA—bismethylamide, *Omniscan*
   10-(2-hydroxypropyl)-1,4,7,10-tetraazacyclododecane-1,4,7-triacetate (GdHPDO3A), *Calteridol* as the chelate molecule.

   As described earlier that the agents currently used are nonspecific extracellular in that they diffuse rapidly from the vascular space into the interstitial space of the tissue. They are eliminated through glomerular filtration and renal excretion. The half life is typically in the order of 90 min with virtually complete elimination of these agents within 24 h. Although they are safe, there were few reported side effects and rare cases of major side effects.

2. The other forms of Gd-chelated contrast agents have both hydrophilic and hydrophobic properties. Two examples are (1) Gd-Benzoxypropionic tetraacetate-Multihance and (2) Gd-Ethoxybenzyl-DTPA as the ligand—Eovist. The presence of a hydrophobic ligand enables hepatocyte fast uptake and eliminated by the biliary system, rather than strictly through renal excretion.
3. Manganese is another metal ion that has been formulated as a T_1 contrast agent for liver imaging. Manganese ion chelated to *N*,*N*-dipyridoxal ethylenediamine-*N*,*N*-diacetate 5,5-bis (phosphate) (*Mangofodipir*) is a T_1-weighted contrast agent that is mainly hepatocyte-selective. It is administered via a slow intravenous injection over 1−2 min. It is used extensively for the determination of the extent of liver metastases in patients under consideration for surgical resection.

3.8.9.1.3.1.2 T_2 Relaxation Agents The second classes of intravenous contrast agents are T_2 relaxation agents. They are typically large molecules containing several iron atoms that collectively form a superparamagnetic center. Because of the large magnetic susceptibility, it distorts the local magnetic field in the vicinity of the agent, causing the nearby water protons to dephase more rapidly than the surrounding tissue. This causes significant signal loss in T_2-weighted SE or GE. The most common T_2 contrast agents are based on superparamagnetic.

1. Superparamagnetic iron oxide
   These agents as particulate molecules called *ferumoxides* are selectively absorbed by the reticuloendothelial cells in the liver, spleen, and bone marrow. Normal tissue in these organ systems take up the agent and generate low signal in T_2- or $T_2^\star$-weighted images. These are of a smaller particle size widely used for liver imaging and are under investigation as contrast agents for the examination of lymph nodes. Normal or hyperplastic lymph nodes take up the agent and lose signal in T_2-weighted images, whereas there is no uptake by malignant lymph nodes and, therefore, have relatively high signal.
2. *Magnetite−dextrin* is the other T_2 agent known as Feridex in clinical use. It is usually administered as a slow drip infusion over 30 min. It leads to selective uptake of the agent by reticuloendothelial cells.

3. *Resovist* (magnetite/maghemite-carboxydextran) is another ferumoxide agent that can be administered in a small dose by bolus injection. The advantage of this agent is ease of administration and less side effect especially no back pain. It is an ultrasmall particulate iron oxide that also increases T_1 contrast immediately following administration and affect both T_1- and T_2-weighted scans. Resovist is currently under consideration for approval by the US Food and Drug Administration.
4. Monoclonal antibodies-labeled iron oxide particles have also been used to targeted specific tissues receptor sites.

3.8.9.1.3.2 Oral Agents Oral MRI contrast agents are typically nonspecific in nature. They are largely used for abdominal and pelvic studies to provide reliable differentiation of bowel from adjacent structures and to provide better delineation of bowel wall processes. They are again categorized as positive or negative agents.

1. Positive agents
   It increases the overall signal intensity within the image, generally by shortening the T_1 or T_2 relaxation times of tissue water. As described earlier, these agents are generally (1) Paramagnetic metal ions or metal—chelate complexes in solution form. Many of these agents are present in naturally occurring products (manganese in green tea and blueberry juice) (2) Other agents have been used such as manganese chloride (Lumenhance), Gd-DTPA (Magnevist) as T_1 relaxation agents and Ferric ammonium citrate (Oncomembrane) as a T_2 relaxation agent.
2. Negative agents
   These eliminate tissue signals from the area of interest because of T_2 shortening. Two approaches are used for negative agents. One method uses suspensions of ferumoxide particle to reduce the T_2 relaxation times and the other approach involves use of agents that contain no protons and, therefore, produces no visible MR signals. The most common agent of this type is Perflubron (perfluorooctylbromide). Barium sulfate, clay, and air have also been employed for intraluminal studies. One disadvantage with many negative contrast agents, especially the iron-based agents, is that they increase the local magnetic susceptibility, which may induce significant dephasing artifacts at high field strengths.

REFERENCES

[1] R.S. Drago, Physical Methods in Inorganic Chemistry, Affiliated East-West Press Pvt. Ltd., New Delhi, 2012.
[2] L. Jackman, Dynamic Nuclear Magnetic Resonance Spectroscopy, Elsevier Inc., 1975.

[3] L.D. Field, S. Sternhell, J.R. Kalman, Organic Structures from Spectra, Wiley Inter Science, New York, 1998.
[4] Physical Organometallic Chemistry, in: M. Gielen, R. Willem, B. Wrackmeyer (Eds.), Fluxional Organometallic and Coordination Compounds, vol. 4, John Wiley & Sons Ltd., Hoboken (NJ, USA), 2004.
[5] M.J. Duer (Ed.), Solid State NMR Spectroscopy: Principles and Applications, Blackwell Science, Malden (USA), 2002.
[6] G. Facey, http://u-of-o-nmr-facility.blogspot.in/2007/11/magic-angle-spinning.html.
[7] D.F. Evans, J. Chem. Soc. (2003) 1959.
[8] K.D. Bartle, B.J. Dale, D.W. Jones, S. Maricic, NMR determination of the magnetic susceptibility of iron proteins: an undergraduate experiment, Biochem. Educ. 3 (3) (1975) 48.
[9] D.L. Rabenstein, NMR shift reagent, Anal. Chem. 43 (1971) 1599.
[10] S. Currie, N. Hoggard, I.J. Craven, M. Hadjivassiliou, I.D. Wilkinson, Understanding MRI: basic MR physics for physicians, Postgrad. Med. J. 89 (2013) 209–223.
[11] M.A. Brown, in: C. Richard, M.D. Semelka (Eds.), MRI Basic Principles and Applications, third ed., John Wiley and Sons, Inc., Hoboken (New Jersey), 2003.
[12] P.K. Mandal, Magnetic resonance spectroscopy (MRS) and its application in Alzheimer's disease, Concepts Magn. Reson. Part A 30A (2007) 40–64.
[13] E. Westman, L.O. Wahlund, C. Foy, M. Poppe, A. Cooper, D. Murphy, C. Spenger, S. Lovestone, A. Simmons, Magnetic resonance imaging and magnetic resonance spectroscopy for detection of early Alzheimer's disease, J. Alzheimers Dis. 3 (2011) 307–319.
[14] D.J. Shah, T.H. Lim, Evaluation of meglumine gadoterate-enhanced MR angiography (MRA) compared with time-of-flight MRA in the diagnosis of clinically significant non-coronary arterial disease: a pooled analysis of data from two clinical trials, Br. J. Radiol. 85 (2012) 596–605.
[15] A. Tartaro, M.T. Maccarone, The utility of gadoteric acid in contrast-enhanced MRI: a review, Rep. Med. Imaging 8 (2015) 25–35.
[16] K. Nael, J.P. Villablanca, R. Saleh, W. Pope, A. Nael, G. Laub, J.P. Finn, Contrast-enhanced MR angiography at 3T in the evaluation of intracranial aneurysms: a comparison with time-of-flight MR angiography, Am. J. Neuroradiol. 27 (10) (2006) 2118–2121.
[17] P. Caravan, J.J. Ellison, T.J. McMurry, R.B. Lauffer, Gadolinium(III) chelates as MRI contrast agents: structure, dynamics, and applications, Chem. Rev. 99 (1999) 2293–2352.

CHAPTER FOUR

Electron Paramagnetic Resonance Spectroscopy

Chandran Karunakaran, Murugesan Balamurugan
VHNSN College (Autonomous), Virudhunagar, India

Contents

Spin Resonance Spectroscopy
ISBN: 978-0-12-813608-9
https://doi.org/10.1016/B978-0-12-813608-9.00004-6

Abstract

Electron paramagnetic resonance (EPR) is the only unambiguous technique to investigate directly the free radicals and paramagnetic compounds viz., transition metal ions, triplet states, etc. So, in this chapter, the principles of EPR, instrumentation (parallel and perpendicular modes), and presentation of spectra are discussed. The Spin Hamiltonian equation used to describe the EPR spectra and parameters are explained. The determination of g-values including factors affecting its magnitude, reference samples, isotropic hyperfine interaction, EPR of hydrogen atom, and free radicals in solution are presented. g- and A-tensor anisotropy, zero-field splitting, Kramer's theory and degeneracy, exchange-coupled EPR, half-field transition, and triplet-state EPR are highlighted. Optically Detected Magnetic Resonance (ODMR), a double resonance technique for studying triplet states is also discussed. The effect Jahn-Teller (JT) coupling/distortion on EPR and optical spectra, its types, consequences, and applications of JT are presented. A brief account and useful list of computer simulation software are also given.

Keywords: Exchange coupled; Half-field transition; JT coupling; Kramer's theory; ODMR; Simulation; Spin Hamiltonian; Triplet; Zero-field.

4.1 INTRODUCTION

The phenomenon of electron spin resonance (ESR), also called electron paramagnetic resonance (EPR)/electron magnetic resonance, was discovered by the Soviet physicist Zavoisky [1a], a year before the discovery of nuclear magnetic resonance (NMR) by the American physicists Bloch et al. [1b] and Purcell and Torrey [1c]. As described in Chapter 1, NMR and EPR are alike in their theoretical principles and are collectively referred to as spin magnetic resonance phenomena. However, the former is observed in radiofrequency (RF) region, while the latter is observed in microwave (MW) region of the electromagnetic spectrum. Although ESR is a highly sensitive technique, it has limited applications because it is observed primarily in systems containing only unpaired electrons. Thus, the systems which can be studied by ESR spectroscopy are organic or inorganic free radicals, triplet states and transition metal ions, their complexes, and metalloproteins with unpaired d- or f-electrons. One of the most important uses of ESR technique lies in the detection of extremely short-lived (transient) free radical intermediate in chemical reactions and in radiation damage [1–5].

4.2 SPIN HAMILTONIAN OF ELECTRON PARAMAGNETIC RESONANCE

The spin Hamiltonian is a mathematical expression of the different interactions that exist in transition metal ions or radicals with unpaired electrons. It can be thought of as a shorthand way of representing the interactions described below. The energy of a paramagnetic species in the ground state is thus described by the static spin Hamiltonian [6].

$$\widehat{H} = \widehat{H}_{EZ} + \widehat{H}_{HF} + \widehat{H}_{ZF} + \widehat{H}_{NQI} + \widehat{H}_{EEI} - \widehat{H}_{NZ} + \widehat{H}_{JT} \quad (4.1)$$

$\widehat{H}_{EZ}$, Electron Zeeman Interaction; $\widehat{H}_{HF}$, Hyperfine Interaction (HF) between electron and nuclear spins; $\widehat{H}_{ZF}$, Zero-Field Interaction; $\widehat{H}_{NQI}$, Nuclear Quadrupole Interaction; $\widehat{H}_{EEI}$, Electron—Electron Interaction (EEI) between two electron spins; $\widehat{H}_{NZ}$, Nuclear Zeeman Interaction; $\widehat{H}_{JT}$, Jahn-Teller (JT) coupling interaction characterizing the splitting due to the interaction of the electronic states with the crystal lattice vibrations.

4.2.1 Electron Zeeman Interaction

For an isolated electron, without any outside forces, it has only an intrinsic angular momentum called "spin." Because an electron is charged, the angular/spinning motion of this charged particle generates a magnetic field. In other words, the electron due to its charge and angular momentum, acts like a little bar magnet, or magnetic dipole, with a magnetic moment (μ).

The electron with spin 1/2, the parallel state (↑) is designated as $m_s = -1/2$ and the antiparallel (↓) state is $m_s = +1/2$ (Fig. 4.1A). The energy of each orientation is the product of μ and B_0.

$$E = \mu B_0 \quad \text{where } \mu = m_s\, g\, \beta \quad (4.2)$$

β is a conversion constant called the Bohr magneton (also called μ_B) and g_e is the spectroscopic g-factor of the free electron and equals 2.0023192778 (≈2.00). Therefore, the energies for an electron with $m_s = +1/2$ and $m_s = -1/2$ are, respectively, as follows:

$$E = m_s\, g\, \beta\, B_0 \quad (4.3)$$

$$E_{1/2} = 1/2\, g\, \beta\, B_0 \quad E_{-1/2} = -1/2\, g\, \beta\, B_0$$

4.2.2 Spin—Orbit Interaction

Consider now an electron placed in an atom or molecule (Fig. 4.1B). Its angular momentum changes because, in addition to the intrinsic spin angular

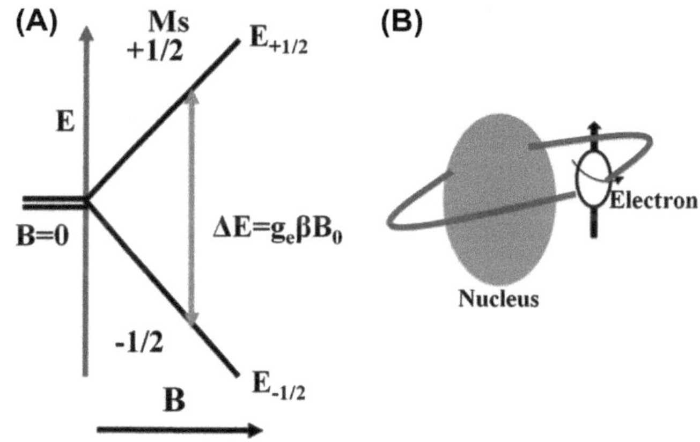

Figure 4.1 (A) Electron Zeeman interaction and (B) Spin—orbit coupling.

momentum, it now also possesses some orbital angular momentum. An electron with orbital angular momentum is, in effect, a circulating current. So, there is also a magnetic moment arising from the orbital angular momentum. These two magnetic moments couple with each other called *spin—orbit interaction* and the energy of this interaction depends on their relative orientations.

In general, the orbital angular momentum is approximately zero for an electron in the ground state (for s electron). However, interaction between the ground and excited states causes admixture of small amounts of orbital angular momentum of excited state to the ground state. The magnitude of the spin—orbit coupling contribution depends on the size of the nucleus containing the unpaired electron. Therefore, organic free radicals, with only H, O, C, and N atoms, will have a small contribution from spin—orbit coupling, producing g-factors very close to g_e. *ESR is the term generally used to study the above system with negligible spin—orbit coupling.* While the g-factors of much larger elements, such as metals and their complexes, may be significantly different from g_e. Hence, *the term EPR is used for these unpaired systems with significant spin—orbit coupling.*

In other words, the unpaired electron experiences the applied magnetic field (B) plus additional magnetic field (δB) including, due to its spin and orbital motions, own nucleus/surrounding ligand nuclei, and also from other core electrons. So, the effective g-value ($g_e + \delta_g$) contains information about the electronic structure of the atom or molecule and also chemical nature of the bond in the molecule. *Thus, the g-value (analogous to chemical shift in NMR) can be used as a fingerprint for the identification and characterization of atoms or molecules.*

4.3 PRESENTATION OF ELECTRON PARAMAGNETIC RESONANCE SPECTRA

In an EPR spectrometer, a paramagnetic sample is placed in a large uniform magnetic field B_0 which, as shown below, splits the energy levels of the ground state by an amount

$$\Delta E = h\nu = g\beta B_0 \tag{4.4}$$

The value of g can be determined from the ΔE, the energy difference between the two spin levels. This can be done by both irradiating the sample either with MWs with a fixed frequency and sweeping the magnetic field or applying a constant magnetic field and scan the frequency of the electromagnetic radiation as in conventional spectroscopy (Fig. 4.2).

In EPR, a klystron MW source is used. This X-band klystron has a narrow spectral band width of about 8.8–9.6 GHz. This makes it impossible to continuously vary the wavelength similar to optical spectroscopy. It is therefore in EPR necessary to vary or sweep the magnetic field, until the quantum of the MWs matches the energy difference between the field-induced energy levels. Thus, a peak (depending on energy levels) due to absorptions with the selection rule $\Delta m_s = \pm 1$ will be observed in the spectrum when the separation of two energy levels is equal to the quantum energy of the incident MW photon. At that instance, the absorption of energies will occur since the condition in Eq. (4.4) is satisfied. This field is called the "field of

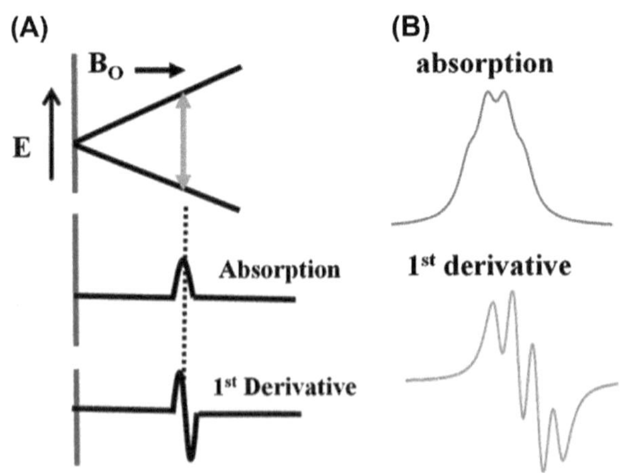

Figure 4.2 Absorption and first derivative electron paramagnetic resonance spectra for single line (A) and multiple lines (B) systems.

resonance," the value of g is then calculated from ν (in GHz) and B_0 (in gauss) using Eq. (4.4).

The direct detection of the absorption signal is possible only for samples containing a high concentration of unpaired electrons. Noise components over a wide range of frequencies appear with the signal, making its detection difficult. This is overcome by field modulation using a phase sensitive detector technique. This gives the signal as a first derivative as shown in Fig. 4.2.

Furthermore, due to short relaxation time of paramagnetic system, the absorption signals are usually broad and hyperfine splittings are buried and not seen (Fig. 4.2B). However, the derivative signal observed using field modulation exhibits well resolved hyperfine features.

4.4 ELECTRON PARAMAGNETIC RESONANCE INSTRUMENTATION

The flow chart and components of an EPR spectrometer are shown in Fig. 4.3.

1. *Source*: A Gunn diode is used to generate monochromatic MW radiation in the 9.75 GHz (known as the X-band) region. Other regions of interest are the L-band (~1.1 GHz), S-band (~3.0 GHz), Q-band (~34.0 GHz), and W-band (~94.0 GHz). In the Gunn tube, the source of MW radiation is a Gunn oscillator (a solid-state device). It also contains those components which control or measure the frequency and the intensity of the MW beam.
2. *The "magnet system"*: It is an electromagnet providing a stable, linear variable and homogeneous magnetic field of arbitrary magnitude.
3. *The cavity system* includes the components which hold the sample and also direct and control the MW beam to and from the sample
4. *The* "detector" *and "modulation systems"* monitor, amplify, and record the signal.

   The MW radiation travels down a waveguide (a type of RF pipe) to the sample, which is held in place in a MW cavity held between the poles of two magnets. The frequency of the MW remains constant, while the impedance of the cavity can be changed mechanically (screw) and electrically. Spectra are obtained by measuring the absorption of the MW radiation while scanning the magnetic field strength. The absorption of this photon by the sample will be indicated by a change in the detector current.

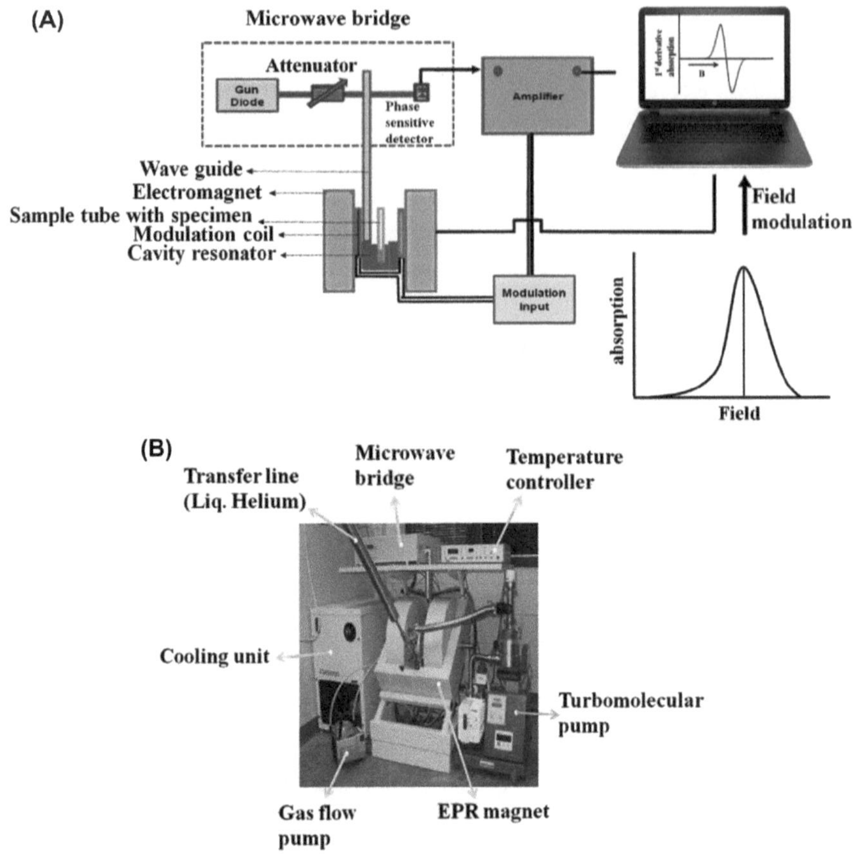

Figure 4.3 (A) Schematic diagram of electron paramagnetic resonance (EPR) spectrometer and (B) Photograph of EPR spectrometer.

5. *Phase sensitive detector*: As mentioned earlier, the direct detection of the absorption signal is possible only for samples containing a high concentration of unpaired electrons; noise components occur over a wide range of frequencies appear with the signal, making its detection difficult. This is overcome by field modulation using a phase sensitive detector technique. It utilizes small-amplitude magnetic field modulation that is used to limit the noise-contributing component to frequency. Modulation at the commonly used frequency of 100 kHz is achieved by placing small Helmholtz coils on each side of the cavity along the axis of the static field. This gives the signal as a first derivative as shown in Fig. 4.4.

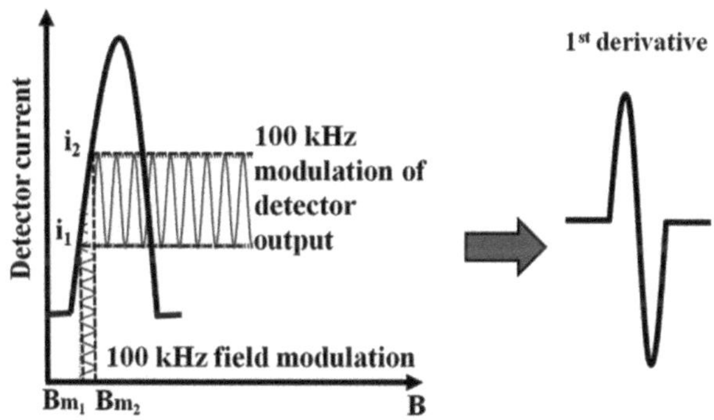

Figure 4.4 Field modulation for obtaining first derivative spectrum.

4.5 PARALLEL AND PERPENDICULAR MODES TRANSITIONS

In general, the MW radiation is (B_1) applied perpendicular ($\perp$) to the magnetic field (B_0) is called normal/ $\perp$ mode EPR. Mostly, EPR spectrum is recorded in this mode. However, certain paramagnetic metal ions such as Fe^{2+}, Ni^{2+}, etc., do not give EPR signals. Furthermore, half-field transitions are weak not observed in normal mode. In such cases, if the MW radiation is applied parallel to magnetic field, these gave well resolved EPR spectra. In this parallel mode, the transitions that are forbidden because the spin has to change by more than ± 1 are well resolved. Also, the transitions in integer spin systems ($S = 1, 2, 3$ etc.), triplets etc., and the "half-field" transitions in interacting exchange spin systems are remarkably enhanced with $B_1 \parallel B_0$. The ER4116DM "dual mode" cavity is designed by Bruker for these studies. It is called dual mode because it operates in one mode with $B_1 \perp B_0$ at about 9.8 GHz, and in another mode with $B_1 \parallel B_0$ at c.9.3 GHz.

4.6 CHOICE OF EXPERIMENTAL CONDITIONS

In principle, the resonance condition $\Delta E = h\nu = g\ \beta\ B_0$ is valid for any frequency (Fig. 4.1). EPR absorption has been detected from magnetic fields of a few gauss ($\nu = 3$ MHz) to fields as large as 120,000 G ($\nu = 360$ GHz). Thus, for a free radical species with $g = 2$, the magnetic field required for resonance absorption is 3400 G ($1\ T = 1 \times 10^4$ G) at about 9.6 GHz. The primary concern in EPR measurement is sensitivity which is governed by Boltzmann distribution of spins. Since, the sensitivity of an EPR spectrometer increases approximately as ν^2, the frequency should be as high as possible.

However, there are three factors which limit the choice of the MW radiation frequency.

1. *Size of the sample*: At high frequency (30–40 GHz), the MW–resonance–cavity dimensions are in order of few millimeters. Although the sensitivity per unit volume is high, the sample volume is limited to about 0.02 cm^3
2. *High frequency* requires also high magnetic field to be homogeneous over the sample volume. Using the conventional magnets, sufficiently homogeneous magnetic fields in excess of 25,000 G are difficult to produce. Therefore, superconducting magnets are required to produce fields of the order of 100,000 G (and higher).
3. *Dielectric absorption* for aqueous samples seriously impairs the sensitivity as the frequency increases. These and other factors resulted in a choice of about 9.5 GHz as the working frequency of most commercial spectrometers.

Sensitivity of EPR is significantly better than the sensitivity of NMR spectroscopy due to the larger magnetic moment of the electron spin compared to the nuclear spins. In particular in high-field (or frequency) EPR spectroscopy, minute sample amounts in the microgram/mole range are sufficient. Concentration for stable organic radicals is in the μmol L^{-1} range. In certain cases, less than a picomolar range of paramagnetic species can be detected.

4.7 PARAMETERS SETTING FOR ELECTRON PARAMAGNETIC RESONANCE SPECTRA

The following parameters must be optimized to obtain a physically meaningful EPR spectrum.

1. *Magnetic field scan range*: If you do not know much about your sample and its absorption region, it is advisable to scan the widest or full magnetic field range available, i.e., from zero up to the maximum. Afterward, the scan range of interest may be selected.
2. *Modulation amplitude*: An oscillating magnetic field at a fixed and stable frequency (mostly 100 kHz) is applied to the sample cavity via coils. The magnetic field is applied continuously throughout the experiment. The amplitude of this modulation is very important because high-modulation amplitudes may distort the individual lines in a spectrum so that valuable information is lost. This is especially important when the lines are sharp and weak.

3. *Sweep time*: Short sweep times may significantly distort the ESR spectrum in that (a) the cross-over point of the first derivative spectrum may be shifted in the direction of the scan; (b) the spectrum becomes asymmetric or (c) the signal intensity reduces.
4. *Sample temperature*: Low concentrations of paramagnetic species may become only visible at low-measuring temperature due to Boltzmann distribution of spins. Furthermore, measurements of the signal intensity as a function of the sample temperature allow one to distinguish different types of magnetic behavior (i.e., paramagnetism, ferromagnetism, antiferromagnetism, etc.).
5. *Microwave power*: At low range of the MW power values, the EPR signal amplitude will increase which is directly proportional to the square root of the MW power received by the sample. This relation, however, is only followed up to a certain MW power level. Above which the signal intensity levels off or even decreases. This is known as *microwave power saturation*. Under such conditions, no quantitative information can be extracted from these saturated spectra. The saturation effects are also more pronounced at low temperatures.

4.8 DETERMINATION OF g-VALUES FROM EXPERIMENTAL ELECTRON PARAMAGNETIC RESONANCE SPECTRA

To determine g-values from EPR spectra, one has to know both the field B_0 and the MW frequency. There are two possible methods: (1) One can use a Gauss meter and a frequency counter, both of which are relatively expensive instruments; (2) the second method is based on the use of a standard with accurately known g-values. A double rectangular cavity (TE_{104}) is often used in that the standard is placed in one half of the cavity and the (unknown) sample (M) in the other. Only one cavity tuning operation is necessary for both measurements. The resonance fields B_R and B_M values are read directly from the EPR spectra as shown in Fig. 4.5. Using diphenylpicrylhydrazine (DPPH) as standard with $g_R = 2.0036$, the g-value is determined from EPR spectra as follows:

$$\begin{aligned} h\nu_R &= g_R \beta B_R \\ h\nu_M &= g_{eff} \beta B_M \\ h\nu_R &= h\nu_M \\ g_{eff} &= g_R (B_R/B_M) \end{aligned} \tag{4.5}$$

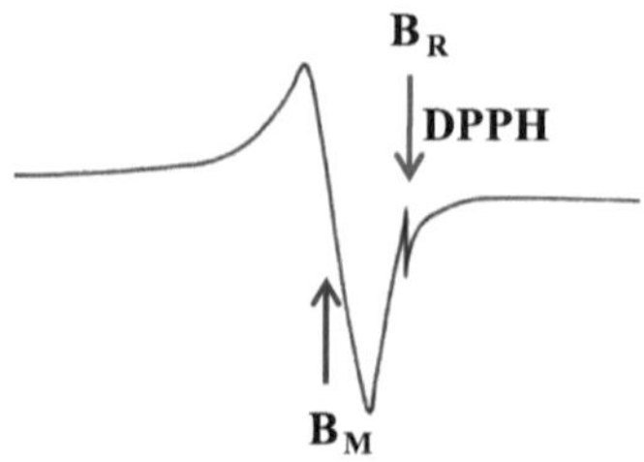

Figure 4.5 Electron paramagnetic resonance spectrum of exchange narrowed Cu(II) complex along with diphenylpicrylhydrazyl radical (DPPH).

4.9 ELECTRON PARAMAGNETIC RESONANCE REFERENCE SAMPLES

1. DPPH has been widely used g-value marker. It has also been used as a standard also for signal intensity. However, the properties of DPPH depend strongly on the solvent from which it was crystallized. Although DPPH is reasonably stable as the solid, in solution the stability is strongly dependent on the solvent, being much more stable in acetonitrile solution. The molar absorptivity of DPPH is $12{,}350 \pm 650$ L M^{-1} cm^{-1} at 520 nm in ethanol used to determine the purity of a sample of DPPH.
2. For transition metal samples, $S = 1/2$ Cu(II) and $S = 5/2$ Mn(II) are commonly used. $CuSO_4 \cdot 5H_2O$, Cu(II)EDTA, and $MnSO_4 \cdot H_2O$ (available with 99% purity) are used as standard.
3. $VOSO_4 \cdot nH_2O$ also has variable number of waters of hydration. So, it was titrated with $KMnO_4$ in the usual volumetric analysis method to determine the V(IV) content and used as standard. The compound was used as the solid, and in some cases ground with NaCl or KCl.
4. Nitroxyl radicals can be synthesized and obtained in sufficient purity that they can be used as quantitative standards for concentrations of organic radicals in fluid and frozen solutions. Nitroxyl radicals can be selected for their solubility in the solvent of interest (ranging from toluene to water) so that the dielectric properties of the known and unknown samples can be matched. It was reported that commercial tempol is purer and easier to handle than tempone, because it is a higher-melting crystalline solid, and it is soluble in both water and alcohols [7].

The g-values of some common EPR standards are as follows: DPPH (2.0037 ± 0.0002), Perylene cation in 98% H_2SO_4 (2.00258 ± 0.00002), *p*-Benzosemiquinone in butanol-KOH at 290 K (2.004679 ± 0.0000129),

Fremy's salt $K_2NO(SO_3)_2$ (2.0057), tetra-cyano-quinoidimethane-lithium salt $C_{12}H_4N_4Li$ ($g = 2.00252$), $CuSO_4 \cdot 5H_2O$ ($g_{||} = 2.27$ and $g_{\perp} = 2.08$), $Cu(acac)_2$ in toluene solution ($g = 2.13$ and 2.126), $VOSO_4 \cdot 5H_2O$ (1.99), $VO(acac)_2 \cdot$ (2.00), VOTPP (tetraphenyl porphyrin) (1.958).

4.9.1 Quantitation of Electron Paramagnetic Resonance Spectra

The area under the absorption spectrum of an EPR signal is, just as in optical spectroscopy, also used for the measurement of the concentration of unpaired electrons. Unlike electronic absorption spectroscopy, there exists no "extinction coefficient" in EPR spectroscopy. It is considered that the $S = 1/2$, $3/2$ etc. systems absorb equally well. To correlate the intensity of the EPR signal with a concentration, a standard or reference is needed. There are different standards available as discussed above and can be used. By comparing the spin concentration of the copper (for example, 10 mM $CuSO_4 \cdot 5H_2O$ solution) as standard with the spin concentration of the signal of interest, the concentration of the sample can be determined. To make a comparison between the EPR signal of the sample and standard, it is essential that the spectra are measured under exactly the identical conditions viz., temperature, MW power, modulation amplitude, conversion time, time constant, total amount of spectra collected, solvent/buffer, the Q factor, etc.

The recorded EPR spectra are usually first derivatives of the normal absorption spectra. So, the spectrum is integrated twice to obtain the intensity (=area under the absorption spectrum).

By determining the intensities for both the unknown sample (I_u) and the copper standard (I_o), one can now determine the concentration of the unknown sample.

$$C_u = I_u C_o / I_o \tag{4.6}$$

where C_u spin concentration of unknown; C_o spin concentration of standard; I_u normalized intensity of unknown; I_o normalized intensity of standard.

If the signal-to-noise is very low, or there is an overlapping spectrum from another species, or the background is large and unavoidable, it may be best to simulate the experimental spectrum and then integrate the computer simulated spectrum to obtain the area or intensity. In addition, a background spectrum can be generated by measuring a tube containing the sample and standard spectra solvent/buffer as the unknown and subtracted from the sample before integration. Furthermore, the equation holds good only if two spectra samples of same spin are compared. So, a standard with the same spin systems should be used.

4.10 ISOTROPIC HYPERFINE INTERACTION

Hyperfine structure (HFS) occurs as a result of the magnetic interaction between the electronic spin S and the nuclear spin I. The magnitude of this interaction is given by A called isotropic hyperfine coupling constant. The hyperfine interaction in ESR spectra is analogous to the fine structure, i.e., nuclear spin–spin interaction in NMR spectra. As a result of this interaction, the ESR signals or peaks are further split into several lines (HFS).

Characteristics of isotropic hyperfine splitting

1. It depends on the magnitude of magnetic moments of nuclear and electron spins.
2. The electron spin density in the immediate vicinity of the nucleus (Fermi contact) (see Section 4.13.1)
3. It is independent of applied magnetic field
4. It obeys $(n + 1)$ rule for $I = 1/2$ i.e., $(2nI + 1)$
5. The intensity ratio is obtained using Pascal triangle (analogous to NMR spin–spin coupling)

Let us consider the hyperfine interaction in the following examples to illustrate and characterize the radicals: (1) hydrogen atom (2) methyl radical (3) (2,2,6,6-tetramethylpiperidin-1-yl)oxyl (4) 1,4-benzosemiquinone radical anion (5) naphthalene negative ion radical (6) *p*-Nitrobenzoate dianion radical, etc.

4.10.1 Electron Paramagnetic Resonance of Hydrogen Atom

It contains an unpaired electron with $S = 1/2$ and a proton with nuclear spin $I = 1/2$ ($m_s = \pm 1/2$ and $m_I = \pm 1/2$). In the absence of magnetic field ($B_0 = 0$), the electron spin energy levels are degenerate, i.e., have the same energy (Fig. 4.6). Application of B_0 lifts the degeneracy between them (the $m_s = -1/2$ sublevel going down and the $m_s = +1/2$ sublevel going up). Each of these electron sublevels further interacts with the nucleus ($I = 1/2$) giving four sublevels designated by the value of m_I. This phenomenon is called *hyperfine splitting.*

The selection rule for hyperfine transitions is given below:

$$\Delta m_s = \pm 1; \quad \Delta m_I = 0$$

This selection rule may be interpreted by considering that the nuclear motion is much slower than the electronic motion. Hence, during the

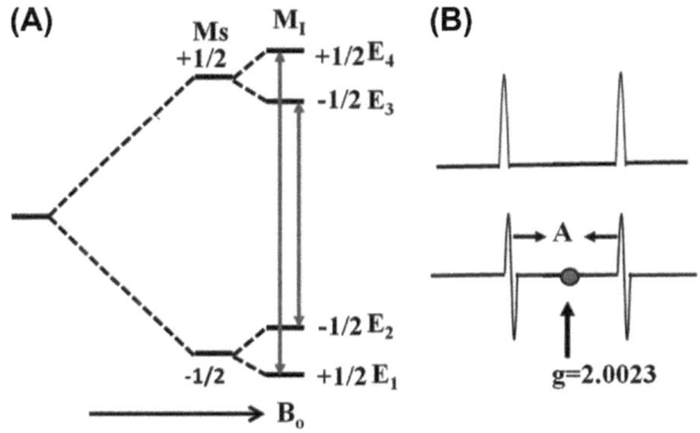

Figure 4.6 The energy level diagram (A) and hyperfine structure (B) in the electron spin resonance spectrum of hydrogen atom.

time electron changes its spin orientation, the nucleus, being heavier, has no time to reorient its spin so that $\Delta m_I = 0$. On using the above selection rule, two ESR lines, i.e., a doublet, are observed, as shown in Fig. 4.6. Both are equally intense, the spacing between them gives the hyperfine coupling constant (A) expressed in units of tesla or millitesla. The required spin Hamiltonian equation and the energy of this interaction are as follows:

$$\widehat{H} = \widehat{H}_{EZ} + \widehat{H}_{HF}$$

where $\widehat{H}_{EZ}$, Electron Zeeman Interaction; $\widehat{H}_{HF}$, Hyperfine Interaction

$$E = g\mu_B B m_s + A m_s m_I$$

where, $A =$ hyperfine coupling constant, $m_I =$ nuclear spin quantum number of the nucleus.

For each value of the electron spin quantum number m_s, the nuclear spin quantum number m_I, can have $-I$, $-I+1$, $\ldots I-1$, I values giving rise to different energy levels.

This effect is illustrated for hydrogen atom ($I = 1/2$ for the proton). The ESR spectrum of a hydrogen atom in a solid matrix consists of two peaks of equal intensity centered at $g = 2.0023$.

The four possible energy states for the hydrogen atom are as follows:

$$E_1 = -1/2 g\mu_B B - 1/4A$$
$$E_2 = -1/2 g\mu_B B + 1/4A$$
$$E_3 = +1/2 g\mu_B B - 1/4A$$
$$E_4 = +1/2 g\mu_B B + 1/4A$$

$$\begin{aligned}\Delta E_1 &= E_3(1/2g\mu_B B - 1/4A) - E_2(-1/2g\mu_B B + 1/4A)\\ &= g\mu_B B - 1/2A\end{aligned} \tag{4.7}$$

$$\begin{aligned}\Delta E_2 &= E_4(1/2g\mu_B B + 1/4A) - E_1(-1/2g\mu_B B - 1/4A)\\ &= g\mu_B B + 1/2A\end{aligned} \tag{4.8}$$

Thus, the separation between the two lines = $(g\mu_B B + 1/2\ A) - (g\mu_B B - 1/2\ A) = A$. Its value is observed to be 506.7 G.

4.10.2 Methyl Radical, $CH_3^{\bullet}$

It contains an unpaired electron with $S = 1/2$ and three equivalent protons. Each proton has a spin $I = 1/2$. Therefore, the total nuclear spin is $I = 3/2$. For each $m_s = \pm 1/2$, the m_I values are $+3/2, +1/2, -1/2, -3/2$ as shown in Fig. 4.7. The interpretation of the ESR spectrum is similar to that of H atom. The ESR spectrum shows four equally spaced lines, $(n + 1)$, i.e., a quartet in the intensity ratio 1:3:3:1. The spacing between any two successive lines represents the isotropic coupling constant (A).

In methyl radical:
1 Proton—splits into 2 lines 1:1
2 Proton—splits into 3 lines 1:2:1
3 Proton—splits into 4 lines 1:3:3:1

4.10.3 (2,2,6,6-Tetramethylpiperidin-1-yl)oxyl (Nitroxide)

(2,2,6,6-Tetramethylpiperidin-1-yl)oxyl is a nitroxide abbreviated as TEMPO. It is a stable free radical with an unpaired electron delocalized

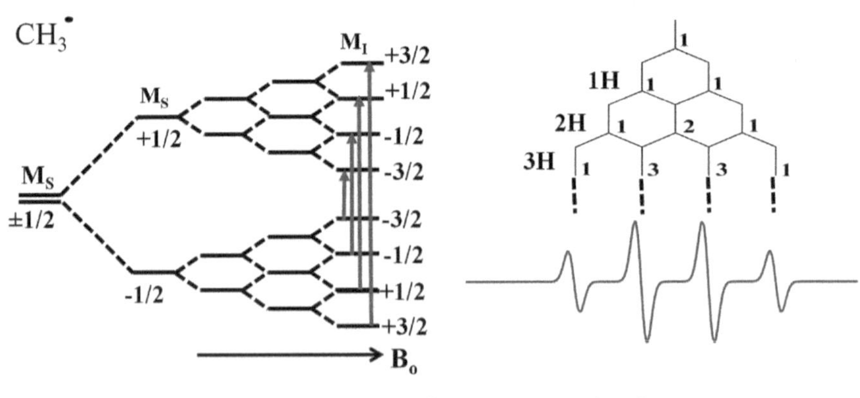

Figure 4.7 The energy diagram and hyperfine structure in the electron spin resonance spectrum of methyl radical $CH_3^{\bullet}$.

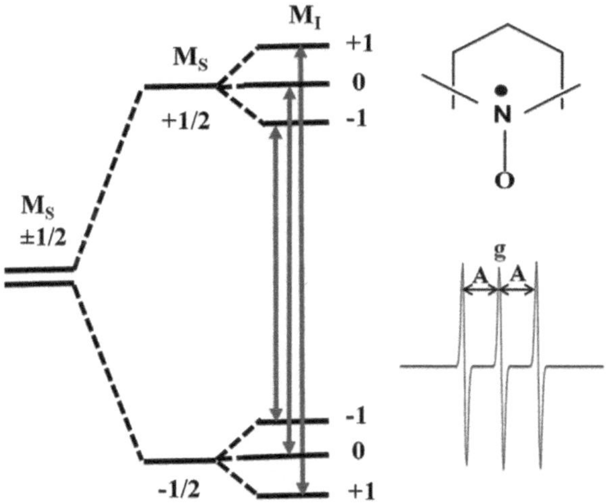

Figure 4.8 The hyperfine structure in the electron spin resonance spectrum of (2,2,6,6-Tetramethylpiperidin-1-yl)oxyl.

between oxygen and nitrogen atoms. As, nitrogen has $I = 1$, each $M_s = \pm 1/2$ splits into three levels $M_I = -1, 0, +1$ giving rise to three peaks of equal intensity as shown in Fig. 4.8.

4.10.4 *p*-Benzosemiquinone Radical Anion

This anion radical has one unpaired electron ($S = 1/2$) interacting with four equivalent protons. The number of peaks observed is calculated as ($2nI + 1 = 2 \times 4 \times 1/2 + 1 = 5$). For four equivalent protons ($I = 2$), M_I values are $+2, 1, 0, -1, -2$. Therefore, the ESR spectrum (Fig. 4.9) contains a quintet, i.e., five equally spaced lines with the intensity ratio 1:4:6:4:6:1.

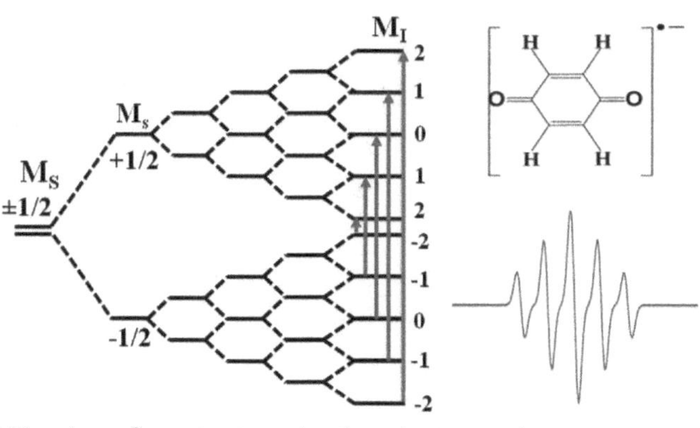

Figure 4.9 The hyperfine structure in the electron spin resonance spectrum of *p*-benzosemiquinone radical anion.

4.10.5 Cyclopentadienyl Radical

This radical has one unpaired electron ($S = 1/2$) interacting with five equivalent protons. Hence, the number of peaks observed is calculated as ($2nI + 1 = 2 \times 5 \times 1/2 + 1 = 6$). Therefore, the ESR spectrum (Fig. 4.10) contains a sextet i.e., six equally spaced lines.

4.10.6 Benzene Anion Radical

This radical has one unpaired electron ($S = 1/2$) interacting with six equivalent protons. Hence, the number of peaks observed is calculated as ($2nI + 1 = 2 \times 6 \times 1/2 + 1 = 7$). Therefore, the ESR spectrum (Fig. 4.11) contains a septet, i.e., seven equally spaced lines.

4.10.7 Naphthalene Anion Radical

This radical anion consists of one unpaired electron interacting with two sets (α and β) of four equivalent protons each as shown in Fig. 4.12. The ESR spectrum would thus show, i.e., $(2nI_1 + 1) \times (2nI_2 + 1) = (4 + 1) \times (4 + 1) = 25$ lines. This radical is formed when naphthalene in solution in 1,2 dimethoxyethane is reduced with potassium metal. A green colored solution is obtained whose EPR spectrum is shown below. By considering the pattern and coupling constants of $a_1 = 4.90$ G and

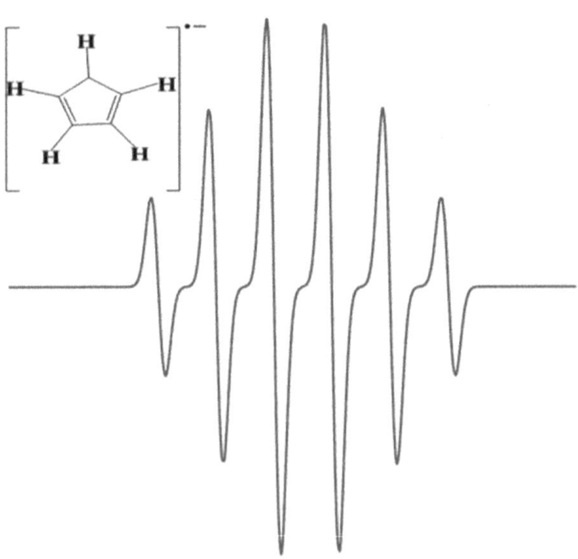

Figure 4.10 Electron spin resonance spectrum of the cyclopentadienyl radical.

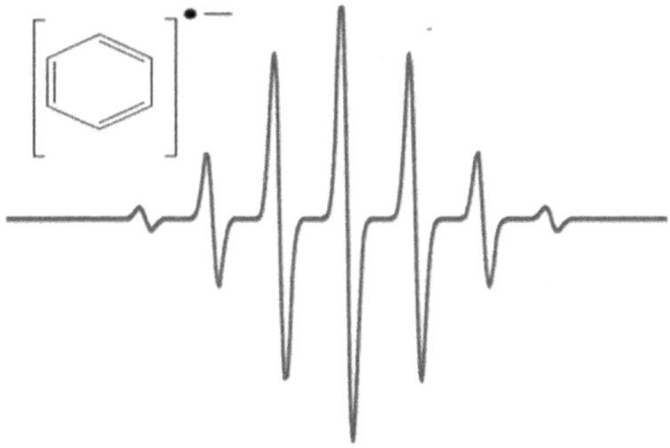

Figure 4.11 The hyperfine structure in the electron spin resonance spectrum of the benzene anion radical.

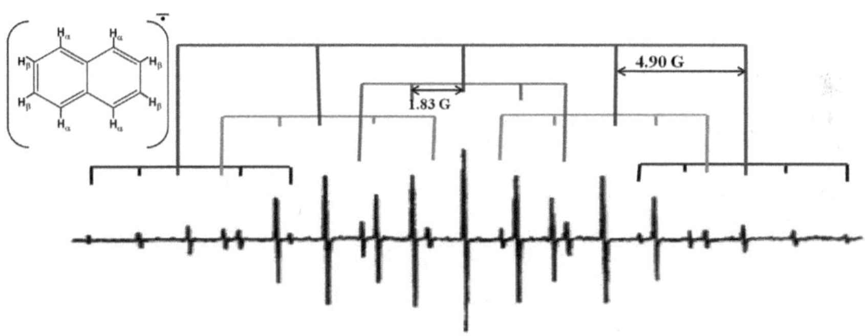

Figure 4.12 Electron spin resonance spectrum of the naphthalene anion radical (K^+ is the counterion) at 298K.

$a_2 = 1.83$ G of hyperfine splittings, the species formed is consistent with the naphthalene radical anion.

4.10.8 McConnel's Relation

McConnel's equation ($a_H = Q\rho_C$) relates the hyperfine coupling interaction with the unpaired electron spin density. The spin density is related to the fraction of time, or to the charge distribution of the electron on a particular *C* atomic orbital (AO). As a first approximation, the spin density on each of the *C* atoms of a π-system radical is given by the square of the coefficient of its AO in the, i.e., singly occupied molecular orbital (SOMO) (probability of finding the unpaired electron on that *C* atom).

$$a_H = Q\rho_C \tag{4.9}$$

a_H = coupling constant of H (attached to carbon atom); ρ_C = unpaired electron density at the C position at which the H is attached; Q = an empirical constant that can range from 20 to 30 G (depending on the type of hydrocarbon).

Thus, the hyperfine couplings can map out electron density distribution of the singly occupied orbital. Moreover, this is usually a frontier orbital and therefore directly related to its chemical reactivity. Therefore, by knowing which sites have the large "*a*" values can tell you which sites are the ones with the most unpaired electron density (donors, nucleophiles). Sites with the smallest hyperfine couplings will be best able to accept electron density (acceptors, electrophiles). For $\rho_C > 0$, we should have an $a_H < 0$ (by means of $\pi - \sigma$ spin polarization). A positive sign of a_H, therefore, must correspond to $\rho_C < 0$. This is due to another type of spin polarization, i.e., the $\pi - \pi$ spin polarization. It arises due to the spin dependent repulsive interaction between the unpaired electron and the electrons of the filled orbitals.

4.10.9 *p*-Nitrobenzoate Dianion Radical

The structure, experimental spectrum and its stick diagram for the reconstruction of the spectrum for the *p*-nitrobenzoate dianion radical are shown in Fig. 4.13. The unpaired electron in the molecule first interacts with the ^{14}N ($I = 1$) with large splitting (A_N) to give a triplet signal of intensity

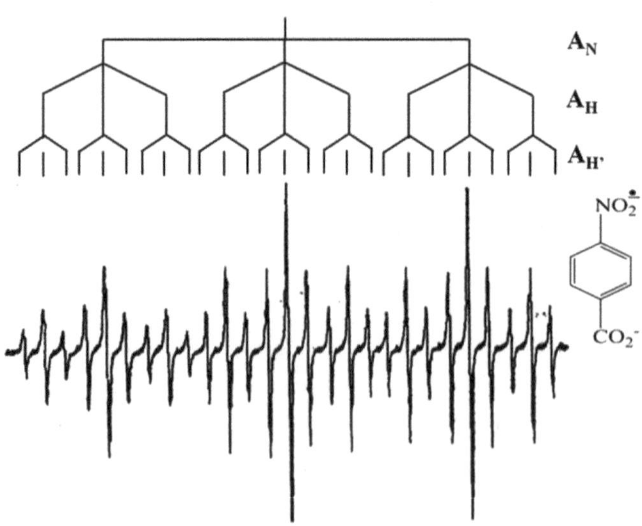

Figure 4.13 The structure stick diagram and electron spin resonance spectrum of *p*-Nitrobenzoate dianion radical.

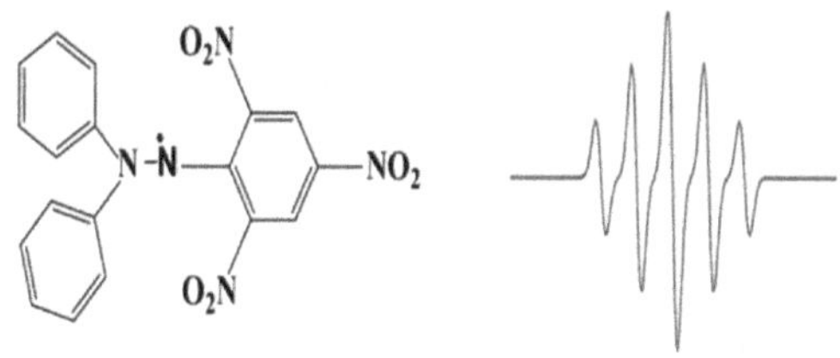

Figure 4.14 The structure of diphenylpicrylhydrazyl radical and its electron paramagnetic resonance spectrum.

ratio 1:1:1. Furthermore, it consists of two sets of two equivalent protons. So, each splits again into a triplet with intensity ratio 1:2:1 and again each signal splits into triplet to give overall 27 lines as shown in Fig. 4.13. However, there is no way based on the spectrum alone to decide which splitting corresponds to which set of protons. A comparison was made with the calculated results and with spectra of similar radicals for which deuterium substituted radicals have been studied. It indicates that the smallest splitting, $A_{H'}$, is associated with the protons closer to the CO_2, while the relatively large splitting, A_H, corresponds to the protons near to the NO_2 group [7].

4.10.10 Diphenylpicrylhydrazyl Radical

In the case of DPPH, the unpaired electron is delocalized on to two nitrogen nuclei ($I = 1$ and $n = 2$). Hence, the two nitrogen nuclei are equivalent and the spectrum (Fig. 4.14) exhibits $2nI + 1 = 5$, five lines with intensity ratio: 1:2:3:2:1. However, in its solid polycrystalline form, it exhibits a single-EPR line that is narrowed due to Heisenberg spin exchange.

4.11 FACTORS AFFECTING THE MAGNITUDE OF g-VALUES

The g-value is given by the expression for an unpaired electron in a gaseous atom or ion, for which Russell-Saunders coupling is applicable as follows [6]:

$$g = 1 + \frac{J(J+1) + S(S+1) - L(L+1)}{2J(J+1)} \tag{4.10}$$

S (total spin angular momentum quantum no.) $= s_1 + s_2 + s_3\ldots$, L (total orbital angular momentum quantum no.) $= l_1 + l_2 + l_3\ldots$, and J (total angular momentum quantum no.) $= |L + S|$.

1. For a free electron ($S = 1/2$, $L = 0$, $J = 1/2$), the g-value, 2.0 is determined from the above equation. However, the actual value for a free electron is 2.0023 where the contribution 0.0023 is due to a relativistic correction.
2. In the case of gaseous halogen atoms, the g-values predicted by the equation were found to agree exactly with the observed values. All the halogen atoms posses the same ground state term ($^2P_{3/2}$, $L = 1$, $S = 1/2, J = 3/2$). On substituting these values in the equation, it yields $g = 4/3$, identical with the experimental value.
3. When the unpaired electron is placed in a chemical/crystal field environment, either in a free radical, or especially in a transition metal ion complex, there is no such agreement observed. This equation is not applicable.

It could be explained due to the perturbation of orbital motion of the electron and the orbital degeneracy, is partly removed or "quenched" due to bonding. For instance, JT distortions (see Section 4.22) serve to remove orbital degeneracy. In addition, a certain amount of orbital degeneracy is sustained due to spin—orbit coupling, i.e., complete removal of the orbital degeneracy is prevented by spin—orbit coupling but higher fold degeneracies are often decreased by this effect. Because of these quenching and sustaining effects, the orbital degeneracy is partly, but not completely, removed and a net orbital magnetic moment results, giving rise to a g-value different from the value of 2.0023 expected if the orbital degeneracy were completely removed.

For most free radicals, spin—orbit coupling is very small and negligible. So, the g-values are nearly equal to the free-electron value. The small deviations (± 0.05 or smaller) often observed are accounted for by a mixing of low-lying excited states with the ground state. Therefore, the g-values are of limited use for free radicals.

4.11.1 g-Values of Transition Metal Ions

On the contrary, the g-values of transition metal ions are much more complicated and interesting. The properties of the transition metals are determined to a large extent by the relative magnitudes of the crystal field (Δ) and spin—orbit coupling (λ). As we discussed above, these two have opposite effects on the orbital degeneracy. Hence, there are three possible cases.

Case (i) $\lambda > \Delta$

The spin—orbit coupling is much larger than that of the crystal-field effect. The rare-earth metal ions (*inner transition*) fall in this class because the f-electrons are well shielded from the crystal-field effects so that LS coupling is not disturbed and J is a good quantum number. Hence, the rare-earth ions and their complexes are more close to free ions and the magnetic moments calculated by the above equation are in close agreement with the experimental values.

Case (ii) $\Delta > \lambda$

The crystal-field effect is strong enough to disrupt the coupling between L and S; hence J is no longer a good quantum number. The splitting of the m_I levels is large (i.e., the orbital degeneracy is removed/quenched) and the EPR transitions are described by the selection rule $\Delta m_s = \pm 1$. The first row transition metals fall into this category. The magnetic moments cannot be calculated by the above equation, but correspond more nearly to the "spin-only" value, g = 2. As noted, the above orbital degeneracy is not removed completely because of the effect of spin–orbit coupling. Consequently a net orbital magnetic moment results giving rise to a g-value different from the free-electron value closer to 2.0023.

For example, ions which have an orbitally nondegenerate ground state such as $Fe^{3+}(^6S)$ and $Mn^{2+}(^6S)$ give g-values nearly equal to the free-electron value, because there is practically no orbital angular momentum. The small deviation from the free-electron value is due to slight spin–orbit coupling.

Case (iii) $\Delta \gg \lambda$

In the strong field case, the effect of the crystal field is very large so that LS coupling is broken down completely. This corresponds to covalent bonding and is applicable to the complexes of the 3d transition metals, such as the cyanides. In many of these cases, a molecular orbital description gives better results than the crystal-field approximation.

These effects can be illustrated using the two specific examples what factors determine the magnitude of the g-values.

1. For *octahedral nickel(II) complexes*, calculations, which include mixing $^3A_{2g}$ ground state with the $^3T_{2g}$ excited state, give the following equation for the g-value.

$$g = 2 - \frac{8\lambda}{10\,Dq} \tag{4.11}$$

where λ is the spin–orbit coupling constant. In hexaquonickel(II), it is found experimentally that $g = 2.25$. Hence $8\lambda/10Dq$ must be equal to -0.25. From the electronic spectrum, one can obtain $10Dq = 8500\ cm^{-1}$. In a molecular orbital description, the extent to which λ is lowered

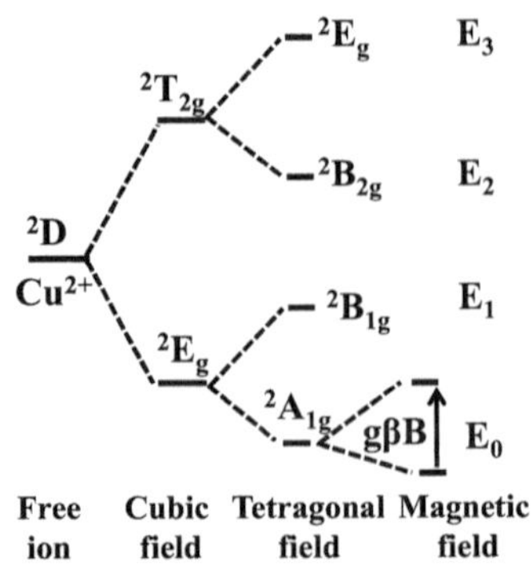

Figure 4.15 Energy level splitting of Cu^{2+} using molecular orbital theory.

from the free-ion value is a measure of the extent of mixing overlapping of metal and ligand orbitals. This example illustrates how both spin–orbit coupling and Dq can affect the magnitude of the g-value.

2. *Cu^{2+} complexes* in a tetragonal field are illustrated in Fig. 4.15. The expression derived for the g-values are,

$$g_{\|} = 2 - \frac{8\lambda}{(E_2 - E_0)} \qquad g_{\perp} = 2 - \frac{2\lambda}{(E_3 - E_0)} \tag{4.12}$$

where E_0, E_1, E_2, and E_3 are the energies illustrated in the Fig. 4.15. If the energy differences $E_2 - E_0$ and $E_3 - E_0$ are obtained or estimated from absorption spectra, whereas $g_{\|}$ and $g_{\perp}$ are measured from the EPR spectrum, the spin–orbit coupling can be evaluated. It can be seen from these equations that the g-value approaches 2.0 as the splitting of the energy levels increases. This illustrates the importance of excited state mixing in distorted complexes. Thus, the magnitude of g-tensor gives considerable information about the ground and excited states in complexes [6].

4.12 g-TENSOR AND ANISOTROPY

The spin Hamiltonian term describing the interaction between an electron spin and the external applied magnetic field (B) is given by

$$\begin{aligned} \widehat{H}_{EZ} &= \mu_B B^T g\, \widehat{S} \\ &= \mu_B \left(B_x B_y B_z\right) \begin{pmatrix} g_{xx} & g_{xy} & g_{xz} \\ g_{yx} & g_{yy} & g_{yz} \\ g_{zx} & g_{zy} & g_{zz} \end{pmatrix} \begin{pmatrix} \widehat{S}_x \\ \widehat{S}_y \\ \widehat{S}_z \end{pmatrix} \end{aligned} \tag{4.13}$$

T is the transpose of the matrix and g-matrix is usually symmetric, in which case it can be transformed into its diagonal form

$$g_{diag} = \begin{pmatrix} g_1 & 0 & 0 \\ 0 & g_2 & 0 \\ 0 & 0 & g_3 \end{pmatrix}$$

via. a rotation R parameterized by three Euler angles (α, β, γ). Here, g_1, g_2, and g_3 are the three principal values of the g matrix. For symmetric isotropic case, $g_1 = g_2 = g_3$, a single-line spectrum is obtained. If g is asymmetric, the diagonalization gives complex principal values.

As stated in the electron Zeeman interaction, the g-value depends on the absolute orientation of the molecule with respect to the external magnetic field. The orientation of the molecule can be characterized by the elevation angle θ between the molecular z-axis and the vector of the magnetic field B. φ denotes the angle between its projection on the molecular xy-plane and the x-axis as shown in Fig. 4.16.

For anisotropic spectrum, the effective g-value depends on both spherical angles and hence the angular dependency of the g-value can be calculated by

$$g(\theta, \varphi) = \left(g_{xx}^2 \sin^2\theta \cos^2\varphi + g_{yy}^2 \sin^2\theta \sin^2\varphi + g_{zz}^2 \cos^2\theta\right)^{1/2} \tag{4.14}$$

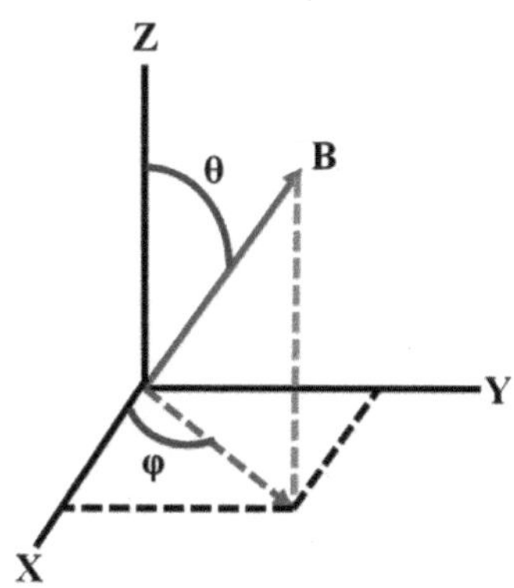

Figure 4.16 Orientation of the magnetic field B with respect to the coordinates (x,y,z) of the molecule.

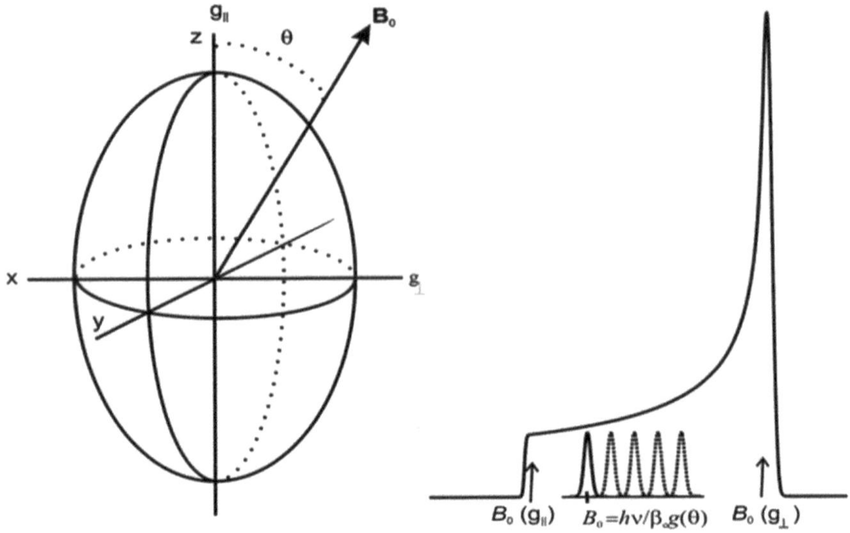

Figure 4.17 The elevation angle θ and the azimuth φ in a unit sphere representing the molecular coordinate system (collinear to the principal g-tensor, g_{xx}, g_{yy}, and g_{zz}) and the magnetic field vector B_0.

In the case of axial symmetry with $g_{zz} = g_{||}$ aligned parallel to the unique axis of the molecular frame and $g_{xx} = g_{yy} = g_{\perp}$ aligned perpendicular to this axis (Fig. 4.17), the g-tensor is given by

$$g = \begin{pmatrix} g_{\perp} & 0 & 0 \\ 0 & g_{\perp} & 0 \\ 0 & 0 & g_{||} \end{pmatrix} \qquad (4.15)$$

Hence, the φ dependence vanishes and the effective g-value is simplified as

$$g(\theta) = \left(g_{\perp}^2 \sin^2\theta + g_{||}^2 \cos^2\theta\right)^{1/2} \qquad (4.16)$$

The isotropic and anisotropy broadened powder spectra for paramagnetic species with axial and orthorhombic symmetries are displayed in Fig. 4.18. Because the intensity scales with the number of contributing orientations, the maximum spectral intensity is observed at effective g-values corresponding to $g_{\perp}$ (g_{xx} and g_{yy}), respectively. In polycrystalline sample, all the orientations are present. Hence, there are more chances that $g_{\perp}$ (g_{xx} and g_{yy}) will be aligned with the field than $g_{||}$ (g_{zz}). As a consequence, the $g_{\perp}$ peak is twice more intense than the $g_{||}$ (Fig. 4.17).

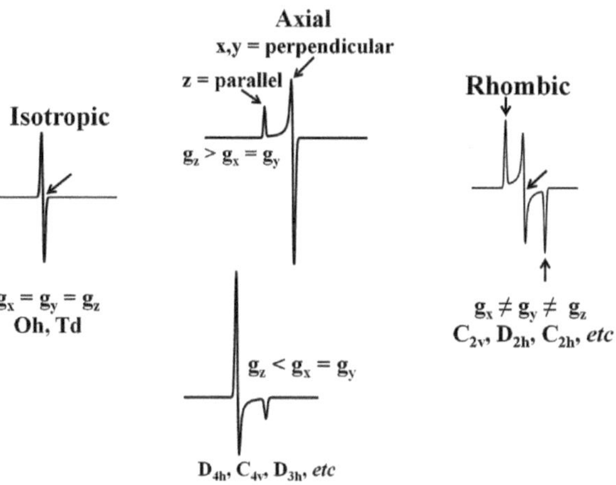

Figure 4.18 Different pattern of electron paramagnetic resonance spectra due to g-tensor values. Isotropic (left), axial (middle), and rhombic (right).

Depending on the magnitude of g-values, there are three classes.

1. The first class is called *isotropic*, where all of the principal g-factors are the same.
2. The second class, called either *axial elongated* (middle top) or *compressed* (middle bottom), there is a unique axis that differs from the other two ($g_x = g_y \neq g_z$) (Fig. 4.18). The g-factor along the unique axis is said to be parallel with it, $g_z = g_{||}$, while the remaining two axes are perpendicular to it, $g_{x,y} = g_{\perp}$.
3. The last class, called *rhombic*, occurs when all the g-factors are different, i.e., $g_x \neq g_y \neq g_z$ (Fig. 4.18, right).

The deviation of the g principal values from the g_e-value of the free-electron spin and its orientation dependence is due to spin–orbit coupling. For a nondegenerate ground state, the orbital angular momentum L is quenched. Hence, only the interaction of excited states with ground state leads to an admixture of the orbital angular momentum to the spin angular momentum. The effective g-tensor can then be expressed as follows:

$$g = g_e + 2\lambda\Lambda_{ij}$$

λ the spin–orbit coupling constant and the symmetric tensor Λ with elements

$$\Lambda_{ij} = \sum\langle\Psi_o|L_i|\Psi_n\rangle\langle\Psi_n|L_i|\Psi_o\rangle/(\varepsilon_o - \varepsilon_n) \qquad (4.17)$$

Each element Λ_{ij} gives the interactions between the SOMO ground state Ψ_o with energy ε_o and the nth excited state Ψ_n with energy ε_n. It is evident that a large deviation from g_e occurs when there is a small energy difference between the SOMO and the lowest excited state leading to a large spin–orbit coupling. For most organic radicals, the excited states are high in energy and therefore $g \approx g_e$. On the other hand, larger deviations are observed for transition metal complexes resulting from the spin–orbit coupling, which is also proportional to the molecular mass of the atom.

In solution, the orientation dependence of the g-tensor is averaged by the fast molecular motion and an isotropic g_{iso} is observed

$$g_{iso} = \frac{(g_{xx} + g_{yy} + g_{zz})}{3} \tag{4.18}$$

4.13 A-TENSOR AND ANISOTROPY

The hyperfine or A-tensor represents the interaction between an unpaired electron and a magnetic nucleus. There are two types of interactions.

1. ***Nuclear hyperfine*** is defined as the interaction of unpaired electron with nuclear magnetic moment and its magnitude is **A.** It results from the nucleus where the unpaired electron originates. For example, the unpaired electron in Cu(II) ion interacts with its own nuclear spin I = 3/2 to give four lines (Fig. 4.19).
2. ***Superhyperfine*** arises from coupling interaction of the unpaired electron with a neighboring nucleus especially ligand atoms. For example, the unpaired electron in the Cu(II) complex also interacts with the ligand "N" nuclear spin I = 1 (see Section 6.1.7.7).

Similar to electron Zeeman term, the Hamiltonian for hyperfine term is given by

$$\frac{\widehat{H}_{HFI}}{h} = \widehat{S}^T A \widehat{I} = \left(\widehat{S}_x \widehat{S}_y \widehat{S}_z\right) \begin{pmatrix} A_{xx} & A_{xy} & A_{xz} \\ A_{yx} & A_{yy} & A_{yz} \\ A_{zx} & A_{zy} & A_{zz} \end{pmatrix} \begin{pmatrix} \widehat{I}_x \\ \widehat{I}_y \\ \widehat{I}_z \end{pmatrix} \tag{4.19}$$

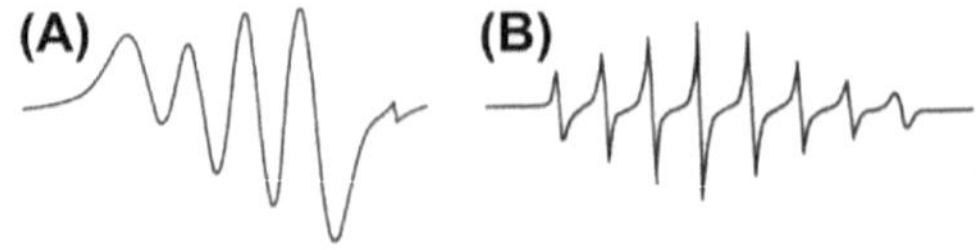

Figure 4.19 Electron paramagnetic resonance spectra of (A) Cu(II) complex and (B) VO(II) complex in solution.

Though the A matrix is asymmetric, it can be transformed to its symmetric diagonal form via a similarity transformation with an orthogonal rotation matrix R.

$$A_{diag} = \begin{pmatrix} A_1 & 0 & 0 \\ 0 & A_2 & 0 \\ 0 & 0 & A_3 \end{pmatrix}$$

4.13.1 Origin of A-Anisotropy

The hyperfine interaction is one of the most important sources of information obtained from EPR spectroscopy. The total Hamiltonian for hyperfine interaction is given by

$$\widehat{H}_{HF} = \widehat{H}_F + \widehat{H}_{DD}$$

Thus, the hyperfine tensor A consists of essentially two factors (1) the isotropic or Fermi contact interaction $\widehat{H}_F$, and (2) the electron–nuclear, dipole–dipole coupling, i.e., $A = A_{iso} + A_{dipolar}$.

The interaction between an electron and a nuclear dipole is the source for the anisotropic component of the hyperfine coupling. The magnitude of it is thus orientation dependent. The dipole–dipole interaction between the magnetic moments of the electron and the nucleus is called electron–nuclear dipolar term and is given by,

$$\widehat{H} = -g_e\mu_B g_n\mu_n\left(\frac{IS}{r^3} - \frac{3(Ir)(Sr)}{r^5}\right) \tag{4.20}$$

here, r is the vector joining the electron and the nuclear spins. The energy of interaction between two magnetic dipoles, electron μ_e and nuclear μ_n, which occurs through space is given by,

$$E = \frac{\mu_e\mu_n}{r^3} - \frac{3(\mu_e r)(\mu_n r)}{r^5} \tag{4.21}$$

The magnitude of the anisotropic hyperfine interaction depends on the orientation of the paramagnetic systems with respect to the external field (B). Mostly, in solid or in frozen solution, both the isotropic and anisotropic hyperfine interaction contributes to the experimental spectrum. Therefore, the A-tensor is split into isotropic and anisotropic parts as follows:

$$A = a_{iso} + \begin{pmatrix} -T & 0 & 0 \\ 0 & -T & 0 \\ 0 & 0 & 2T \end{pmatrix}$$

$$a_{iso} = \frac{A_1 + A_2 + A_3}{3}$$

T is the dipolar coupling constant as a function of g_n and the electron–nuclear distance "r." Because T is traceless tensor, the dipolar (anisotropic) part of the hyperfine interaction is averaged to zero by fast rapid tumbling motion in low-viscosity solution. Under this condition, only the isotropic part of the hyperfine interaction exists. Therefore, the solution spectrum is isotropic in nature. The isotropic Fermi contact term is represented by,

$$\widehat{H}_F = a_{iso} SI$$

The isotropic hyperfine coupling constant, a_{iso} is given by,

$$a_{iso} = (2/3)\{\mu_o/\hbar\} g_e g_n \beta_e \beta_n |\Psi_o(0)|^2 \tag{4.22}$$

$|\Psi_o(0)|^2$ is the electron spin density at the nucleus. Assuming 100% spin density at the nucleus, the $a_{iso}(^1H) = 50.68$ mT and $a_{iso}(^{19}F) = 1886.53$ mT are obtained. The Fermi contact interaction is significant only if there is finite (unpaired) electron spin density at the nucleus, i.e., in s-orbital or s-hybridized orbital where it possesses a finite electron spin density at the nucleus. There will be no isotropic interaction with a nucleus if the unpaired electron is present in an orbital with a node at the site of the nucleus (p, d, f, … orbitals). However, the isotropic interaction a_{iso} can be also significant when the unpaired electron mainly resides in a p- or a d-orbital. In such cases, the spin density at the nucleus is induced by configuration interaction or spin polarization, which arise due to the spin density transfer from the SOMO to the nucleus via s-orbital.

4.14 MEASUREMENT OF ELECTRON PARAMAGNETIC RESONANCE SPECTRA

The EPR spectra can be recorded in solution, powder (frozen), and single-crystal samples and their characteristics are as follows.

4.14.1 Solution or Isotropic Electron Paramagnetic Resonance

Isotropic spectra are largely observed in solution of the sample at room temperature because of the rapid tumbling motion of the molecules. For example, the solution EPR of (A) Cu(II) complex and (B) VO^{2+} complex are shown in Fig. 4.19.

Although four hyperfine lines [Cu^{2+}(I = 3/2)] and eight lines [VO^{2+}(I = 7/2)] of equal intensities are expected, the observed isotropic spectra as shown in Fig. 4.19 reveal incomplete averaging of anisotropies in the g-value and hyperfine value due to variations in line width (LW) from one hyperfine component to another.

$$LW = \alpha + \sum \left(\beta_i m_i + \gamma_i m_i^2 + \cdots \right) \quad (4.23)$$

$$\alpha = k_1 (\Delta g)^2 T_R$$
$$\beta_i = k_2 (\Delta g)(\Delta A_i) T_R$$
$$\gamma_i = k_3 (\Delta A_i)^2 T_R$$
$$\text{where } \Delta g = g_{\parallel} - g_{\perp}, \Delta A_i = A_{i,\parallel} - A_{i,\perp}$$

k_1, k_2, and k_3 are proportionality constants. T_R, rotational correlation time. Since these terms are proportional to T_R, they increase with decreasing temperature.

4.14.2 Powder Electron Paramagnetic Resonance

As described in g-anisotropy section, in powders, frozen solutions, and glasses, the observed spectrum is a result of superposition of all possible orientations of single crystals giving rise to statistically weighted average. In certain cases, the principal g- and A-values are obtained from powder data. For example, the powder EPR spectra of axially symmetric Cu(II) and VO(II) complexes [8] and extraction of their principal values of g- and A-tensors are represented in Fig. 4.20. However, powder lines shapes become complicated when more than one type of species is present and/or when hyperfine lines overlap especially when the g- and A-tensors do not coincide. These complications are overcome by recording the spectra at

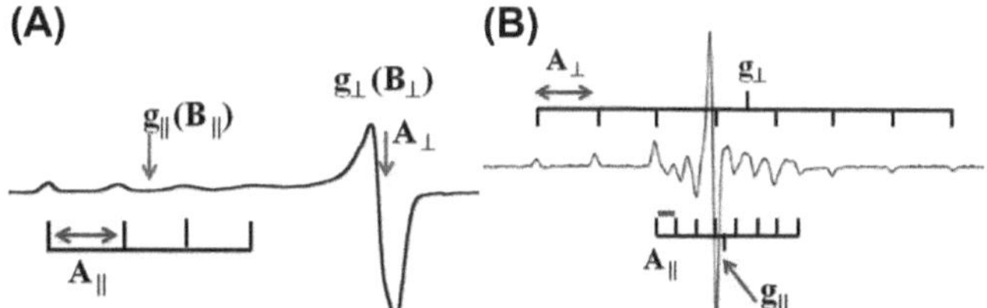

Figure 4.20 Powder electron paramagnetic resonance spectra of (A) Cu(II) and (B) VO(II) complexes and their principal values. *Reprinted with permission from Elsevier.*

two different frequencies viz., X- and Q-bands to sort out the field-dependent (g) and field-independent (A) terms in the Hamiltonian (see Section 5.1). Furthermore, MW power saturation techniques and temperature variation will also be helpful to distinguish the two different species.

However, the orientation of the magnetic tensors relative to crystallographic axes and hence with respect to molecular framework cannot be obtained from powder spectra. In such a case, measurements need to be done on single crystals. Nevertheless, the principal values from powder data are obtained easily by a single measurement. It thus serves as a useful check for the interpretation of single-crystal data. There is also almost a parallel analogy between powder and single-crystal data in EPR and X-ray crystallography.

4.14.3 Single-Crystal Electron Paramagnetic Resonance

Single-crystal measurements have been used for evaluation of the principal values of magnetic tensors. This method consists of measuring the variation of g- and A-magnetic tensors for rotations about three mutually perpendicular axes in the crystal, which may coincide with the crystallographic axes or are related to the crystallographic axes by a simple rotation or transformation. The experimental set up for single-crystal measurement is shown in Fig. 4.21. For example, the angular g-value variations for Cu(II) complex containing two geometries square pyramidal and octahedral for X-band (one site) and Q-band (two sites) are shown in Fig. 4.22 [9].

From the maxima and minima of g-values obtained in the three orthogonal planes (axes), the matrix elements of the g^2-tensor are obtained. On Jacobi diagonalization of this matrix, it gives the Eigen values corresponding to the principal values of the g-tensor. The transformation matrix, which diagonalizes the experimental g^2-matrix, provides the direction cosines of these tensors with respect to the three orthogonal rotations. However,

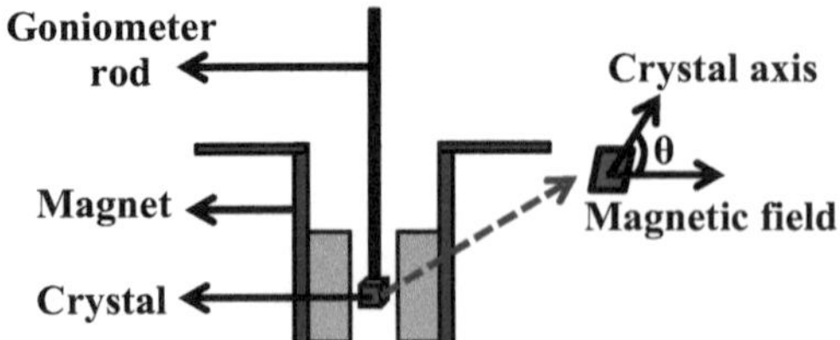

Figure 4.21 Single-crystal rotation for electron paramagnetic resonance using goniometer rod.

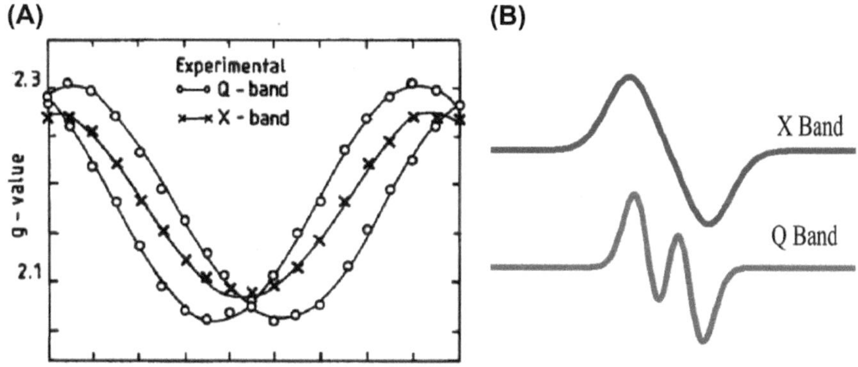

Figure 4.22 Single-crystal electron paramagnetic resonance of the Cu(II) complex experimental angular variations of g for rotation about a axis for Q-band (*open circles*) and X-band (*crosses*) (left) (A) X-band shows exchange-averaged signal (B) Q-band exhibiting two well-resolved resonances for two distinct geometries (right). *Courtesy Dr. C. Karunakaran.*

complications arise when more than one magnetically distinct site per unit cell are present because there is no a priori predictability of the relations between the sites and spectra in the three planes. This leads to several possible permutations leading to many g^2-tensors. For example, when a system has n sites, then one can end up with $2n^3$ tensors, including for not performing a proper rotation, i.e., clockwise or anticlockwise. A careful and close examination, however, invariably leads to the identification of the proper combinations and the corresponding direction cosines. Similarly, for hyperfine tensor, when g is not highly anisotropic, the same procedure as above is adopted. If this is not the case, it is necessary to follow the variation of $g^2A(\theta)$ in the three principal planes.

The spin Hamiltonian for a paramagnetic system, considering only the electronic Zeeman and hyperfine terms, is expressed as follows:

$$\widehat{H} = \beta(g_{11}B_1S_1 + g_{22}B_2S_2 + g_{33}B_3S_3) + (A_{11}S_1I_1 + A_{22}S_2I_2 + A_{33}S_3I_3) \tag{4.24}$$

Let (n_1, n_2, n_3) be the direction cosines of the magnetic field B with respect to the axes of the g- and hyperfine tensors. It is assumed here that g- and hyperfine tensors are coincident.

$$\begin{aligned} g &= \left(g_{11}^2 n_1^2 + g g_{22}^2 n_2^2 + g_{33}^2 n_3^2\right)^{1/2} \text{ and} \\ K &= 1/g\left[g_{11}^2 A_{11}^2 n_1^2 + g_{22}^2 A_{22}^2 n_2^2 + g_{22}^2 A_{33}^2 n_3^2\right]^{1/2} \end{aligned} \tag{4.25}$$

To evaluate the matrix elements of the hyperfine tensor, one has to consider the angular variation of $(gK)^2$, because (gK) has a linear angular dependence on g.

Therefore,

$$(gK)^2 = g^4A^2$$

From this equation, the matrix elements of the hyperfine tensor matrix are evaluated using the same procedure above used to derive g-tensor matrix.

4.14.4 Direction Cosines of the Substitutional Sites

The single-crystal X-ray analysis data provide the positional parameters p, q, r and the unit cell dimensions a, b, c and α, β, γ. For crystal system with nonorthogonal crystal axes, the positional parameters p, q, r of the various atoms can be changed over to an orthogonal framework and the Cartesian coordinates X, Y, Z could be calculated using the relation

$$\begin{pmatrix} X \\ Y \\ Z \end{pmatrix} = \begin{bmatrix} a & b & c\cos\beta \\ 0 & b\sin\gamma & (c/\sin\gamma)(\cos\alpha - \cos\beta\cos\gamma) \\ 0 & 0 & d \end{bmatrix} \begin{bmatrix} p \\ q \\ r \end{bmatrix} \tag{4.26}$$

where, $d = [c^2 - c^2\cos^2\beta - (c^2/\sin^2\gamma)(\cos\alpha - \cos\beta\cos\gamma)^2]^{1/2}$.

By setting the metal as the origin, the coordinates of various atoms in the crystal surrounding the metal are calculated. The normalized Cartesian coordinates of these atoms give the direction cosines of the metal–ligand bond of the coordination polyhedron. The direction cosines of these metal–ligand bonds can be compared with the direction cosines of the g- and A-tensors, obtained by the procedure described in the previous section. Sometimes, it is found that the magnetic tensor directions may coincide with some of the bond directions, which may not be so in low-symmetry cases [10].

4.15 ZERO-FIELD SPLITTING, KRAMER'S THEORY, AND DEGENERACY

So far, we have considered a species with only one unpaired electron. Let us now consider the species or transition metal ions that contain more than one unpaired electrons. In such cases, the spin degeneracy can be also resolved by the crystal field. Thus, the spin levels may be split even in the absence of a magnetic field due to the crystal field. This phenomenon

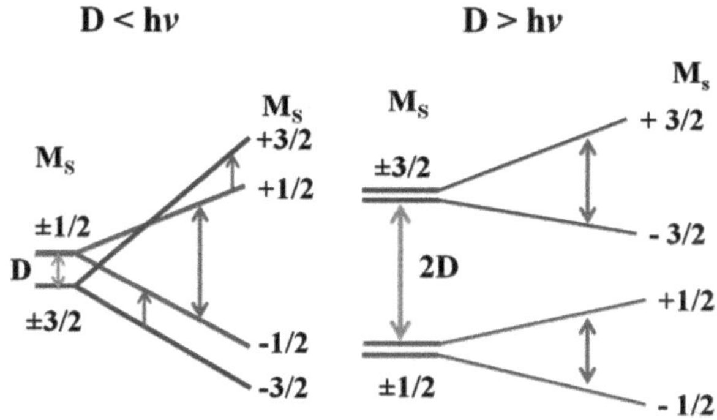

Figure 4.23 Zero-field splitting of Cr^{3+} ion (d^3) and Co^{2+} (d^3) high spin ($s = 3/2$).

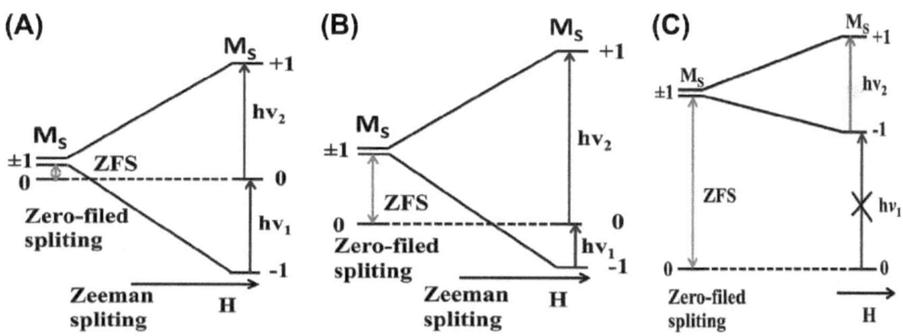

Figure 4.24 The splitting of triplet state due to different magnitude of zero-field splitting (ZFS). (A) $D = 0$, (B) $D > 0$, (C) $D >> 0$.

is called *zero-field splitting* (ZFS) (D). There are two cases of species having either odd or even number of unpaired electrons. For the species containing an odd number of unpaired electrons, the spin degeneracy of every level remains doubly degenerate (Fig. 4.23) this is known as *Kramers' degeneracy*. However, for an even number of unpaired electrons, the spin degeneracy of ground state may be removed completely by the crystal field (Fig. 4.24). These concepts and the consequences of ZFS are used to predict how many EPR peaks are expected for transition metal ion complexes and illustrated below with selected examples [6].

4.15.1 Even Number of Unpaired Electrons

Consider molecules (in triplet state) or ions with two unpaired electrons (Ni^{II}), i.e., $S = s_1 + s_2 = 1$ and hence $M_s = -1, 0, +1$. The total quantum number S is the sum of the spin quantum numbers of the unpaired electrons. There are three possible cases.

Case (i) $D = 0$
In the absence of ZFS, the two possible transitions are expected, $0 \rightarrow +1$, $-1 \rightarrow 0$ with the selection rule ($\Delta m_s = \pm 1$). These two are degenerate and hence only one signal is observed in the EPR spectrum (Fig. 4.24A).
Case (ii) $D > 0$
Here, the zero-splitting removes the degeneracy in M_s as shown in Fig. 4.24B. Furthermore, splitting occurs by the applied magnetic field. The two resulting transitions are not now degenerate. Hence, two peaks are observed in the spectrum when ZFS exists.
Case (iii) $D \gg 0$
When spin–orbit coupling is large resulting in strong ZFS, the energy level splitting diagram for the triplet state, $^3A_{2g}$ of Ni^{2+} (d^8 ion with $t_{2g}^0\ e_g^1$) in an octahedral field is shown in Fig. 4.24C. There are no EPR transitions observed under normal conditions because the transition ($h\nu_1$) energy is very much greater than MW energy. The peak ($h\nu_2$) is also forbidden transition ($\Delta M_s = \pm 2$).

4.15.2 Odd Number of Unpaired Electrons

Consider, for example, Mn(II) (d^5) where there are odd number of five unpaired electrons and hence it exhibits Kramer's degeneracy. The term symbol for the free ion ground state is ^6S. The ZFS produces three doubly degenerate spin states ($M_s = \pm 5/2$, $\pm 3/2$, $\pm 1/2$) (Kramer's degeneracy). Each of these is split into two singlets by the applied field, producing six levels. Hence, five transitions ($-5/2 \rightarrow -3/2$, $-3/2 \rightarrow -1/2$, $-1/2 \rightarrow 1/2$, $1/2 \rightarrow 3/2$, $3/2 \rightarrow 5/2$) are expected and called *fine splitting/structure*. The spectrum is further complicated by the hyperfine splitting due to the manganese nucleus ($I = 5/2$). Thus, five peaks each further split into six hyperfine components with equal intensities are expected and called *hyperfine splitting* (see Section 6.1.3). In contrast to hyperfine splitting, the term fine splitting is used when an absorption band is split because of nondegeneracy arising from ZFS. Components of fine splitting have varying intensities. The intensity is greatest for the central lines and smallest for the outermost lines.

The magnitude of the ZFS in transition metal ions can generally be described as arising from the crystal field. However, we have seen above that Mn^{2+} (d^5) which have a spherically symmetric electron distribution, so that the ^6S ground state is not largely split by the crystal field, but mainly due to ZFS. Spin–orbit coupling mixes excited states which are split by the crystal field with the ground state. This mixing gives rise to a small ZFS in Mn^{2+}. The spin–spin interaction has a negligible effect. However, for triplets (S = 1) the spin–spin interaction term is often solely responsible for the ZFS where there is no crystal-field effect. For example, the triplet state of naphthalene will be discussed later (see Section 4.17).

4.15.3 Effective Spin, S′

For an odd number of unpaired electrons, Kramer's degeneracy leaves the lowest spin state doubly degenerate ($M_s = \pm 1/2$). If the ZFS is large, this lowest doublet will be isolated from the higher lying doublets (Fig. 4.23 right). Transitions will then be observed only in the low-lying doublet, and hence the effective spin will appear to have a value of only 1/2 ($S' = 1/2$).

An example is high-spin Co^{2+} (d^7) with three unpaired electrons, i.e., $S = 3/2$ ($M_s = \pm 3/2, \pm 1/2$). The cubic field leaves a ^4F ground state which, as a result of lower symmetry crystal fields and spin–orbit coupling, gives rise to two doublets. The lowest doublet ($M_s = \pm 1/2$) is separated from the next ($M_s = \pm 3/2$) by about 200 cm^{-1}. Thus, the effective spin has a value of 1/2 ($S' = 1/2$) instead of 3/2 ($S = 3/2$) (see Section 4.15.1). Furthermore, the application of these concepts to various transition metal ions in different crystal fields will be treated in detail in Chapter 6 and the predicted splitting correlated with the experimental results.

4.15.4 Forbidden Transitions

In V^{3+} (d^2) triplet state the magnitude of the ZFS exceeds the energies of usual EPR transitions (Fig. 4.24C). When this happens, it is not possible to observe EPR transitions with $\Delta m_s = \pm 1$, but it has been possible to observe a weak transition corresponding to $\Delta m_s = 2$, i.e., between the $m_s = \pm 1$ levels. This line was further split into eight components due to hyperfine splitting because $I = 7/2$ for V^{51}.

4.16 EXCHANGE COUPLED ELECTRON PARAMAGNETIC RESONANCE

Consider a system of two weakly coupled unpaired electrons conveniently described by two single spin Hamiltonians and two additional terms [11].

$$\widehat{H}_o(S_1, S_2) = \widehat{H}_o(S_1) + \widehat{H}_o(S_2) + \widehat{H}_{exch} + \widehat{H}_{dd} \quad (4.27)$$

These excess terms characterize the contributions due to spin exchange interaction $\left(\widehat{H}_{exch}\right)$ and dipole–dipole coupling $\left(\widehat{H}_{dd}\right)$.

The exchange interaction is a true quantum mechanical phenomenon and is, thus, not understandable in terms of a classical analogue. It is rather a consequence of the Pauli principle, which states that the total wave function of a many-particle system consisting of fermions viz., electrons has to be antisymmetric on exchange of particles. In general, the exchange interaction is written as sum of two parts an isotropic and an anisotropic component. The latter arises from magnetic dipolar interactions or pseudodipolar exchange interactions. For the exchange coupled complexes, only isotropic part exists and the anisotropic part is negligible. Two contributions to the isotropic part are to be considered.

1. *Direct (Heisenberg-van Vleck)* exchange arises from the direct overlap of magnetic orbitals centered at the two interacting paramagnets (see Section 4.16.3). The exchange interaction is called antiferromagnetic if the total spin of the ground state is minimum (i.e., spins are paired up) (Fig. 4.25B) and ferromagnetic (Fig. 4.25A) if the spin is maximum (i.e., for parallel spins). The most frequently used spin Hamiltonians are

$$\widehat{H} = \overrightarrow{\widehat{S}_1} J \overrightarrow{\widehat{S}_2} \quad (\text{for } J < 0)$$

$$\widehat{H} = -2J \overrightarrow{\widehat{S}_1} \overrightarrow{\widehat{S}_2} \quad (\text{for } J > 0)$$

   for which the exchange is antiferromagnetic if $J < 0$, and ferromagnetic if $J > 0$, respectively.

2. *Indirect exchange or superexchange*, this second contribution is mediated via bridging ligand orbitals located between the two interacting centers, so that the ligand forms a bond with either center.

In many cases, the direct exchange is antiferromagnetic because an electronic configuration with mostly paired spins is energetically favorable. However, the superexchange may be ferro- or antiferromagnetic depending

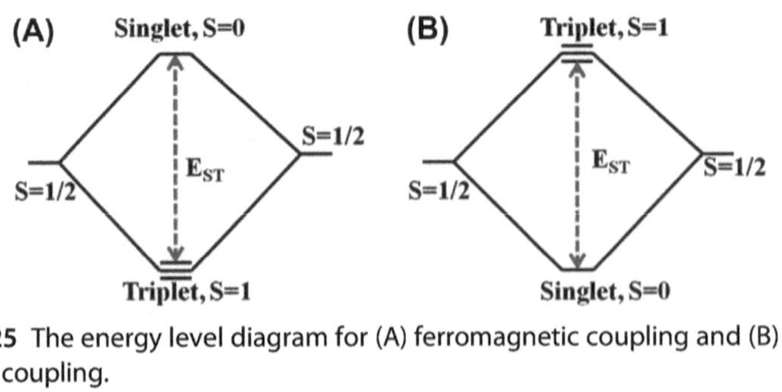

Figure 4.25 The energy level diagram for (A) ferromagnetic coupling and (B) Antiferromagnetic coupling.

on the extent of overlap of the magnetic orbitals and on the particular bridging situation. In most exchange cases, the antiferromagnetic contributions are dominant.

The single spins S_1 and S_2 couple to give a total spin S, which is

$$|S_1 - S_2| \leq S \leq |S_1 + S_2|$$

The energies and energy differences between adjacent energy levels are given by,

$$E_S = -J[S(S+1) - S_1(S_1+1) - S_2(S_2+1)] \quad (4.28)$$

$$E_S - E_T = 2JS$$

For a system $S_1 = S_2 = 1/2$, the isotropic part can be described in terms of chemical bonding. Ferromagnetically coupled spins form a weakly bonded triplet state (with $S = 1$), while a singlet state ($S = 0$) is formed in case of antiferromagnetic coupling. The singlet triplet energy gap, ΔE_{ST}, is given by,

$$\Delta E_{ST} = E_S - E_T = (3J/2) - (-J/2) = 2J \quad (4.29)$$

In the case of diradical system, the exchange interaction between two electrons 1 and 2 according to the isotropic Heisenberg Hamiltonian $\hat{H} = -2J\hat{S}_1\hat{S}_2$, where the Eigen functions are the singlet and triplet functions. The Eigen values are separated by 2J as shown in Fig. 4.25. For an exchange coupled diradical system, the spin–spin interaction for the populated triplet state is detectable by EPR.

4.16.1 Determination of Exchange Coupling (J)

Although the paramagnetism of the triplet obeys the Curie Law, the concentration of triplet state is dependent on temperature and is given by the Boltzmann distribution,

$$\frac{n_T}{n_T + n_S} = [\text{TRIPLET}]_{rel} = \frac{3\exp\left(\frac{-\varepsilon_T}{RT}\right)}{3\exp\left(\frac{-\varepsilon_T}{RT}\right) + \exp\left(\frac{-\varepsilon_S}{RT}\right)}$$

On multiplication of the numerator and denominator by, $\exp\left(\frac{\varepsilon_S}{RT}\right)$, and $\varepsilon_T - \varepsilon_S = \Delta E_{ST}$ gives,

$$[\text{TRIPLET}]_{rel} = \frac{3\exp\left(\frac{-\Delta E_{ST}}{RT}\right)}{1 + 3\exp\left(\frac{-\Delta E_{ST}}{RT}\right)}$$

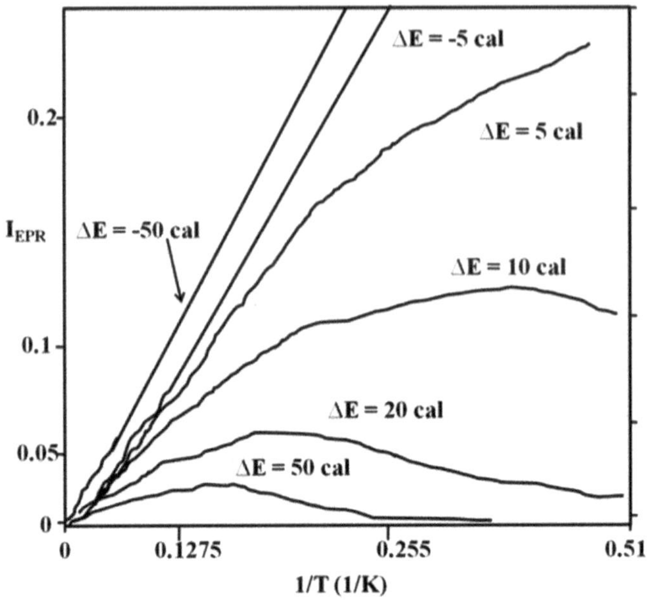

Figure 4.26 Electron paramagnetic resonance signal intensity versus 1/T for different values of ΔE_{ST}.

Thus, the EPR signal intensity is given by,

$$I_{EPR} = \frac{C}{T}[\mathrm{TRIPLET}]_{rel} = \frac{C}{T}\left[\frac{3\exp\left(\frac{-\Delta E_{ST}}{RT}\right)}{1+3\exp\left(\frac{-\Delta E_{ST}}{RT}\right)}\right] \tag{4.30}$$

The constant C is called curie constant (material specific). For different singlet—triplet gaps, a plot of EPR signal intensity versus 1/T is obtained in Fig. 4.26. It is evident from figure that very little change in the slope of the plot occurs as the triplet becomes more stable. From the ΔE_{ST} obtained from the graph, the J value is readily calculated.

4.16.2 Half-Field Transition

An important component of the EPR spectra of exchange coupled binuclear units viz., copper and vanadyl complex, is the (forbidden) transition between the components $M_s = \pm 1$ of the triplet state. This transition is called *Half-field transition* because it occurs at half the field (i.e., at $g = 4$ of the field 160 mT) of the more intense high-field transition at 330 mT, $g = 2$.

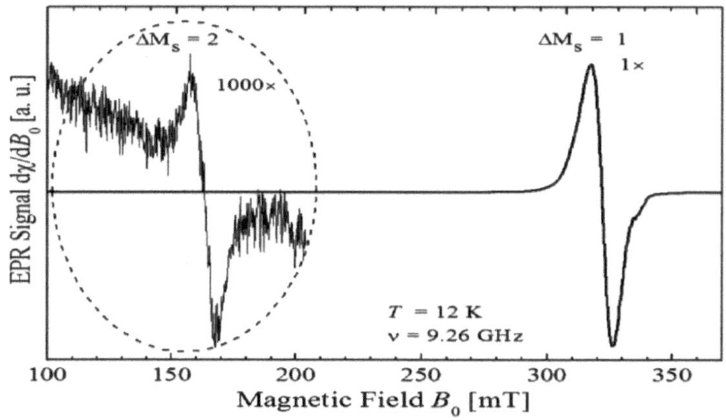

Figure 4.27 Electron paramagnetic resonance (EPR) spectrum of half-filled transition for binuclear Cu(II) compound. *Reproduced with permission from Elsevier.*

This forbidden transition, $\Delta M_s = 2$, for a spin pair in a triplet state between the levels with $m_s = -1$ and $m_s = +1$ conventionally appear to be possible for close spins (r ~ 4–5 Å). This transition has been thus used as fingerprint for the binuclear system with two spins each, 1/2. It is further used to calculate the value of D and the distance between the spins from powder and single-crystal data. The interspin distance evaluation is not a relevant issue when the structure is already determined. But other unknown parameters can be calculated from the study of the half-field resonance observed in EPR spectra of binuclear Cu(II) compound with pairs of Cu(II) ions bridged by one pyrophosphate tetra anion in Fig. 4.27 [12]. The inset displays with a vertical amplification ×1000 the signal observed around 160 mT. It indicates that the ratio of the areas of the large peak (I_o) observed around g = 2, to the small half-field peak are ~1000. The intensity of this forbidden line due to the transitions $I_{\pm 1}$ is thus very much lower than the peak at g ~ 2. Nevertheless, the equation, $I_{\pm 1}/I_o \sim 1/r^6$, allows us to estimate r. Due to exchange of spin states rapidly, it averages out the HFS and fine structure so that only a single "exchange narrowed" resonance line remains.

4.16.3 Direct Exchange Coupled Binuclear System

In several crystals, the magnetic ions are strongly coupled in pairs. For example, copper acetate dimer (Fig. 4.28A) where the d^9 Cu^{++} ion has a spin, S = 1/2, and two spins are directly exchange coupled into a singlet ground state and an excited triplet 300 cm^{-1} above it. At room temperature, there is a high concentration of triplet states exhibiting a single resonance

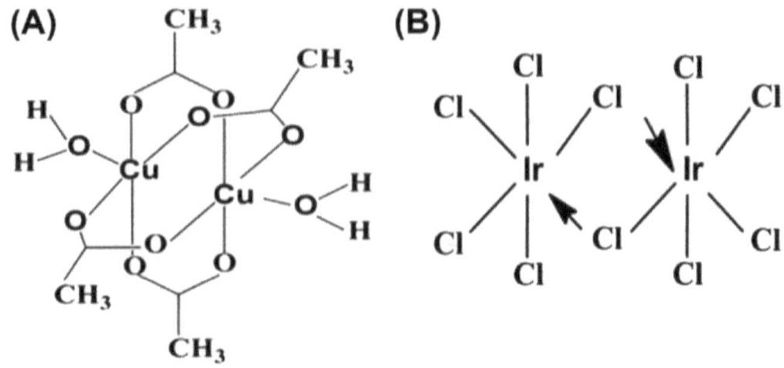

Figure 4.28 Structure of (A) copper acetate dimer and (B) K_2IrCl_6 dimer.

line with no HFS resolved. But on cooling to 77°K, the triplet concentration decreases greatly and seven hyperfine splitting ($2nI + 1 = 2 \times 2 \times 3/2 + 1 = 7$) from two equivalent copper atoms appears. Finally, at 20 K all the spins are paired (antiferromagnetic) and the EPR spectrum disappears.

4.16.4 Superexchange Coupling

Another antiferromagnetic example is K_2IrCl_6 dimeric salt where exchange couplings occur through superexchange mechanism detectable by EPR. It was grown magnetically to dilute single crystals of about 5% K_2IrCl_6 in a diamagnetic K_2PtCl_6 host lattice. There is a fairly high concentration of paramagnetic ion-pairs occupying adjacent lattice sites. Each d^5 Ir^{4+} low-spin ion has $S = 1/2$ and the pairs interactions to give a triplet EPR spectrum with a small ZFS. On decreasing the temperature, the spectrum weakened as the thermal population of the triplet state becomes lower, and the J (~ 7.5 cm^{-1}) is determined. Thus, EPR gives information about the antiferromagnetism of pure K_2IrCl_6 dimer. The exchange interaction here between the ions is not due to direct overlap of their d-orbitals as in Cu(II) dimer. Instead the spin is transmitted through the intervening chlorine ligands resulting in superexchange coupling (Fig. 4.28B).

4.17 TRIPLET STATE ELECTRON PARAMAGNETIC RESONANCE (DIPOLE–DIPOLE INTERACTION)

Another spin–spin interaction namely dipole–dipole that is also detectable by EPR is described by the ZFS parameters **D** and **E** and is also shown in Fig. 4.29. The energy of this interaction is given by the spin–spin Hamiltonian

$$\hat{H} = \hat{S}D\hat{S}$$

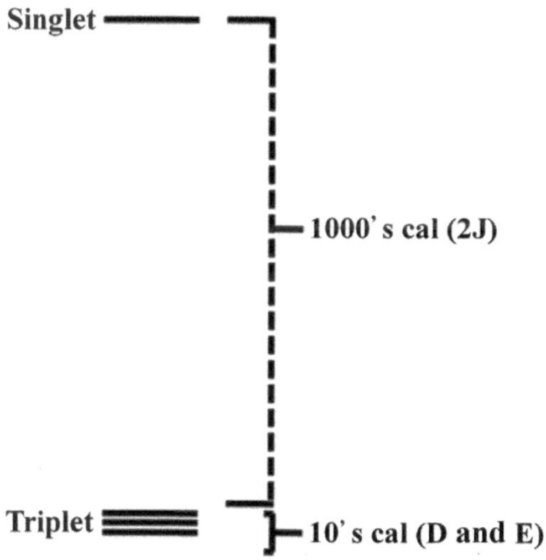

Figure 4.29 Triplet–singlet energy level splitting diagram showing J, D, and E.

It gives the energy separation within the triplet manifold. The exchange and dipolar interactions are the interactions which lift the degeneracy of the four spin states (the one singlet excited state and the three substates of the triplet ground state). Here, the exchange interaction lifts the degeneracy between the singlet and triplet states and *the dipolar interaction lifts the degeneracy of the three triplet substates.*

When two unpaired electrons present in a system are aligned antiparallel, the multiplicity is a singlet [2(0) + 1 = 1]. There is only one spin state and thus no EPR transition is observed. If the spins are aligned parallel (S = 1), the multiplicity is a triplet [2(1) + 1 = 3]. The three substates, the allowed transitions and the EPR spectra are shown in Fig. 4.30. According to the

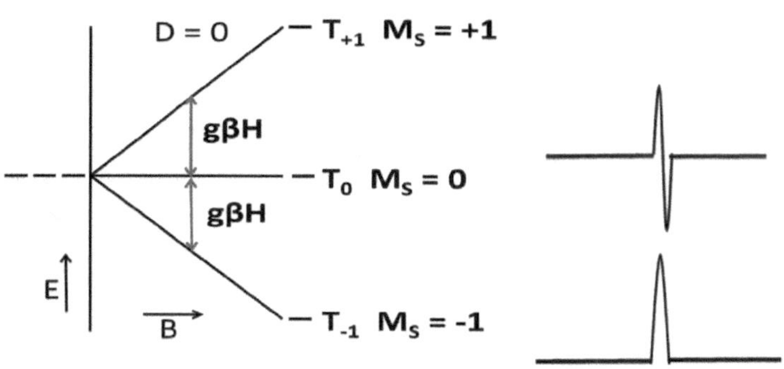

Figure 4.30 Energy level diagram and electron spin resonance spectra of S = 1 triplet system (D = 0).

$\Delta M_s = \pm 1$ selection rule, there are two allowed transitions. However, it exhibits only one resonance as the two transitions are degenerate.

4.17.1 Determination of D and E From Triplet Electron Paramagnetic Resonance

In the case of benzene and naphthalene, the splitting of the spin substates does not result from an externally applied field, but rather from the molecular geometry. This effect is due to the anisotropic electron distribution and is described by the ZFS parameters **D** and **E**. **D** is defined as the energy difference between the lower energy state and the average of higher degenerate energy states as in (Fig. 4.31A) or between the higher energy state and the average of the lower two energy states as in (Fig. 4.31B). The parameter **E** is 1/2 times the energy difference between the higher energy states.

There are three cases based on D (1) **(D > 0) D** is positive for an oblate spin distribution—a flattening in one direction as shown in Fig. 4.31A (2) **(D < 0) D** can also be negative when the electron distribution is prolate—an elongation in one direction (Fig. 4.31B). Thus, the geometrical shape of the spin distribution can be readily estimated by EPR from **D** and **E** parameters. (3) $\mathbf{D} = \mathbf{0}$ For spherical electron distribution (cubic symmetry) as in Fig. 4.30, there is no zero-field energy state splitting.

The different EPR spectral patterns obtained for triplet states in the above different symmetry environments are described as follows.

1. *Cubic symmetry*: The simplest triplet EPR spectrum is obtained for triplet system with *cubic symmetry*. As shown in Fig. 4.30. There are two allowed transitions. Because these two are degenerate, only one signal is detected. Due to the spherical symmetry, the splitting pattern will be identical for all three canonical orientations, i.e., the spectra observed for H_X, H_Y, and H_Z applied fields will be identical to each other.
2. *Axially symmetric*: This system is generally characterized by a four line spectrum with the distance between the outermost peaks equal to 2**D** (Fig. 4.32).
3. *Rhombic symmetric*: For a rhombic system where $|\mathbf{D}| \neq 3|\mathbf{E}|$ each canonical orientation will result in a different spectrum giving a six line spectrum in Fig. 4.33. From the EPR spectrum, the energy values of **D** and **E** by the relations given in Fig. 4.33 are readily determined.

4.17.2 Triplet Electron Paramagnetic Resonance of Naphthalene and C_{60}

1. The triplet state of naphthalene is formed on UV irradiation. The single-crystal EPR spectrum was studied by doping it into durene. This dilution

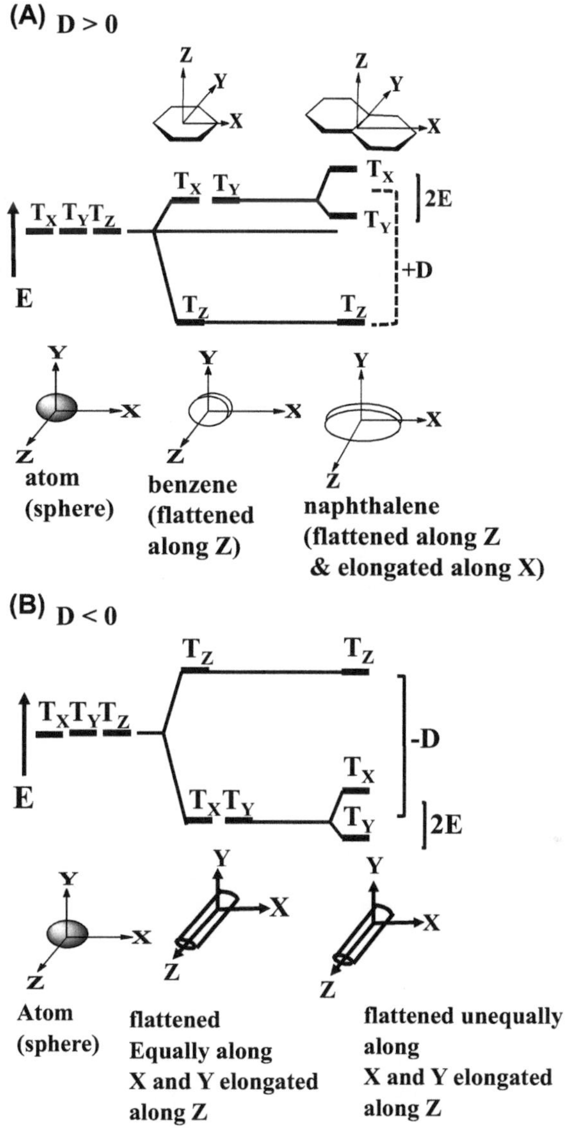

Figure 4.31 Energy level splitting for (A) oblate (D > 0) and (B) prolate (D < 0) molecules.

remarkably increases the lifetime of the triplet state. The single-crystal spectrum consists of three peaks which changes resonance fields with orientation of the crystal. Two of this peak corresponds to $\Delta M_s = \pm 1$ and the third transition is $\Delta M_s = \pm 2$. Although it is forbidden, it is allowed here when the ZFS is small compared to the MW frequency [11].

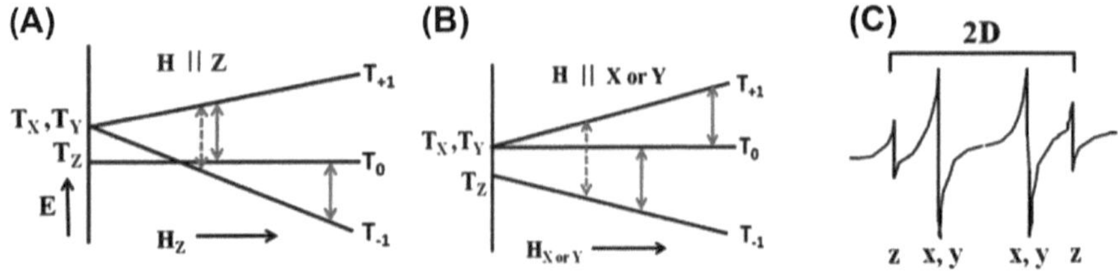

Figure 4.32 The energy diagram and electron paramagnetic resonance spectra of $S = 1$ for $D > 0$ axial system. (A) Applied field aligned with the molecular Z-axis. (B) Applied field aligned with the molecular X- or Y-axis. (C) Simulated spectrum of powder sample for all orientation.

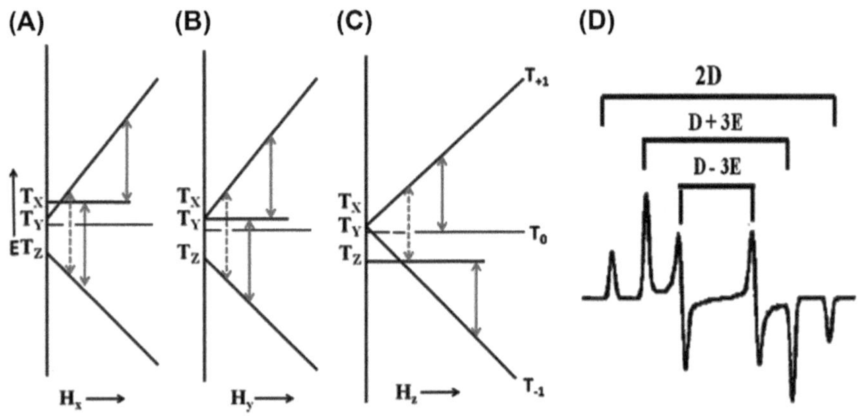

Figure 4.33 The energy diagram and absorption curves for $S = 1$, $D > 0$ and $|D| \neq 3|E|$ rhombic system. (A) Applied field aligned with the molecular Z-axis. (B) Applied field aligned with the molecular Y-axis. (C) Applied field aligned with the molecular X-axis. (D) Simulated electron paramagnetic resonance powder spectrum.

2. The EPR spectra of the triplet state of C_{60} and its derivative phenyl-C_{61}-butyric acid methyl ester in toluene solution are shown in Fig. 4.34, generated by using laser pulse with $\lambda = 532$ nm. The derivative of C_{60} exhibited much more higher intensity than C_{60}. The schematic energy level diagram and the transitions, which can be either absorptive or emissive, are shown in Fig. 4.34. S_0 and S_1 are the ground and first excited singlet state of the system; T_{-1}, T_0, and T_{+1} are Zeeman levels of the triplet state in the presence of a strong (compared to the Zero-field interaction) magnetic field; P_X, P_Y, and P_Z are the populations of the corresponding zero-field triplet sublevels T_X, T_Y, and T_Z. The triplet signal has an absorption line (positive) followed by emission line (negative) [13].

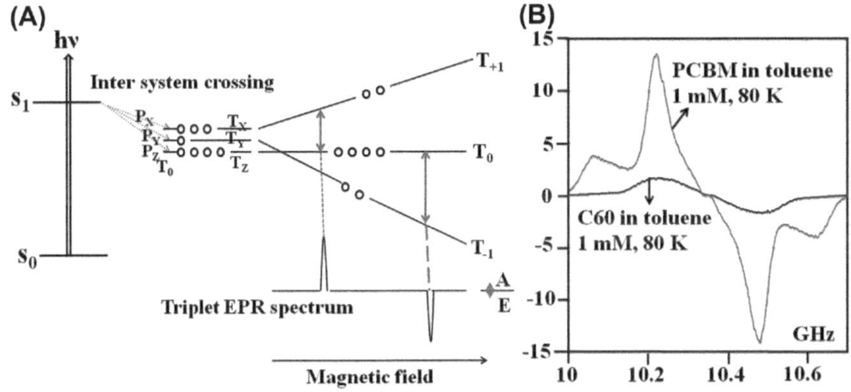

Figure 4.34 The energy level diagram (A) and triplet electron paramagnetic resonance (EPR) spectrum of C_{60} and its derivative (B). *PCBM*, phenyl-C_{61}-butyric acid methyl ester. *Reproduced with permission from Elsevier.*

4.18 OPTICALLY DETECTED MAGNETIC RESONANCE (ODMR)

It is a double resonance technique that combines optical measurements (fluorescence, phosphorescence, absorption) with ESR spectroscopy. After the first triplet state optically detected magnetic resonance (ODMR) experiments on organic molecules in zero-magnetic field reported in 1968 by Schmidt and van der Waals, the number of double resonance studies on excited triplet states viz., chlorophyll, carotenoids, fullerenes, etc., grew rapidly. ODMR spectroscopy is mainly applied to triplet states, in the absence of external magnetic fields (the so-called *zero-field ODMR* or *ZF-ODMR* spectroscopy). Electronic triplet states can be seen as intrinsic molecular "spin probes" for a system, because they carry electronic paramagnetism and are endogenous probes of molecular structure and molecular interactions with the environment.

4.18.1 Mechanism of Triplet State Formation

A triplet state is an electronic state in which two electrons in different molecular orbitals have parallel spins, as shown in Fig. 4.35. The name "triplet" reflects that there are three triplet sublevels as discussed earlier (see Section 4.15.1). These sublevels are degenerate only for spherical molecular symmetry. For lower symmetry, the threefold degeneracy is removed, even in the absence of an applied magnetic field [14].

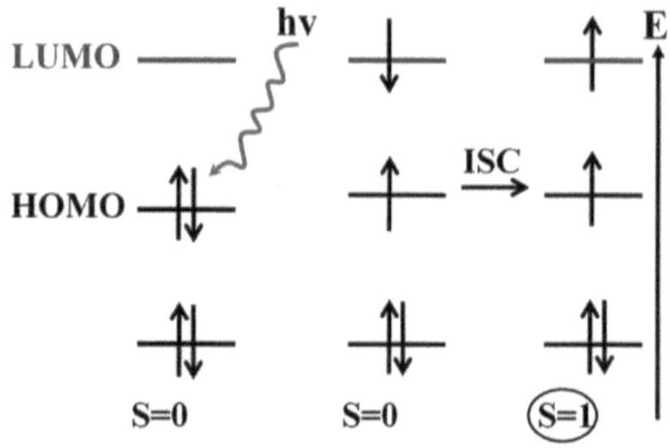

Figure 4.35 Molecular orbitals of ground, excited singlet (S = 0), and triplet (S = 1) states. *HOMO*, highest-occupied molecular orbital; *ISC*, intersystem crossing; *LUMO*, lowest-unoccupied molecular orbital.

Electronic triplet states in molecules are rare but during chemical and biological systems they are often formed thermally by photo-excitation or, in some cases, by chemical reactions, starting from singlet states. The Jablonski diagram in Fig. 4.36 illustrates the relevant molecular electronic states and their population and decay routes. The ground state of a typical organic molecule is a singlet state (S = 0), and absorption of light leads to excited singlet states. The unpairing of the electron spins is possible only if there is some degree of spin–orbit coupling (in which the magnetic field

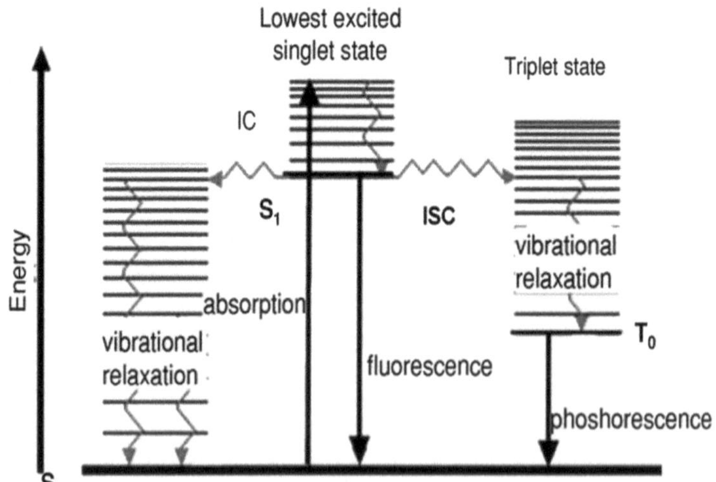

Figure 4.36 Jablonski diagram showing the mechanism of formation of triplet state. *ISC*, intersystem crossing.

arising from the orbital motion of the two electrons interacts with the spin magnetic moments). If this happens, then the molecule undergoes intersystem crossing (ISC) and becomes a triplet state. The atomic spin–orbit coupling is proportional to the nuclear charge. Often, in organic molecules, only light atoms (C, N, O) contribute to the molecular structure; thus the spin–orbit interaction is weak, and the ISC process is slow.

After an excited molecule crosses into a triplet state ($S = 1$), it may remain there for a long time because the deexcitation is spin-forbidden. The triplet state may decay to the ground singlet state with (phosphorescence) or without emission of radiation. The energy of the lowest triplet state, T_0, is usually lower than that of the first excited singlet state, S_1. It is because the two electrons are farther apart in the triplet state than in the singlet state, hence the Coulombic repulsion is less intense and the triplet state energy is lower.

4.18.2 Determination of Zero-Field Splitting Parameters

Those molecular paramagnetic triplet states, for which the spin–orbit coupling, i.e., the interaction between the spin angular momentum and the orbital angular momentum of the electron is not large, have less energy splitting of the three sublevels (X,Y,Z). However, there is a magnetic dipole moment associated with the spin angular momentum of each unpaired electron, the splitting is largely due to the magnetic dipole–dipole interaction between the two unpaired spins. This interaction is given by the spin Hamiltonian, $\widehat{H}_{SS}$, which depends on the molecular symmetry and average distance vector r between the spins:

$$\widehat{H}_{SS} = \frac{3}{4}\left(\frac{g^2\beta^2\mu_0}{4\pi}\right)\left[\left(\frac{S_1S_2}{r^3} - \frac{(S_1r)(S_2r)}{r^5}\right)\right] \tag{4.31}$$

g is the electronic g-value, β is the electronic Bohr magneton, $\widehat{S}_1$ and $\widehat{S}_2$ are the spin angular moments of the two electrons, and r is the distance between the two electrons.

The energies of the three triplet sublevels in zero field in the principal axes system are X, Y, Z Fig. 4.37. Because $X + Y + Z = 0$ (the trace of the tensor is zero), the energies may be written in terms of two parameters, D and E, called ZFS parameters.

$$\begin{aligned} D &= -3Z/2 \\ E &= -1/2(X - Y) \\ |D| &\geq 3\,|E| \end{aligned} \tag{4.32}$$

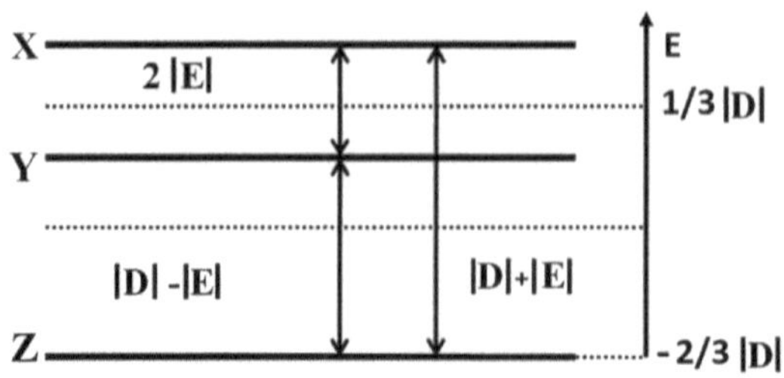

Figure 4.37 The energy levels of the triplet spin sublevels together with the three possible transitions.

D and E reflect the structural and symmetry information that is the spatial distribution of the electrons. D is sensitive to the asymmetry of electronic distribution along the Z-axis, while E reflects the distribution in the X,Y plane. The ZFS parameters are thus the fingerprints of the molecular species. The ZFS parameters are also sensitive to electric fields.

The basic apparatus required for ODMR in zero field is not expensive; however, it is not commercial. The equipment essentially consists of an optical excitation source (lamp or continuous laser beam) and an optical detector (photomultiplier or photodiode). Furthermore, it includes MW source sweep generator perpendicular to the optical source and detector for observing EPR transitions and spectra.

For example, ODMR spectrum of chlorophyll triplet states in photosystem I formed under illumination exhibited three EPR transitions $|D| + |E|$, $|D| - |E|$, and $2|E|$ by sweeping MW radiation (MHz) in the absence of magnetic field [14].

4.18.3 Advantages and Limitations of Zero-Field Optically Detected Magnetic Resonance

ODMR is a quite sensitive method compared to conventional EPR. This is due to the reason that the energies of optical quanta (detected by ODMR) are much higher than the MWs quanta (detected by EPR), and thus enhances the sensitivity of the detector (the quantum-up conversion factor is about 105). Furthermore, the lines are much narrower, because there is no anisotropy associated with an applied magnetic field. A unique feature of ODMR is the possibility to probe the EPR resonance at different optical wavelengths, which may help to distinguish among different triplet states present in a sample. Thus, it allows correlating the optical and magnetic properties of the system.

However, the major limitations arise from the need to operate at low temperature. The application at physiological temperatures is precluded except for crystals. For triplet state transitions which do not saturate at low MW power, the use of high power often induces MW reflections and artifacts in the detection; however, saturation is required. ODMR with respect to EPR is that it is restricted to triplet states, while radicals are not detectable [13].

4.18.4 Applications of Optically Detected Magnetic Resonance

All biological materials viz., chlorophyll, carotenoids, etc., contain molecules such that their triplet states can be used as important probes of their molecular structure, of their structural environment, of their binding sites, etc. Furthermore, all proteins contain as naturally occurring potential triplet probes their aromatic amino acids (phenylalanine, tryptophan, and tyrosine) which can be studied by ODMR [14]. Aromatic residues close to active sites of enzymes may be sensitive to substrate or dye binding. In other cases, the triplet states of endogenous cofactors, in biological systems, are potential tools for studying the dynamics and the function of the macromolecules [15]. The application of ODMR in photosynthesis research has been extensively studied and reported [13].

4.19 NUCLEAR ZEEMAN INTERACTION

Analogous to the electron Zeeman interaction, nuclear spins also couple to the external magnetic field. This contribution is described by the nuclear Zeeman term.

$$\widehat{H}_{NZI} = -\mu_n g_n B^T \bar{I} \tag{4.33}$$

The spin quantum number I and the g_n factor are inherent properties of a nucleus. In most experiments, the nuclear Zeeman interaction can be considered isotropic. So, it hardly influences EPR spectra, however, it affects the nuclear frequency spectra measured by EPR techniques viz., electron spin echo envelope modulation (ESEEM), electron nuclear double resonance (ENDOR), HYSCORE etc. In EPR, the chemical shifts and chemical shift anisotropy are neglected.

4.20 NUCLEAR QUADRUPOLE INTERACTION

Nuclei with spin $I > 1/2$ have an electric quadrupole moment that can interact with the local electric field gradient (EFG) at the nucleus. The spin Hamiltonian term describing this nuclear quadrupole interaction is

$$\hat{H}_{NQI} = \hat{I}^T Q \hat{I} = \left(\hat{I}_x \hat{I}_y \hat{I}_z\right) \begin{pmatrix} Q_{xx} & Q_{xy} & Q_{xz} \\ Q_{yx} & Q_{yy} & Q_{yz} \\ Q_{zx} & Q_{zy} & Q_{zz} \end{pmatrix} \begin{pmatrix} \hat{I}_x \\ \hat{I}_y \\ \hat{I}_z \end{pmatrix} \tag{4.34}$$

where Q is the nuclear quadrupole moment. The Q matrix can be transformed into diagonal form

$$Q = R(\alpha, \beta, \gamma) \begin{pmatrix} Q_1 & 0 & 0 \\ 0 & Q_2 & 0 \\ 0 & 0 & Q_3 \end{pmatrix} R(\alpha, \beta, \gamma)^T \tag{4.35}$$

where Q_1, Q_2, and Q_3 are the three principal values.

Because Q is traceless, it gives $Q_{xx} + Q_{yy} + Q_{zz} = Q_1 + Q_2 + Q_3 = 0$. The relations between the diagonal matrix and the nuclear quadrupole parameters e^2Qq/h and η (asymmetric parameter) are as follows:

$$Q = \begin{pmatrix} Q_1 & 0 & 0 \\ 0 & Q_2 & 0 \\ 0 & 0 & Q_3 \end{pmatrix} = \frac{e^2Qq/h}{4I(2I-1)} \begin{pmatrix} -(1-\eta) & 0 & 0 \\ 0 & -(1+\eta) & 0 \\ 0 & 0 & 2 \end{pmatrix}$$

The product e^2Qq is called nuclear quadrupole coupling constant.

$$e^2Qq/h = 2I(2I-1)Q_3$$

$$\eta = \frac{Q_1 - Q_2}{Q_3} \tag{4.36}$$

e^2Qq/h can be positive or negative, and η is between 0 and 1. The EFG is denoted by Eq. (4.34) arising due to asymmetric charge distribution. It's SI unit is V/m^2. The EFG is symmetrical if $q_{xx} = q_{yy} = q_{zz}$ then there is no quadrupole effect. In EPR spectra, nuclear quadrupole interactions usually result in resonance shifts and also the appearance of forbidden transitions. These small second-order effects are difficult to observe in EPR. However, in nuclear frequency spectra recorded by EPR viz., ENDOR, ESEEM, the quadrupole couplings manifest themselves as first-order splittings.

4.21 JAHN-TELLER SPLITTING

Analogous to electron and electron/nuclear spin interactions, the EPR and optical spectra are further affected by JT coupling/splitting owing to the interaction of the electronic states with the crystal lattice vibrations called electron–phonon coupling ($10^2 - 10^3$ cm^{-1}) [16]. This effect could be explained by *Jahn-Teller theorem* as follows:

> *In a nonlinear molecule, if degenerate orbitals are asymmetrically occupied, a distortion will occur to remove the degeneracy to lower energy (or) in an electronically degenerate state, a nonlinear molecule undergoes distortion to remove the degeneracy by lowering the symmetry and thereby lowering the energy.*

An electronically degenerate state means the availability of more than one degenerate orbital for an electron (Fig. 4.38). In this condition, the degenerate orbitals are asymmetrically occupied. For example, in octahedral symmetry, the d^1 and d^2 configurations are said to be electronically degenerate because the three t_{2g} orbitals with same energy are available for the electrons to occupy. The degenerate orbitals are also said to be asymmetrically occupied by electrons. The lone (or) two electrons may occupy any of the degenerate t_{2g} orbitals as shown in Fig. 4.38.

On the other hand, the d^3 and d^6 (low spin) in octahedral geometry are symmetric and nondegenerate. It is not possible to place two electrons in one orbital, which is against the Hund's rule of maximum multiplicity. Hence, only one arrangement for d^3/d^6 configurations is possible according to Hund's rule. So, they are JT inactive.

4.22 JAHN-TELLER DISTORTION

Likewise, octahedral d^9 configuration, based on crystal field theory (CFT) the last unpaired electron may occupy either d_{z^2} or $d_{x^2y^2}$ orbitals

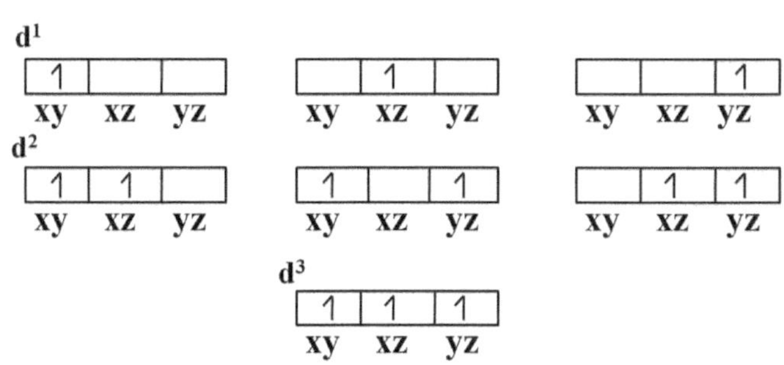

Figure 4.38 Octahedral symmetry of d^1, d^2, and d^3 configurations.

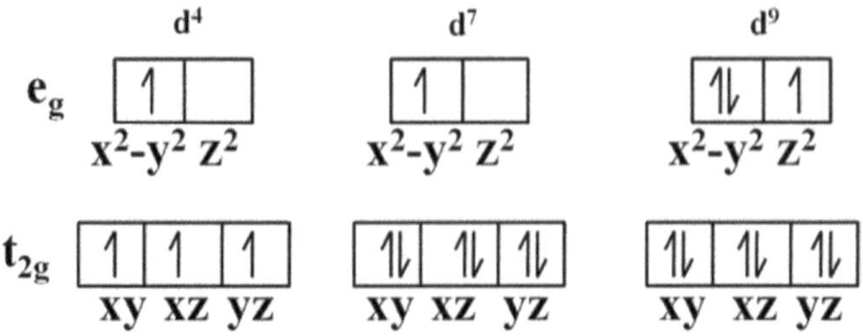

Figure 4.39 Symmetry of high- and low-spin d^4, d^7, and d^9 configurations.

of e_g set (Fig. 4.39). If it occupies d_{z^2} orbital, most of the electron density will be concentrated between the metal and the two ligands on the z-axis. Thus, there will be greater electrostatic repulsion associated with these ligands according to CFT than with the other four ligands on xy-plane. This asymmetric distribution of the electron density may increase the overall energy of the system. To get rid of this, the Cu(II) complex suffers elongation of bonds on z-axis and lowers the symmetry. Conversely, the occupation of electron in the $d_{x^2-y^2}$ orbital would result in elongation of bonds along the x and y axes. This effect is called JT distortion [17].

There is no JT distortion in pure octahedral symmetry. Theoretically, the electronic degeneracy is possible in all the configurations except d^3, d^8, d^{10}, high-spin d^5, and low-spin d^6 configurations.

1. **Strong JT Distortions**
   Distortions are usually observed in **high-spin d^4, low-spin d^7,** and **d^9** configurations in the octahedral environment. The JT distortion is usually significant here because of asymmetrically occupied e_g orbitals as they are directed toward the ligands and the increase in energy is considerably more.
2. **Weak JT Distortions**
   In case of unevenly occupied t_{2g} orbitals, the JT distortion is very weak because the t_{2g} set orbitals do not point directly at the ligands. Therefore, the energy increase is much less exhibiting weak JT effect. Weak JT distortions are largely observed in d^1, d^2, low-spin d^4 and d^5, and high-spin d^7 and d^7 configurations. The tetrahedral complexes also do not exhibit JT geometry distortion. Again, in this case also the ligands are not pointing toward the orbitals directly and hence there is less energy stabilization on distortion (Fig. 4.40).

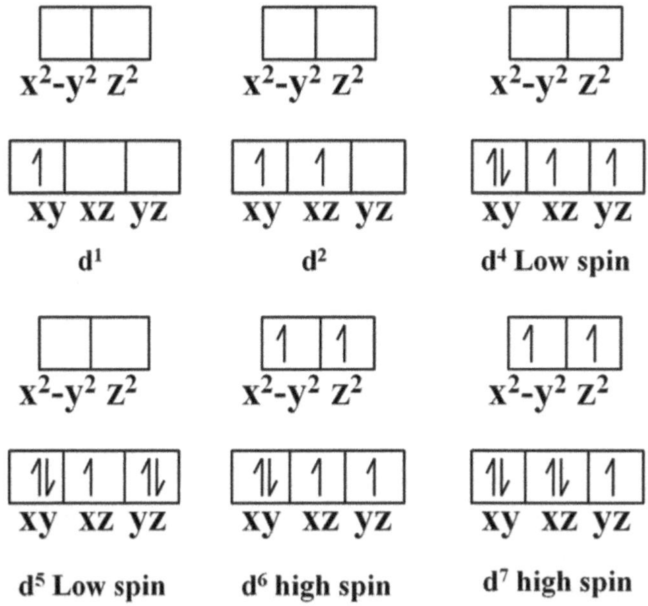

Figure 4.40 Symmetry of low- and high-spin d^1, d^2, d^4, d^5, d^6, and d^7 configurations.

4.22.1 Types of Jahn-Teller Distortions

The degeneracy of orbitals in octahedral symmetry can be removed by lowering the symmetry of molecule. This occurs due to either elongation of bonds along the z-axis (Z-out distortion) or by shortening of the bonds along the z-axis (Z-in distortion). Thus an octahedrally symmetrical molecule is distorted to either tetragonal elongation or compression.

4.22.1.1 Tetragonal Elongation (Z-out Distortion)

In this case, the energies of d-orbitals with z-factor (i.e., d_{z^2}, d_{xz}, d_{yz}) are lowered if the bonds along the z-axis are elongated. This is the most preferred distortion and occurs in most of the cases, especially when the degeneracy occurs in e_g level.

For example, the octahedral d^2, d^4 (high-spin), d^7 (low-spin), d^8 (low-spin), and d^9 configurations orbital the z-out distortion. For example, $[Cu\ (en)_2\ (H_2O)_2]^{2+}$.

4.22.1.2 Tetragonal Compression (Z-in Distortion)

In this case, the energies of orbitals with z-factor are *increased* when the bonds along the z-axis are shortened. This type of distortion is observed in case of octahedral d^1 configuration. Here, the lone electron will now occupy the d_{xy} orbital with lower energy.

For example, the octahedral d^1 configurations such as Ti(III) in $[Ti(H_2O)_6]^{3+}$ and VO(II) complexes can show Z-in distortion.

4.22.2 Consequences and Applications of Jahn-Teller Effect

The following are examples given to illustrate the consequence of JT effect and its applications to interpret the observed structural and spectroscopic phenomena. The extraordinary stability of Cu(II) ion can be explained by taking into account of strong JT distortion in this d^9 ion.

4.22.2.1 Jahn-Teller Distortion in Cu(II)

1. $[Cu\ (H_2O)_6]^{2+}$
   The Cu(II) ion in the complex is surrounded by six water molecules in tetragonal geometry, i.e., four of which are at the corners of square plane and are at shorter distances with stronger interactions. The remaining two are weakly interacting with the metal ion at distant axial positions.
2. $CuSO_4 \cdot 5H_2O$
   In the solid $CuSO_4 \cdot 5H_2O$, the copper ion is surrounded by four water molecules in square planar geometry. There are also weak interactions with the sulfate ions in the axial positions. The fifth water molecule is hydrogen bonded to the sulfate ion. That means the two water molecules on the axial positions are completely eliminated and substituted by weak sulfate ions while forming the solid.

   Due to interactions with these ligands, the d-orbitals are split into different energy levels, which make d—d transitions possible. It absorbs red color and emits blue color during the d—d transition. But on heating, all the water molecules are lost, making all the d-orbitals degenerate again. Hence the anhydrous $CuSO_4$ formed is colorless.
3. Extra stability of Cu^{2+} ions
   The relative stabilities of complexes formed by high-spin divalent first row transition metal ions are given by Irving—Williams series as follows:

$$Mn(II) < Fe(II) < Co(II) < Ni(II) < Cu(II) > Zn(II)$$

4.22.2.2 Effect of Jahn-Teller on Optical and Electron Paramagnetic Resonance Spectra

The absorption band observed in the electronic spectrum of aqueous Ti(III), a d^1 octahedral system, is not symmetric but rather shows a distinct broad shoulder due to JT distortion.

The JT distortion is negligible in the case of degenerate t_{2g} orbitals in the ground state. Hence, no distortion in the ground state. But when the electron gets excited, the configuration now becomes $t_{2g}^0\ e_g^1$, which is again degenerate. Hence in the excited state, the JT distortion exists. So, the promotion of electron may occur to either of the two nondegenerate e_g orbitals, the d_{z^2}, and $d_{x^2-y^2}$. Thus, two transitions are possible. But a shoulder appears because the energy difference between two transitions is small distortion. The effect of JT on EPR spectra is described in detail in Chapter 5 (see Section 6.1.7.4).

4.22.2.3 Coordination Lability in $[Cr(H_2O)_6]^{2+}$ and $[Co(NH_3)_6]^{2+}$

The $[Cr(H_2O)_6]^{2+}$ undergoes substitution reaction easily because the Cr(II) ion is a high-spin d^4 system with one electron in the e_g orbital. Hence, it is electronically degenerate and shows strong JT distortion. So, the hydrated Cr(II) ion is coordinatively labile.

Similarly, the ease of substitution reaction in $[Co(NH_3)_6]^{2+}$ by water molecules can also be explained. In this case, the Co(II) ion is coordinatively labile because it is a low-spin octahedral d^7 ion which is degenerate in e_g set. Hence, it undergoes JT distortion and is labile.

4.22.2.4 Disproportionation Reaction of Au(II)

Au(II) ion is less stable and undergoes disproportionation to Au(I) and Au(III). Whereas the Cu(II) and Ag(II) ions in the same groups are comparatively more stable. One may expect same stability because all are d^9 systems and undergo the JT distortion. However, the crystal-field energy Δ value increases down the group Cu(II) < Ag(II) < Au(II). Hence, in Au(II) ion, it reaches a maximum, which causes high destabilization of the last electron, which is now occupying the $d_{x^2-y^2}$. This makes Au(II) highly reactive, which may undergo either oxidation to Au(III), a d^8 system or reduction to Au(I), a d^{10} system.

The d^8 system, Au(III) is stable as the electron from the $d_{x^2-y^2}$ is completely removed. So, it prefers square planar geometry and more stable than both Au(II) and Au(I). The d^{10} system, Au(I) favors mostly linear geometry with coordination number = 2.

4.22.2.5 $[Cr(III)(acac)_3]$ and $[Mn(III)(acac)_3]$

The complex $Cr(acac)_3$ with Cr(III) d^3 ion does not have degeneracy (JT inactive ion). Thus the geometry of $Cr(acac)_3$ is a regular octahedron. However, octahedral Mn(III) complexes (d^4) are prone to JT distortion and exhibits dynamic JT distortion.

4.23 COMPUTER SIMULATION AND SOFTWARES

The simulation of the EPR spectrum is usually carried out to calculate the various spin Hamiltonian parameters viz., g, A, D, E, etc. It is done by fitting these parameters in the total spin Hamiltonian with the experimental spectrum. There are two kinds of methods available viz., perturbation and direct diagonalization methods. For example, *SimFonia* program for powder spectral simulation is based on perturbation theory. This method is preferred because of the speed of calculation and the intuitiveness of the results. However, it is an approximate method for finding the energy Eigen values and vectors of the spin Hamiltonian. Here, the dominant interaction especially electronic Zeeman, i.e., larger while the other interactions are perturbation terms. The various interactions considered are $\widehat{H}_{EZI}$, $\widehat{H}_{ZFS}$, $\widehat{H}_{NHI}$, $\widehat{H}_{NQI}$, and $\widehat{H}_{HF}$ (see Section 4.2). The transition energies thus obtained are fitted with Lorentzian and Gaussian as well as combination of the two line shapes to get the simulated spectrum.

On the other hand, for instance, *EPRNMR/Xsophe* programs set up spin Hamiltonian matrices and determine their Eigen values (energies) using exact diagonalization. This program not only calculates the energy levels and spectral simulation but also carry out comparison with the observed and parameters optimization.

The National Institute of Environmental Health Sciences (NIEHS) maintains a database of EPR simulation programs (ESR software database) [18]. Here a list of various other softwares available and their application to various cases is listed [19] as follows:

1. Bruker Instruments offers (1) *SimFonia* based on perturbation method (2) *Xsophe*—allows choice between perturbation and diagonalization methods. Recently, *Molecular Sophe* (MoSophe) an integrated computer simulation software suite for the analysis of continuous wave (CW) and pulsed EPR spectra has been developed.
2. *WinSIM* is dedicated to solution spectra of spin traps and was developed at the NIEHS.
3. *Easy Spin* is a versatile program for solution and solid-state applications.
4. *SpinCount* was developed by Hendrich, a program that emphasizes on spin quantitation.
5. *SLE solvers*—Slow-motion spectra of nitroxide radicals can be simulated and fitted using the suite of highly optimized *SLE solvers* developed by the Freed group at the ACERT center at Cornell.

6. *E-SpiReS* is a program again for slow motion simulation that also interfaces to quantum chemistry programs.
7. *WinMOMD* is a program for simulation of slow-motional nitroxide spectra based on the MOMD (microscopic order, macroscopic disorder) model.
8. *EWVoigt* supports nitroxide spectra in the fast-motion regime using convolution methods.
9. *EPRSIM-C* implements wide variety of models for nitroxide spectra using evolutionary fitting algorithms.
10. *DIPFIT*—Dipolar broadening of CW EPR spectra of nitroxides can be simulated and analyzed using *DIPFIT*.
11. *EPRNMR* program developed by Weil is designed for solid-state EPR and also used for single-crystal spectra.
12. *QPOW* and *SIMPOW6* for powder and single-EPR were developed at the University of Illinois.
13. *DDPOW* is a program for mononuclear and binuclear complexes and also supports LW equation, a program by Weihe.
14. *SPIN* a program developed at the National High Magnetic Field Laboratory is largely applicable for high-spin systems.
15. *Xemr* is a general-purpose EPR simulation program for wide applications.
16. *EPRsim32* is a powder CW EPR program that includes genetic fitting algorithms.
17. *MAGRES* from Nijmegen was an early one reported specifically useful for ENDOR and ESEEM simulations. *GENDOR* is an ENDOR simulation program developed by Hoffman at Northwestern.
18. HYSCORE (hyperfine sublevel correlation) simulation programs were pioneered by Goldfarb and Schweiger.
19. *SimBud* simple program by Astashkin's is useful for simulation of pulse EPR spectra.
20. *OPTESIM* (OPTimization of ESEEM SIMulation)—The OPTESIM toolbox consists of 26 Matlab functions and 8 Java classes that can be divided into the following four categories: experimental data filtering, numerical simulation, simulation parameter optimization, and distributed computation framework.

For NMR spectra, several simulation programs have been developed and reviewed. *SIMPSON*, *SPINEVOLUTION*, spin works, dynamic NMR, iNMR, gNMR and *Spinach* are widely used. Among the many programs,

Spinach is a very versatile and efficient program that is useful for large NMR spin systems, and also supports EPR studies.

REFERENCES

[1] N.M. Atherton, Principles of Electron Spin Resonance, Horwood, Chichester, 1993; (a) E.K. Zavoisky, Paramagnetic relaxation of liquid solutions for perpendicular fields, Zhur. Eksperiment. i Theoret. Fiz. 15 (1945) 344–350;
(b) F. Bloch, W.W. Hansen, M. Packard, Nuclear induction, Phys Rev 69 (1946) 127;
(c) E.M. Purcell, H.C. Torrey, R.V. Pound, Resonance absorption by nuclear moments in a solid, Phys Rev 69 (1946) 37–38.
[2] J.A. Weil, J.R. Bolton, J.E. Wertz, Electron Paramagnetic Resonance, Elementary Theory and Practical Applications, John Wiley & Sons, Inc., New York, 1994.
[3] C.P. Slichter, Principles of magnetic resonances, in: Springer Series in Solid-State Sciences 1, third ed., Springer, Berlin, Heidelberg, New York, 1990.
[4] A. Carrington, A.D. McLachlan, Introduction to Magnetic Resonance, Harper & Row, New York, 1969.
[5] C.P. Poole, Electron Spin Resonance. A Comprehensive Treatise on Experimental Techniques, John Wiley, New York, 1983.
[6] R.S. Drago, Physical Methods in Inorganic Chemistry, Affiliated East-West press Pvt Ltd, New Delhi, 2012.
[7] J.E. Harriman, Theoretical Foundations of Electron Spin Resonance, Academic Press Inc. (London) Ltd, 1978.
[8] J.W. Van Schijndel, E.G. Vollen Broek, R. Wever, The vanadium peroxidase from the fungus curvulania in equalis, a novel vanadium enzyme, Bio. Chem. Biophys. Acta 1161 (1993) 249–256.
[9] C. Karunakaran, et al., EPR of an exchange-coupled, helical hydrogen-bridged one dimensional Cu(II) complex containing both octahedral and square pyramidal geomentrices in the same unit cell, Mol. Phys. 100 (2002) 287.
[10] C. Karunakaran, K.R. Justin Thomas, A. Shunmugasundaram, R. Murugesan, X-ray crystal structure and spectroscopy of a pseudosquare pyramidal Cu(II) complex, trans-dinitratotetrakis (trans-4-styrlpyridine) copper (II), J. Chem. Crystallogr. 29 (1999) 413–420.
[11] R.S. Drago, Physical Methods for Chemists, second ed., Surfside Scientific Publishers, Gainesville, USA, 1992.
[12] R.P. Sartoris, O.R. Nascimento, R.C. Santana, M. Perec, R.F. Baggio, R. Calvo, Structure and magnetism of a binuclear Cu(II) pyrophosphate: transition to a 3D magnetic behavior studied by single crystal EPR, Dalton Trans. 44 (2015) 4732–4743.
[13] D. Carbonera, Optically detected magnetic resonance (ODMR) of photoexcited triplet states, Photosynth. Res. 102 (2009) 403–414.
[14] A.H. Maki, Optically detected magnetic resonance of photoexcited triplet states in biochemical spectroscopy, Methods Enzymol. 246 (1995) 611–638.
[15] R.H. Clake, Triplet State ODMR Spectroscopy, John Wiley and Sons, New York, 1982.
[16] J.D. Lee, Concise Inorganic Chemistry, fourth ed., Chapman & Hall, 1991.
[17] M.C.M. O'Brien, C.C. Chancey, Jahn-Teller effect: an introduction and current review, Am. J. Phys. 61 (1993) 688–697.
[18] R. Cammack, Y.C. Fann, R.P. Mason, Electron Spin Resonance Software Database. http://tools.niehs.nih.gov/stdb/index.cfm/spintrap/epr_home, accessed 22 August 2013.
[19] Stefan stall, Computational Modeling and Least-Squares Fitting of EPR Spectra, Wiley-VCH Verlag GmbH & Co., KGaA, 2014.

CHAPTER FIVE

Advances in Electron Paramagnetic Resonance

Chandran Karunakaran, Murugesan Balamurugan
VHNSN College (Autonomous), Virudhunagar, India

Contents

Spin Resonance Spectroscopy
ISBN: 978-0-12-813608-9
https://doi.org/10.1016/B978-0-12-813608-9.00005-8

Abstract

This chapter deals with the recent advances in electron paramagnetic resonance (EPR) methods especially using multifrequency including low- and high-field EPR on Zero-Field Splitting, g-anisotropy, and role in in vivo EPR spectroscopy. It also includes advances in pulsed-EPR using electron spin echo, field-swept, echo-detected, saturation recovery EPR for measurement of electron spin—relaxation times. Time-resolved and rapid-scan EPRs are discussed to study the unstable free radicals intermediates. The theory and application of multipulse EPR, viz., electron spin echo envelope modulation (ESEEM) including two-pulse ESEEM, three-pulse ESEEM, and two-dimensional hyperfine sublevel correlation (HYSCORE) are discussed. Furthermore, multiresonance EPR techniques viz., electron—nuclear double resonance (ENDOR), continuous wave (CW) ENDOR, Pulsed ENDOR and its types, Davies ENDOR, and Mims ENDOR are described. In addition, electron—electron double resonance (ELDOR) CW and pulsed ELDOR or double electron—electron resonance for distance measurements are presented.

Keywords: DEER; ELDOR; ENDOR; ESEEM; HYSCORE; Multifrequency EPR; Time resolved; TRIPLE.

5.1 MULTIFREQUENCY ELECTRON PARAMAGNETIC RESONANCE

Multifrequency electron paramagnetic resonance (EPR) involves the use of high- and low-microwave (MW) frequencies in studying of paramagnetic samples (Fig 5.1). High frequency is defined as MW frequencies above X-band (~10 GHz) and Q-band (~34 GHz) also called high-field EPR. Low frequency is defined as MW frequencies below X-band [1–3]. Frequencies are often identified by the waveguide size known as frequency bands. Table 5.1 lists the most commonly used bands in EPR.

So, EPR measurements can be performed at a range of different frequencies as shown in Table 5.1 and commercial available spectrometers are covering the range 1–94 GHz. There are several obvious reasons for

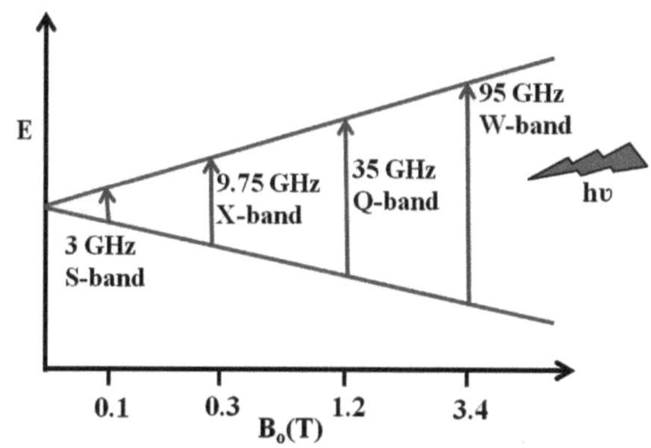

Figure 5.1 Low- and high-microwave frequency.

Table 5.1 Different frequency band used for electron paramagnetic resonance (EPR) measurements

Frequency band	MW frequencies range (GHz)	Common MW frequency for EPR
L	1–2	~1
S	2–4	~4
C	4–8	
X	8.2–12.4	9.2–9.9
K	18–26.5	~24
Q	26.5–40	~34
V	40–75	
W	75–110	~94
D	110–170	

MW, microwave.

going to higher frequencies, including (1) enhanced g resolution, (2) improved sensitivity,(3) simplification of spectra (particularly for systems with large zero-field splitting (ZFS)), (4) improved resolution of fine structure, (5) spectra from integer spin systems, (6) sign of ZFS, (7) Orientation, i.e., anisotropy of spin label in membrane, (8) Molecular dynamics on the different time scales using different MW frequencies.

5.1.1 Enhanced Resolution of g-Anisotropies

Enhanced resolution of small g-value differences will undoubtedly be achieved going from low to high frequencies. For example, the single-crystal EPR spectra of Cu(II) complex existing in square pyramidal and octahedral geometries in the crystal lattice for a given orientation gave one broad band at X-band, whereas two resolved (as shown in Fig. 5.2) signals at Q-band [4]. Furthermore, the powder EPR at X-band of the same Cu(II) complex gives a broad signal at X-band, however, the g-anisotropies ($g_\perp$ and $g_{||}$ values) are well resolved at Q-band [4].

In another example, the powder EPR for a Cu(II) ion (in a square planar environment) at two different frequencies (9.5 and 34 GHz) are recorded as shown in Fig. 5.3. At 9.5 GHz, the g-anisotropy is not sufficiently resolved; the parallel hyperfine (HF) component overlaps with the perpendicular component. This situation complicates the analysis of the spectrum, particularly with respect to the exact determination of the g-value. In such cases, it is difficult to decide whether the spectrum is indeed axial ($g_1 \neq g_2 = g_3$) or slightly rhombic ($g_1 \neq g_2 \approx g_3$). The g-anisotropy can be easily resolved at

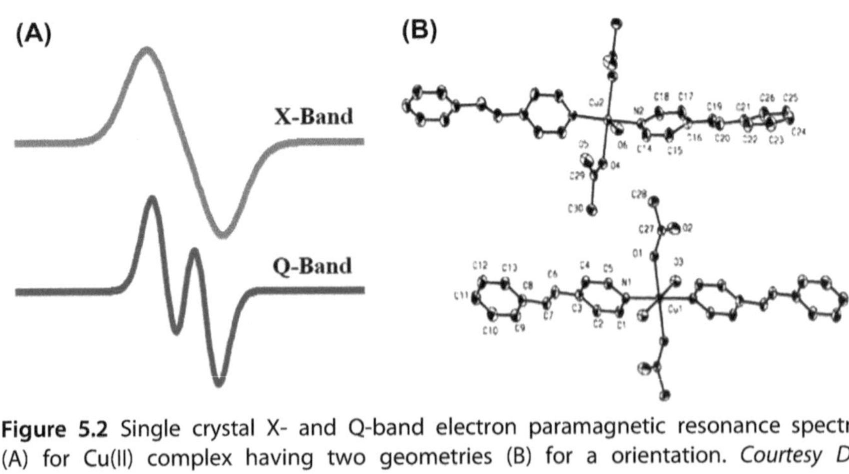

Figure 5.2 Single crystal X- and Q-band electron paramagnetic resonance spectra (A) for Cu(II) complex having two geometries (B) for a orientation. *Courtesy Dr. C. Karunakaran.*

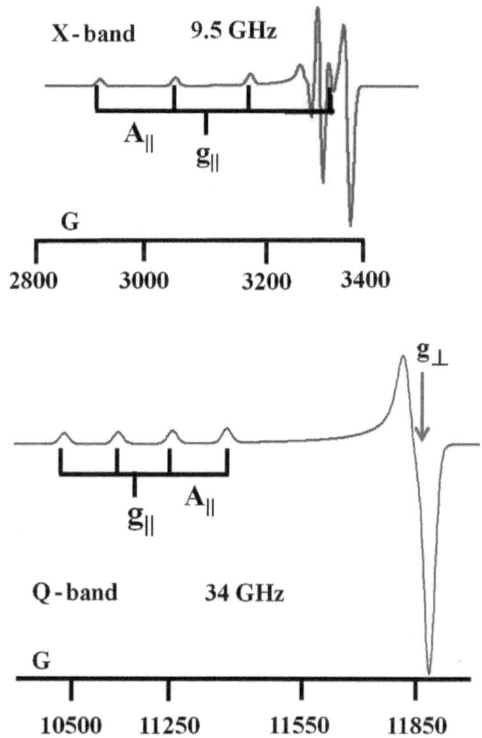

Figure 5.3 Multifrequency electron paramagnetic resonance of Cu(II) ion in square planar complex ($g_{||} = 2.18$, $g_{\perp} = 2.05$, $A_{||} = 140$ G, $A_{\perp} = 15$ G). *Courtesy Dr. C. Karunakaran.*

higher frequencies (34 or 94 GHz). Here, Q-band (34 GHz) is quiet sufficient to provide a clear assignment on exact g-values.

5.1.2 Effect on Zero-Field Splittings

The spin Hamiltonian equation consists of magnetic field–dependent interaction (first term) and magnetic field–independent interaction (second and third terms) [1–3].

$$\hat{H} = \underbrace{\beta_e SgB}_{\text{Magnetic field dependent}} + \underbrace{SDS + SAI}_{\text{Magnetic field independent}} \tag{5.1}$$

The first term in the spin Hamiltonian is the electronic Zeeman interaction. Owing to the B variable, this is the magnetic field–dependent term. The second two terms, the ZFS and nuclear HF interactions, are the magnetic field independent terms. The ZFS term only contributes for spin

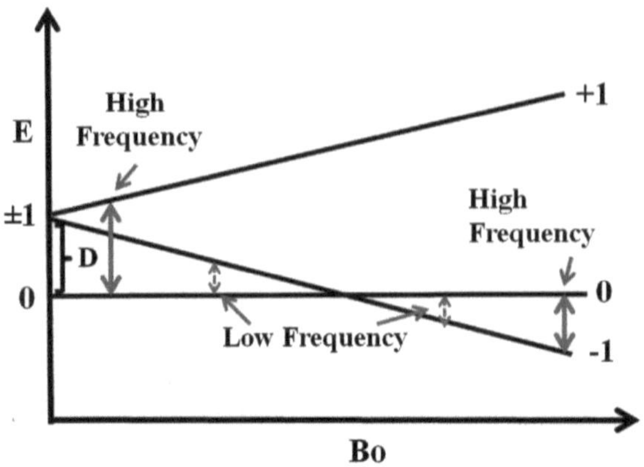

Figure 5.4 Magnetic fields for resonance: $h\nu > ZFS$ (zero-field splitting) or $h\nu < ZFS$.

systems with $S > 1/2$. Large ZFSs that are greater than $h\nu$ may make observation of EPR impossible. In such instances, MW frequencies employed at high-field EPR and therefore greater than the ZFS can bring the field for resonance down to achievable values (Fig.5.4).

If spectra are observed, for samples with $h\nu < ZFS$, the spectra will be difficult to interpret because of second-order splittings. At higher frequency, it often results in simplified spectra. The relative values of the ZFS and HF interactions thus exert a strong influence on the spectral shape.

5.1.3 Improved Resolution of Fine Structure

For Cu(II) and other complexes, it is observed that higher frequencies usually lead to broader lines owing to g-anisotropy and g-strain. Whereas for Mn^{2+} ethylenediaminetetraacetic acid in H_2O at room temperature, the lines become narrower with increasing frequency. For this high spin system with ^6S ground state, there is no g-anisotropy. ZFSs can often lead to very broad lines at low frequency owing to anisotropy. Second-order perturbation theory treatment of the spin Hamiltonian predicts that the anisotropy of the EPR spectrum decreases with increasing frequency. Hence, at higher frequencies the spectrum exhibits remarkable resolution enhancement as shown in Fig. 5.5.

5.1.4 High-Field Electron Paramagnetic Resonance of Integer Spin Systems

EPR using conventional X- and Q-bands frequencies have been widely applied to $S = 1/2$ spin systems. However, this technique is less useful to

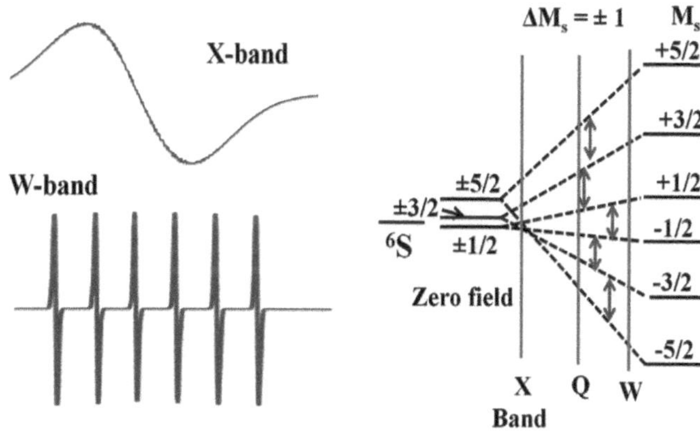

Figure 5.5 Increased resolution for high spin Mn^{2+} system.

high-spin especially integer S = 2 for Fe(II) system [2]. It is EPR inactive because of high ZFS. The recently developed high-field EPR and frequency EPR have been used to obtain EPR spectra and their zero field parameters. For example, $[Fe(H_2O)_6]ClO_4$ with $S = 2$, $D = 11.2\ cm^{-1}$, and $E = 0.70\ cm^{-1}$.

5.1.5 Molecular Dynamics From Multifrequency Electron Paramagnetic Resonance

In time domain experiments, EPR's time scale is nanoseconds, whereas nuclear magnetic resonance (NMR) is milliseconds. EPR is thus sensitive to the fast tumbling motion of the probe thereby providing the local fluidity. Furthermore, multifrequency approach provides snapshots at different time scales. For instance, HF tensors are resolved at lower frequencies and g-tensor at higher frequencies. Protein global slow motion is provided by X-band and fast internal motion of spin label by high-frequency EPR. In other words, the slow overall and collective motion will show up at lower frequencies. However, the fast motions will show up at higher frequencies [5].

5.1.6 Role in In Vivo Spectroscopy

If we have large sample *viz.,* mice, we need a large resonator. In general, as the MW frequency increases, the size of the resonator usually decreases. For the large sample size viz., mice to fit inside the resonator, we need to use lower frequencies (see Section 5.11). Applications where you may encounter large samples, for instance, in vivo spectroscopy and imaging,

Figure 5.6 Bruker ER 6502 resonator using L-band for in vivo spectroscopy. *Courtesy Bruker.*

samples exhibit high dielectric losses such as aqueous samples. Thus, it is quite difficult to measure at higher frequencies because of absorption, dielectric loss, and penetration problems. So, lower frequencies are useful for large biological samples (Fig. 5.6) [6].

5.1.7 Low-Frequency Electron Paramagnetic Resonance

Generally, Cu(II) complexes at higher frequencies lead to broader lines owing to g-anisotropy and strain and superhyperfine line not resolved. So, frequencies lower than X-band were used to give better-resolved continuous wave (CW)—EPR spectra. Not only better-resolved superhyperfine spectra were obtained but also a new type of copper complex in proteins was observed while studied at S-band using loop-gap resonator. In the $g_{||}$ region of the spectrum, there is narrowing of the line widths for negative values of M_I with minimum line widths near 3 GHz [7].

For example, the S-band EPR (3.5 GHz) of the octarepeat (PHGGGWGQ) prion protein segment in complex with Cu(II) is shown in Fig. 5.7. The Cu(II) HF (4 lines) are indicated as ($M_I = -3/2$, $-1/2$, $+1/2$, $+3/2$) in the $g_{||}$ region. Furthermore, the superhyperfine eight-line pattern is clearly resolved in $M_I = -1/2$ region. The computer simulated spectrum of this region is shown in Fig. 5.7 using three equivalent nitrogen ($a_H = 13$ G) and one proton ($a_H = 10$ G). Based on S-band low-frequency EPR, the structure was thus determined much before the X-ray

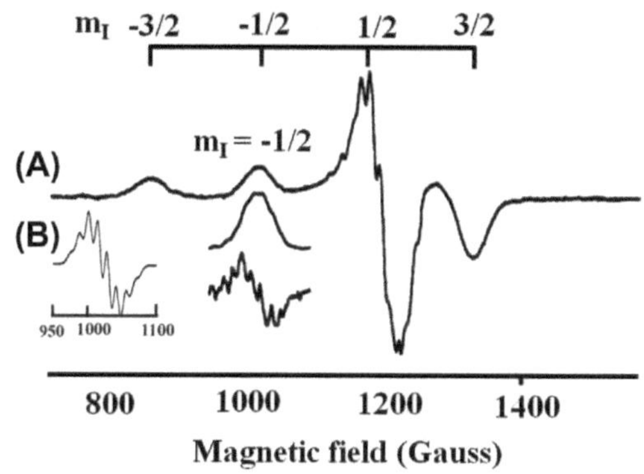

Figure 5.7 S-band electron paramagnetic resonance of Cu(II) complex of prion protein segment (A) and its simulation (B). *Reprinted with permission from American Chemical Society.*

report both agreeing each other. Hence, the EPR method will likely be accurate for cases where the structure has not yet been determined by X-ray.

Similarly, for Ni(I) d^9 complex in NiN_4 environment, the X- and W-band spectra measured are shown in Fig. 5.8. At lower frequency, the superhyperfine for four nitrogens ($2nI + 1 = 2 \times 4 \times 1+1 = 9$ lines) in the $g_\perp$ are well resolved, whereas, at higher frequency, these superfine lines are not resolved at W-band [8].

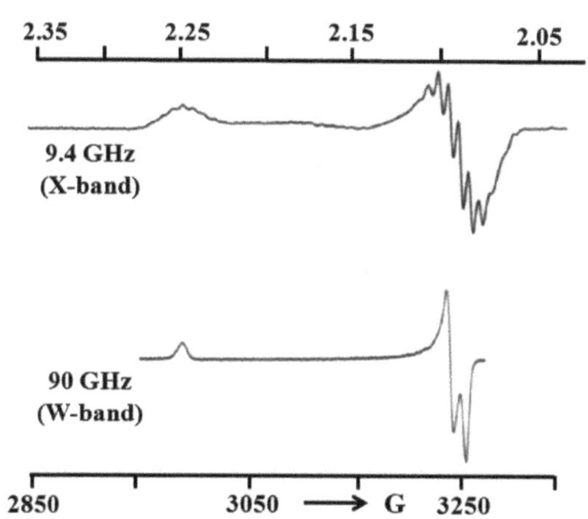

Figure 5.8 Superhyperfine splitting in electron paramagnetic resonance Ni(I) complex. *Reproduced with permission from Springer.*

5.1.7.1 Half-Field Transitions From Cu(II) Dimers

Both $\Delta m_s = 1$ and $\Delta m_s = 2$ signals are detected from cupric dimers. Often the half-field transitions are too weak to observe at X-band, and even more difficult to observe at higher frequencies because the transition probability is inversely proportional to the square of the MW frequency, i.e., the transition probability should be about nine times more at 3 GHz [7].

5.1.7.2 Low Frequency Over High Frequency

The general rule is that low MW frequencies are better for resolution of HF parameters and high frequencies are better for g-values [7].

5.2 PULSED ELECTRON PARAMAGNETIC RESONANCE

There are several ways to investigate a spin system by EPR. In CW EPR as discussed earlier, one measures the response of continuous MW irradiation of sample, while sweeping the magnetic field. This is equivalent to the sweep of MW frequency at fixed B_0. On resonance, the MW energy is partially absorbed. Another possibility is pulsed-EPR method similar to NMR to excite the electron spin system with a fixed range of frequency, high-power MW pulse at a constant magnetic field and to monitor the response. The shorter the pulse length is the greater the excitation bandwidth. One observes a rotating magnetization in the XY plane after 90°pulse (Fig. 5.9) and its decay due to dephasing of the individual spins

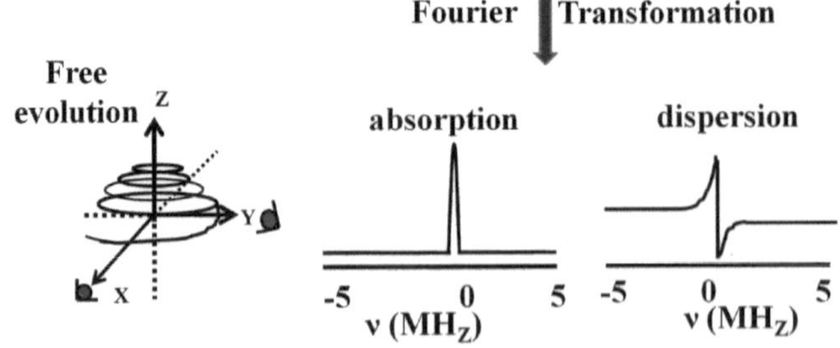

Figure 5.9 Free induction decay and Fourier transform electron paramagnetic resonance.

(T_2 relaxation) and due to relaxation to thermal equilibrium (T_1 relaxation). This is called free induction decay (FID). Fourier transform (FT) of the FID response of the sample gives a spectrum in the frequency domain. FID signals can be monitored, or further pulses applied to give rise to spin echoes (electron spin echo (ESE) spectrum) which are the basis of many important techniques. The FT of the FID signal gives spectrum equivalent to the CW-EPR spectrum [9—12].

5.2.1 Electron Spin Echo—Electron Paramagnetic Resonance

The FT-EPR technique is rarely used as it suffers from two major problems. First, the excitation bandwidth of the shortest available MW pulse (4 ns) does not allow for a complete excitation of spin systems with even moderately broad spectral widths. Second, the strong excitation pulse causes a ringing of the resonator, which prevents the recording of a signal for a certain time after the pulse. The dead time for X-band is ~100 ns. To circumvent this especially second problem, recent EPR experiments are based on the detection of ESE rather than FID. This spin echo phenomenon was first observed by Erwin Hahn in 1950. Using more than one pulse, because of the nonlinearity of spin systems, ESE phenomena are produced. The two-pulse sequence (Hahn echo) and the diagram describing the motion of the magnetization vectors in a primary echo experiment is shown in Fig. 5.10. Before the experiment, at thermal equilibrium, the magnetization vector is oriented along the z-axis (Fig 5.9). On applying the first $\pi/2$ pulse

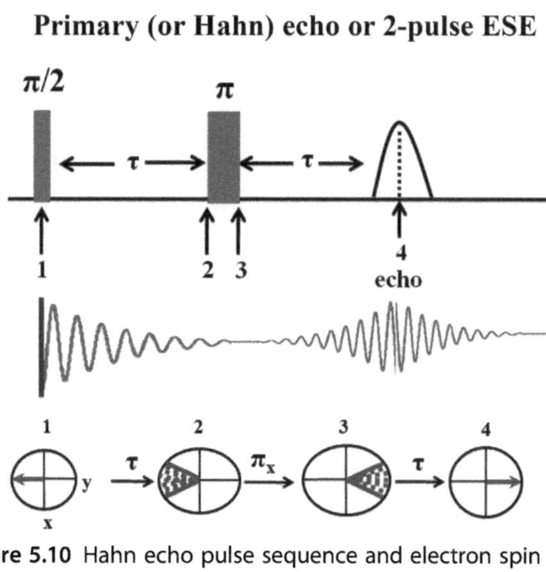

Figure 5.10 Hahn echo pulse sequence and electron spin echo.

along the x-axis, the magnetization rotates to the −y-axis (1). After the pulse, the different spin packets start to process with their individual Larmor frequencies around the z-axis, resulting in a dephasing of the transverse magnetization. After time τ (2), a second π pulse again along the x-axis turns all the magnetization vectors through 180 degrees about this axis (3). As the directions of rotation of the individual spin packets are not changed by the refocusing pulse (π), after another time τ, all the vectors are aligned along the xy-axis, giving rise to a net y-magnetization called ESE (4).

5.2.1.1 Advantages

1. Pulsed FT methods offer the advantage of higher sensitivity.
2. Moreover, sophisticated methods using several pulses help to disentangle the various interaction parameters hidden in CW-EPR spectra. This leads to the field of multidimensional magnetic resonance.
3. The importance of spin echo experiments in pulse EPR is explained by the broad lines encountered in solid-state EPR, which make the recording of FID often impossible. Anisotropic interaction such as the g-tensor or the B_0 in homogeneities leads to an inhomogeneous broadening of the EPR line. Unlike the homogeneous line width, the inhomogeneous line width is not related to any relaxation time. The dephasing of the magnetic moments under the action of the magnetic field distribution leads to a rapid vanishing of the transversal magnetization, which often occurs already during the spectrometer dead time.
4. Particularly interesting is the combination of high magnetic fields/high MW frequencies provides high gauss and high time resolution using pulsed methods. However, the high frequency leads to limitations of MW technology and restriction to very basic (field swept) pulsed methods.

5.2.2 Field-Swept, Echo-Detected Electron Paramagnetic Resonance

Though the ESE spectra circumvent the dead time problem, it does not overcome the insufficient excitation bandwidth. This problem can be solved by sweeping of the magnetic field analogous to the CW EPR. The field-swept, echo-detected EPR is one of the simplest pulse EPR techniques. Here again the $\pi/2-\tau-\pi$ pulse sequence is used to generate a spin echo signal. However, the MW frequency and the pulse sequence are kept constant and echo intensity is recorded as a function of the magnetic field.

This pulse experiment has some advantages over conventional CW EPR, such as the absence of MW irradiation during signal detection, and that the measured EPR response is proportional to the magnetization vector amplitude at the moment of MW excitation. Thus, the noise level is defined by only spectrometer limits, without additional noise from the MW source, and spectral analysis is simplified due to the absence of MW excitation in the detected signal. In particular, CW EPR is insensitive to very broad spectra due to the field modulation when collecting data. Echo-detected EPR does not suffer from this problem. However, echo-detected EPR does suffer from some serious limitations, including time resolution and sensitivity. The time resolution is limited by two factors: the length of the pulse sequence and the dead time of the spectrometer. Dead time is defined as the time between switching off the MW power and when the ringing intensity in the cavity decreases to the same order of magnitude as the echo intensity, which is when it becomes possible to measure the response signal.

5.2.2.1 Pulsed Electron Paramagnetic Resonance Instrumentation

The basic setup of a pulsed EPR spectrometer is straightforward. One needs a magnet, a transmitter, which provides the pulse MW field, a MW resonator, and a receiver, which has to have the required time resolution. The optimum setup depends on the desired application and purpose. In recent years, a lot of different high-field/high-frequency spectrometers have been developed pulsed as well as CW. The spectrometers vary in frequency, in time resolution, in sensitivity, and in MW output power, which determines the $\pi/2$ pulse length.

5.2.3 Comparison of Pulsed Electron Paramagnetic Resonance Versus Nuclear Magnetic Resonance

Because of the much shorter electron (cf. nuclear) spin relaxation times, much shorter pulse lengths are required in EPR than in NMR, and hence it is necessary to measure at low temperatures (e.g., for transition ions). Moreover, the excitation bandwidth is rarely sufficient to excite the entire EPR spectrum. Nevertheless, pulsed EPR has developed into a range of robust and hugely powerful techniques, exploiting the advantages of time domain and multidimensional methods. Such experiments give access to information that is not accessible by CW methods, including information on weakly coupled nuclei, weakly coupled electron spins, and spin dynamics (relaxation) etc.

5.3 SATURATION RECOVERY ELECTRON PARAMAGNETIC RESONANCE

This technique is used to measure electron spin—relaxation times to understand the relaxation mechanism. It is applied to oximetry, Heisenberg exchange, and spin—spin distance measurements [13].

A qualitative understanding of relaxation times is essential for selection of parameters for CW EPR and electron—nuclear double resonance (ENDOR) experiments and for prediction of feasibility of pulsed-EPR experiments. Quantitatively, the relaxation times provide insight into electronic structure, motion dynamics, and other processes that contribute to relaxation.

There are three methods to measure electron spin—lattice relaxation times (1) Saturation recovery (SR) (2) Inversion recovery, and (3) Echo-detected SR.

The measurement of T_1 by CW SR was first demonstrated for nuclear spins by Bloembergen [13a]. Subsequently, there have been many applications to electron spins. He measured the change in the z magnetization following a MW pulse by using a pickup coil outside the resonant cavity. This method is known as SR. The pulse sequence for SR EPR is sketched in Fig. 5.11A. A high-power pulse is first applied to saturate the EPR transition. The length of this pulse defines whether the experiment is viewed as long- or short-pulse SR.

In a short-pulse SR experiment, there frequently will be contributions to the recovery curve from spectral diffusion processes including molecular tumbling (in fluid solution), nuclear spin relaxation, or cross relaxation. In these cases, a T_{1e} measurement produces a recovery curve that is "the sum of all possible relaxation pathways rather than the relaxation between only the observed levels."

5.3.1 Comparison of Saturation Recovery With Inversion Recovery and Echo-Detected Saturation Recovery

Inversion recovery and echo-detected SR are two other pulse techniques that have been used to measure T_{1e} (Fig. 5.11B and C). In an inversion recovery experiment, the first pulse is a 180 degrees pulse that inverts the spin magnetization. The time between the inverting pulse and the two-pulse spin echo sequence that is used to monitor the net magnetization is varied. The disadvantage of the inversion recovery sequence is that spectral diffusion processes with time constants that are shorter than T_{1e} contribute to the recovery curve. In echo-detected SR (Fig. 5.11C) a longer lower-power pulse

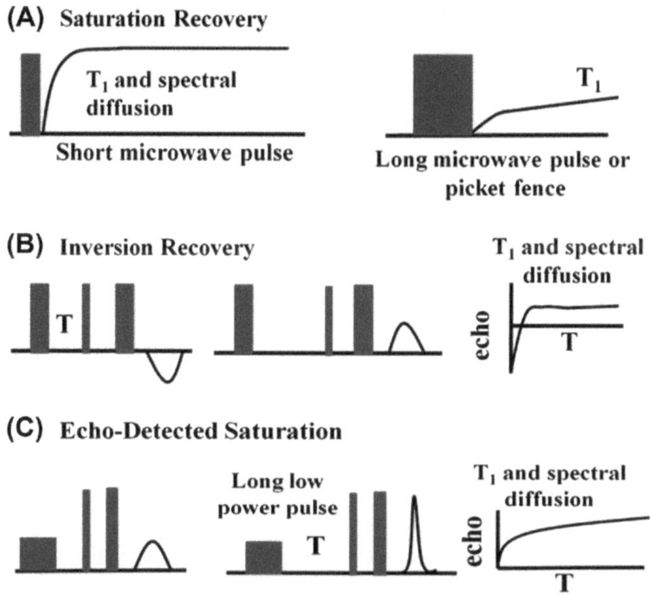

Figure 5.11 Pulse sequences for measurement of T_1. The time constant measured is an "effective" T_1 that includes spectral diffusion contributions. (A) Saturation recovery, (B) inversion recovery and (C) echo-detected saturation recovery.

is used to saturate the EPR transition and the net magnetization as a function of time is monitored with two-pulse spin echoes. In principle, the recovery curve obtained using echo-detected SR method should have the same time constant as obtained by SR, if the lengths and powers of the saturating pulses are comparable [13].

An advantage of SR is that the saturating MW pulse is produced by a MW source. It can, in principle, be made as long as required to saturate the spectral diffusion processes.

5.4 TIME-RESOLVED ELECTRON PARAMAGNETIC RESONANCE

Time-resolved EPR (TR-EPR) has proven to be a valuable and powerful technique for the characterization of the structure and dynamics of radicals produced in photochemical reactions of organic molecules [14,15]. In addition, the technique provides information on the interactions of radical pairs, spin polarization mechanisms, spin relaxation, molecular motion and the selectivity of triplet sublevel populations. There is one serious problem, in TR-EPR spectroscopy. For commercial CW-EPR spectrometers, a magnetic field modulation of 100 kHz is routinely used to improve the sensitivity of the spectrometers. The response time of such spectrometers is limited to tens of microseconds. This time resolution is usually inadequate for the study of photo-induced processes in solution. Two methods have been devised for the improvement of the response time, eliminating the field modulation with a loss of signal sensitivity.

1. The first method is the direct CW detection and
2. The second is the pulsed-MW ESE or FT method.

The time resolution reachable with the first method is about 50 ns and with the second one about 10 ns. The FT methods may be further classified in terms of the process of analyzing the signal as FID or as an ESE. The CW and FT methods are expected to differ somewhat in terms of sensitivity and resolution due to differences in the technical aspects of data acquisition and analysis. The CW method depends mainly on the spin—lattice relaxation time T_1, and the FT method depends mainly on the phase memory time T_M (which is a function of all processes that lead to the relaxation of the spin system, such as the spin—lattice relaxation time T_1, spin—spin relaxation times T_2, and chemical reactions). So, there is a fundamental difference between the two methods that could result in significant differences in the experimental spectra.

Although TR-CW and TR-FT EPR spectra of radicals produced by laser flash photolysis are typically experimentally equivalent except for signal to noise and resolution of the EPR signals, there are cases in which specific effects can influence T_2 for certain lines in the TR-FT spectrum and thereby complicate the identification of the radicals responsible for the experimental spectrum. It was shown that certain lines could be reduced in intensity in the TR-FT spectra as a result of the shortening of T_2 due to hindered rotation. Although the "*missing or diminished signals*" in a TR-FT spectra caused by a short T_M complicates the interpretation of TR-FT EPR spectra, a combination of TR-CW and TR-FT and multiple methods of generation of radicals provides a means of resolving the complications and also leads to insights into the radical structure being investigated.

5.4.1 Time-Resolved Electron Paramagnetic Resonance Spectrometer

For the measurements of a TR-CW EPR, a Bruker spectrometer ESP 300E equipped with a MW amplifier and a TE_{102} MW cavity (MW power 9 mW, response time 40 ns) is commercially available operating without field modulation. Radicals possessing electron spin polarization are produced

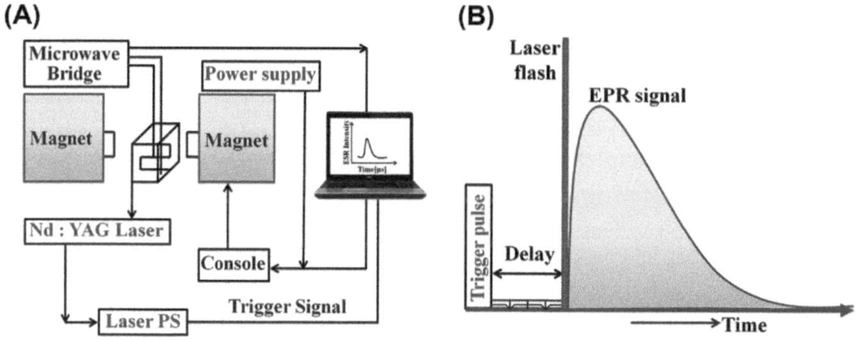

Figure 5.12 (A) Block diagram of the time-resolved electron paramagnetic resonance (TR-EPR) spectrometer and (B) TR-EPR signal.

in the cavity of the spectrometer by a short-laser pulse (Fig. 5.12). The EPR transitions are observed with nanosecond time resolution in a fixed external magnetic field B_0. The resonance of the EPR spectrometer after a light pulse at a precise, fixed value of the static magnetic field was measured and the data displayed on computer. A block diagram of the TR-EPR experiment is shown in Fig. 5.12A. The timing sequence for the TR-EPR experiment is shown in Fig. 5.12B.

Although, the sensitivity of the TR-EPR spectrometers is much lower than that of the conventional ones, the discovery of extremely large anomalies in the EPR intensities of many transient radicals has improved the sensitivity of their detection.

5.4.2 Rapid-Scan Electron Paramagnetic Resonance

Rapid-scan EPR is a new EPR method in which a direct-detected signal is (i.e., without magnetic field modulation and phase-sensitive detection) similar to TR-EPR, whereas the magnetic field is being scanned through the spectrum rapidly relative to electron spin—relaxation times. Rapid-scan EPR encompasses the regime in which the magnetic field sweep is fast relative to relaxation times, which is a newly developed intermediate regime between CW and pulsed EPR.

5.5 ELECTRON SPIN ECHO ENVELOPE MODULATION (ESEEM)

In 1965, Rowan, Hahn, and Mims observed that for certain samples the decays of primary ESE were modulated with frequencies that correspond to nuclear frequencies and their combinations (differences and sums). This electron spin echo envelope modulation (ESEEM) effect is caused by the HF/superhyperfine and nuclear quadrupole interaction (NQI). Thus, it allows for a detection of nuclear spectra without directly exciting the nuclear spins. The basic principles of many pulse EPR (ESEEM) experiments can be explained by considering a spin system consisting either of one electron spin $S = 1/2$ and one nuclear spin $I = 1/2$ or of one electron spin $S = 1/2$ and one nuclear spin $I = 1$. The treatment here is restricted to these two cases for simplicity [9—12].

5.5.1 Spin System S = 1/2, I = 1/2

The energy-level diagram for an $S = 1/2$, $I = 1/2$ system, along with the EPR and nuclear frequency spectra, are shown in Fig. 5.13. There are two cases depending on the interaction (1) the weak coupling $|A| < 2|\omega_I|$ and (2) strong coupling $|A| > 2|\omega_I|$.

$\omega_s = g_e\beta_e B_0/\hbar$ and $\omega_I = g_n\beta_n B_0/\hbar$ are the electron and the nuclear Zeeman (NZ) frequency, respectively, expressed in angular frequency units. The effective quantization vector, experienced by the nucleus is obtained by addition of the HF and NZ interactions (NZI). In the case of isotropic, the NZI and the HF field act in the same direction z for both weak- and strong coupling in the $m_s = -1/2$ manifold. The splitting between the NZ states is thus increased by both interactions. However, in the $m_s = 1/2$ manifold, the NZ and HF field, act in the opposite direction. At exact cancellation ($|A| = 2|\omega_I|$), these two terms cancel and the nuclear frequency

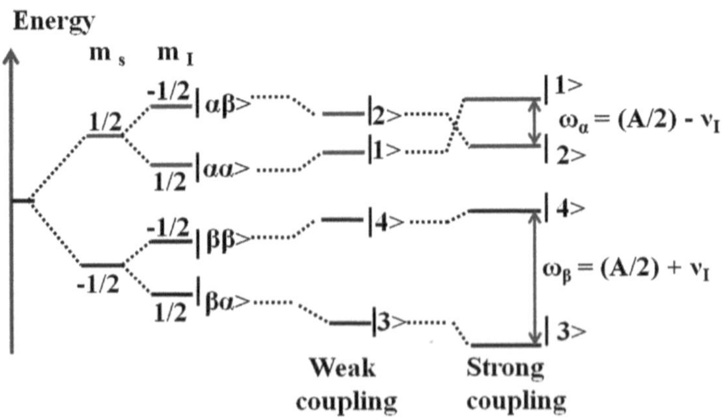

Figure 5.13 Energy-level diagram for an $S = 1/2$, $I = 1/2$ system in the weak- and strong-coupling cases showing nuclear transitions.

is zero in this manifold. In the case of strong coupling, where HF > NZ, the energy of the $|\alpha\beta\rangle$ and $|\alpha\alpha\rangle$ states are therefore interchanged in their order.

5.5.1.1 Two-Pulse Electron Spin Echo Envelope Modulation

In the two-pulse ESEEM (2p) experiment (Fig. 5.14), the intensity of the primary echo is recorded as a function of the time interval τ between the $\pi/2$ and π pulses. The echo amplitude modulation formula for an $S = 1/2$, $I = 1/2$ system is given as follows.

$$V_{2p} = 1 - k[2 - 2\cos(\omega_\alpha \tau) - 2\cos(\omega_\alpha \tau) + \cos(\omega_- \tau) + \cos(\omega_+ \tau)]/4 \quad (5.2)$$

where k is the orientation dependent modulation depth parameter given by

$$k(\theta) = \sin^2 2\eta = \left(B\omega_I/\omega_\alpha\omega_\beta\right)^2 \quad (5.3)$$

where 2η is the angle between the nuclear quantization axes in the 2 m_s manifold and ω_I is the nuclear Larmor frequency, $\omega_{13} = (\omega_S + \omega_-)/2$, $\omega_{24} = (\omega_S - \omega_-)/2$ and $\omega_{14} = (\omega_S + \omega_+)/2$, $\omega_{23} = (\omega_S - \omega_+)/2$, respectively, with $\omega_+ = \omega_{12} + \omega_{34}$ and $\omega_- = \omega_{12} - \omega_{34}$.

ESEEM techniques rely on the excitation of both allowed- and forbidden-EPR transitions using MW radiation. There is no ESEEM effect if $I = 0$. The energy level–splitting diagram for $S = 1/2$, $I = 1/2$ weakly

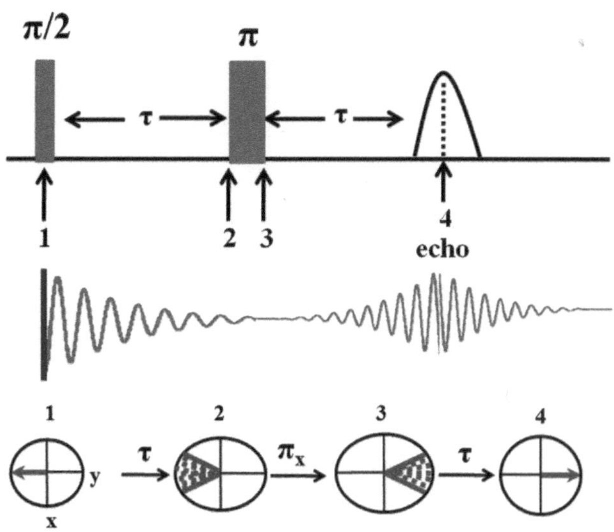

Figure 5.14 Pulse sequence for two-pulse electron spin echo envelope modulation (A) and the diagram (B) showing the motion of the magnetization vectors.

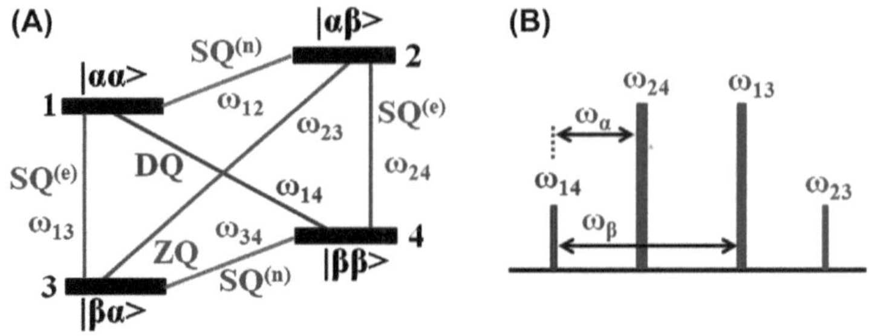

Figure 5.15 For a weakly coupled S = 1/2; I = 1/2 spin system ($|\omega_I| > |A/2|$). (A) Energy-level diagram with allowed single quantum (SQ) transitions of the electron and nucleus and forbidden electron paramagnetic resonance double quantum (DQ) and zero quantum (ZQ) transitions and (B) corresponding stick electron spin echo envelope modulation spectrum.

coupled system in Fig. 5.15. It further, shows the allowed single-quantum electron (SQe) transitions (ω_{13} and ω_{24}), single-quantum nuclear (SQn) transitions (ω_{12} and ω_{34}), and forbidden EPR double transitions (ω_{14}) zero-quantum transition (ω_{23}).

In the case of an isotropic HF interaction or B_0 oriented along one of the principal axis of the HF tensor ($\theta = 0$ or $\theta = \pi/2$), the echo modulation disappears because B in Eq. (5.3) is zero. The modulation depth is maximum when $|A| \approx 2|\omega_I|$. Here, A is the HF coupling at an arbitrary orientation.

The disadvantage of the two-pulse experiment is that the primary echo decays with the phase memory time, T_M, which is often very short. This can be overcome with the three-pulse ESEEM experiment.

5.5.2 Spin System S = 1/2, I = 1

Consider now nuclei I = 1, with the HF interaction (A) approximately twice the NZI. The energy-level splitting of one m_s manifold is primarily determined by the NQI. There are two cases as follows (1) $A = 2|\omega_I|$ and (2) A > NZ and NQI

Case (i) At exact cancellation

When the HF and the NZ terms match, the effective field of nucleus vanishes in the $m_s = 1/2$ manifold. The nuclear frequencies within this manifold thus correspond to the nuclear quadrupole resonance frequencies $\omega_0 = 2K\eta$, $\omega_- = K(3 - \eta)$, and $\omega_+ = K(3 + \eta)$ with $K = e^2qQ/4\hbar$. This condition leads to a line narrowing and consequently the intensity becomes high. However, the manifold $m_s = -1/2$ where the NZ and HF interactions are additive giving rise to much

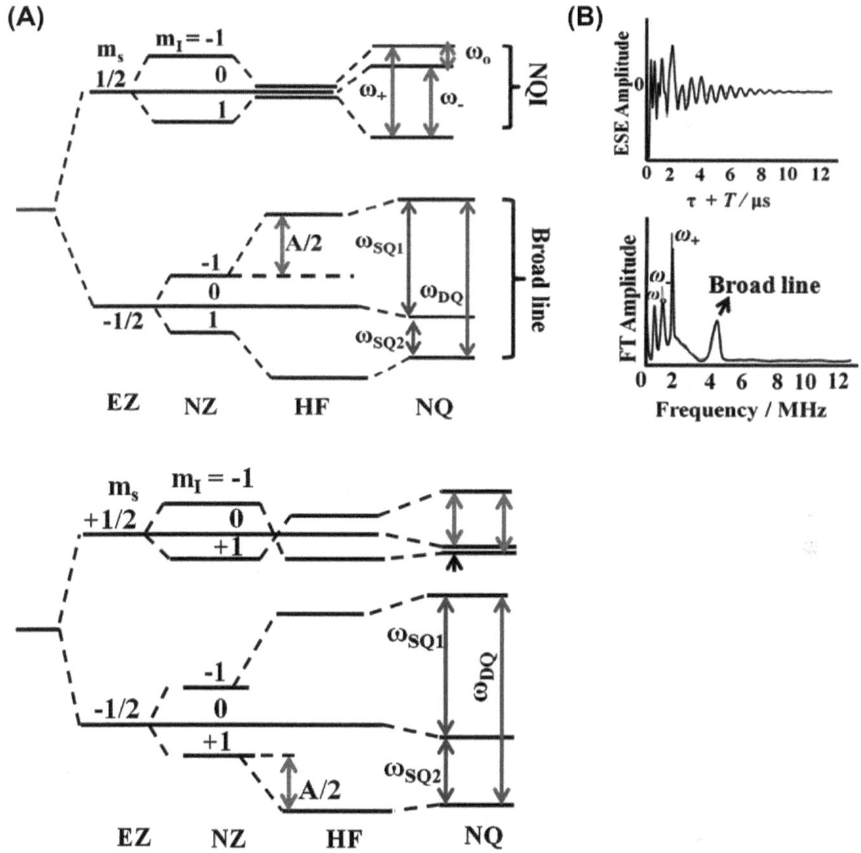

Figure 5.16 (A) Energy level—splitting diagram for S = 1/2, I = 1 spin system under the condition of exact cancellation (A) $|\omega_I| = |A/2|$ weak coupling, (B) time and frequency domain electron spin echo envelope modulation spectra for Cu(II) complex with weak coupling and (bottom) shows $|A| > 2|\omega_I|$ strong-coupling case. *DQ*, double quantum; *ESE*, electron spin echo; *EZ*, electron Zeeman; *FT*, Fourier transform; *HF*, Hyperfine; *NZ*, nuclear Zeeman; *NQ*, nuclear quadrupole; *SQ*, single quantum. *Reproduced with permission from Wiley-VCH Verlag GmbH & Co. KGaA, Weinheim.*

broad resonances and often the only distinguishable feature is the double-quantum (DQ) transition, $\Delta m_I = 2$. This is because the DQ frequencies are free from quadrupole broadenings to first order. The energy level—splitting diagram illustrating this case is given in Fig. 5.16. If the anisotropic HF interaction is small compared to the isotropic and the NQI, the DQ frequency is given by

$$\omega_{dq} = 2\left[(m_s a_{iso} + \omega_I)^2 + K^2(3 + \eta)^2\right]^{1/2} \tag{5.4}$$

For example, the ESEEM spectra of Cu(II) complex are shown in Fig. 5.16B. It exhibits ω_0, ω_- and ω_+, and DQ transitions (broad line) corresponding to weak-coupling case [16].

Case (ii)

When the HF energy with coupling constant A exceeds both the NZI and the NQI for an axially symmetric system with $I = 1$, the energy level—splitting diagram given in Fig. 5.16 (bottom) is obtained. In ESEEM experiments, usually both single-quantum (SQ) and DQ nuclear transitions can be observed, while in ENDOR spectroscopy only the SQ nuclear transitions are observed, and the corresponding frequencies are given in general by the first order equations

$$\omega_{sq} = A/2 \pm \omega_I \pm 3P_z/2 \tag{5.5}$$

The four ENDOR lines predicted by Eq. (5.5) are often not resolved. If the P (nuclear quadrupole tensor) is smaller than the ENDOR line width, 0.1—0.5 MHz for frozen solution of transition metal ions, only two lines centered at A/2 and separated by $2\omega_I$ are observed.

5.5.2.1 Three-Pulse Electron Spin Echo Envelope Modulation

Although two pulse ESEEM is very fast method, it has the following major drawbacks

1. Overlapping of signals due to line broadening.
2. It depends on electron spin—spin relaxation that is often very short, so that one has broad signals.
3. Nuclear combination frequencies are observed. However, no such combination frequencies are obtained in 3p ESEEM. This significantly simplifies spectral interpretation.
4. Furthermore, an increased spectral resolution is achieved using 3p ESEEM beause the echo decay is related to the phase memory time of the nuclear spins, which is longer than electron-phase memory time.
5. In contrast to 2-pulse ESEEM, modulations due to couplings with several nuclei are only multiplicative in 3p in ESEEM (but summative in 2p) with respect to the electron spin manifold.

In the three-pulse ESEEM experiment (Fig. 5.17), an envelope modulation due to the stimulated echo (instead of primary echo using Hahn echo sequence) is observed when the time delay T is incremented. The first $\pi/2$ pulse creates allowed and forbidden electron coherence evolving during τ. The second $\pi/2$ pulse partially transfers this electron coherence to nuclear coherence, which evolves during the evolution time T and decays

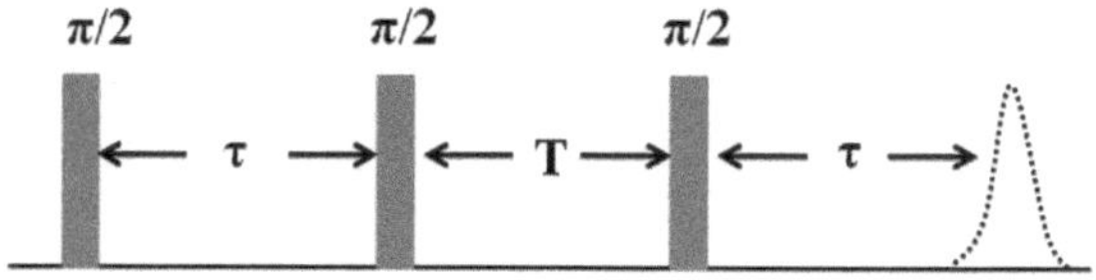

Figure 5.17 Pulse sequence for three-pulse electron spin echo envelope modulation experiment.

with the transverse nuclear relaxation time T_{2n}, which is usually much longer than the corresponding relaxation time T_M of the electrons. The third $\pi/2$ pulse transfers the nuclear coherence back to observable electron coherence.

For a two-spin system, with the contribution from the $\alpha = 1/2 (\beta = -1/2)$ electron spin manifold the modulation of the stimulated echo is given by

$$V^{\alpha(\beta)}(\tau, T) = 1 - k\left[1 - \cos\left(\omega_{\beta(\alpha)}\tau\right)\right]\left[1 - \cos\left(\omega_{\alpha(\beta)}(\tau + T)\right)\right]\Big/2 \quad (5.6)$$

When T is varied, the echo envelope is modulated only by the two basic frequencies ω_α and ω_β. In contrast to the two-pulse ESEEM experiment, the sum ω_+ and the difference ω_- frequencies do not appear. The three-pulse ESEEM amplitudes depend on τ, which results in blind spots at $\omega_{\alpha(\beta)}$ when $\tau = 2\pi n/\omega_{\beta(\alpha)}$ ($n = 1, 2, \ldots$). As a consequence, the experiment often has to be performed for different τ values and spectra have to be added to avoid blind spots artefacts.

5.6 HYPERFINE SUBLEVEL CORRELATION (HYSCORE)

Hyperfine sublevel correlation (HYSCORE) is a four-pulse ESEEM and two-dimensional (2D) technique (Fig. 5.18) where an additional π pulse is introduced between the second and third $\pi/2$ pulse of the 3-pulse ESEEM [9–12,17]. It interchanges the nuclear coherences between α and β manifolds of the electron spin. The two interpulse times t_1 and t_2 are varied independently. 2D Fourier transformation leads to frequency-domain spectra where the nuclear frequencies belonging to different m_s manifold are correlated. This 2D EPR spectrum was invented in 1986 by Hofer et al. [9–12,12a]. Since the spectral density is distributed over two dimensions, HYSCORE spectra provide an improved resolution and contain information that cannot be easily be retrieved by one-dimensional (1D) ESEEM methods.

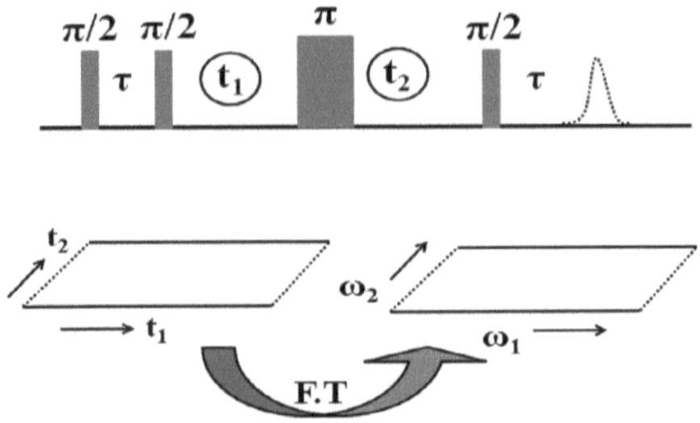

Figure 5.18 Hyperfine sublevel correlation (four-pulse electron spin echo envelope modulation). *FT*, Fourier transform.

For the $S = 1/2$, $I = 1/2$ system described earlier, the HYSCORE spectrum is characterized by a correlation pattern as shown in Fig. 5.19A. For the weak-coupling case, the cross peaks are found in the (++) quadrant, whereas they appear in the (−+) quadrant for the strong-coupling case. As can be seen from Fig. 5.19A the HF value can easily be estimated from the peak positions. Furthermore, the correlation pattern considerably facilitates spectral assignment for complex systems where the electron is interacting with many nuclei. For disordered systems, e.g., frozen solutions of proteins, each magnetic field setting corresponds to a selection of molecular orientations (so-called orientation selection) and the HYSCORE signals will no longer be sharp cross peaks but extended ridges representing the sum of the cross peaks corresponding to each of these orientations. This is shown in Fig. 5.19B, where a proton HYSCORE spectrum of a frozen solution of a Cu(II)-bound prion protein (PrP(23−231)) is depicted [18]. The intersection of the diagonal and the antidiagonal (dashed line) represents the proton Larmor frequency for this field. This is a typical example of the weak-coupling case (signals in (++) quadrant). Analysis of the HYSCORE spectra of disordered systems, however, requires spectral simulation and is nontrivial.

For $S = 1/2$, $I = 1$ system (e.g., an unpaired electron interacting with a ^{14}N nucleus), the correlation pattern becomes more complicated. In cases where only one molecular orientation is selected by the magnetic field setting (single-crystal case), six nuclear frequencies can be detected: Four SQ frequencies ($|\Delta m_I| = 1$) and two DQ frequencies ($|\Delta m_I| = 2$) (Fig. 5.19C). The corresponding HYSCORE spectrum will then reveal information about the nine correlations linking the nuclear frequencies of the

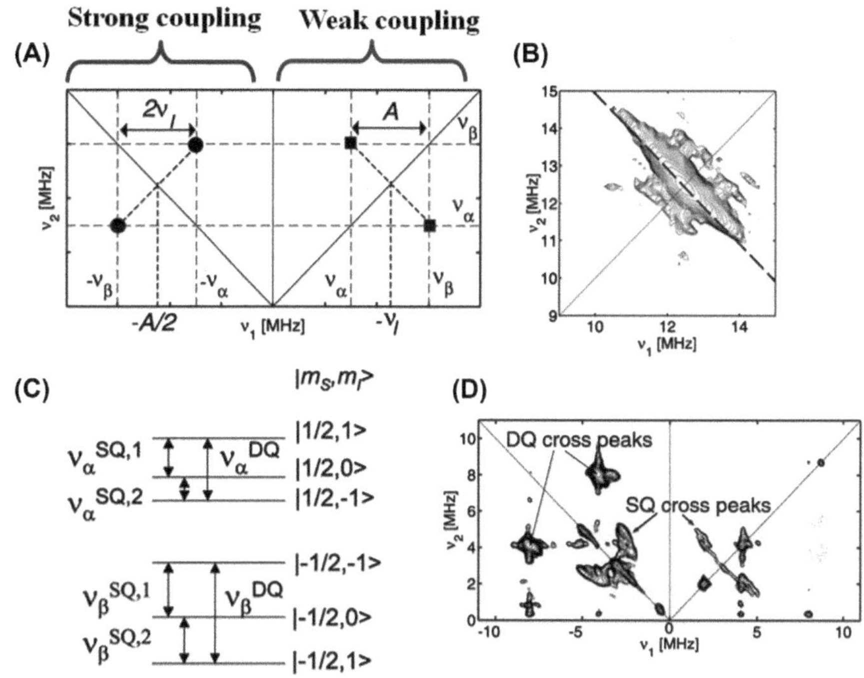

Figure 5.19 (A) Schematic representation of an hyperfine sublevel correlation (HYSCORE) spectrum of an S = 1/2, I = 1/2 system, (B) experimental proton HYSCORE spectrum of a frozen solution of Cu(II)-bound PrP(23–231) taken at an observer position corresponding to g = 2.37. (C) Schematic representation of the energy levels for an S = 1/2, I = 1 system indicating the double-quantum (DQ) and single-quantum (SQ) nuclear frequencies. The example represents a strong coupling case. (D) Experimental ^{14}N HYSCORE spectrum of a frozen toluene solution of vanadyl (II) complex at g = 2.056 (strong-coupling case). *Reproduced with permission from Royal Society Chemistry.*

two m_s manifolds. For cases with a considerable nuclear quadrupole contribution (which is generally the case for ^{14}N nuclei), the HYSCORE spectra are dominated by the DQ cross peaks, linking the frequencies [18].

$$\nu_{\alpha,\beta}^{DQ} = 2\sqrt{\left(\nu_I \pm \frac{A}{2}\right)^2 + \left(\frac{e^2qQ}{4h}\right)^2(3+\eta^2)} \tag{5.7}$$

where A is the HF coupling at the particular observer position.

These DQ cross peaks will appear in the (++) quadrant for a weak-coupling case and in the (−+) quadrant for a strong-coupling case. The other cross peaks can be found spread over both quadrants. This is exemplified in Fig. 5.19D showing the ^{14}N HYSCORE spectrum of a VO(II) salen-type molecule taken at an observer position known to excite all molecular orientations in the ligand plane [18]. This example was chosen because it nicely represents some of the essential characteristics of nitrogen HYSCORE spectra of disordered systems. Similar spectra have been reported for VO(II) model systems of the enzyme bromoperoxidase [18].

There are two factors affecting the ESEEM/HYSCORE spectra (1) Blind spots (2) Unwanted echoes along with the stimulated echo removed by using phase-cycling method.

1. Blind spots

   A feature of the 3-pulse ESEEM and the HYSCORE technique is a consequence of the factors $[1\text{-}\cos(\omega_{(\alpha,\beta)}\tau)]$. For $\tau = 2\pi n/\omega_{\beta(\alpha)}$ ($n = 1, 2, \ldots$), the modulation at frequency $\omega_{\beta(\alpha)}$ vanishes. This leads to so-called blind spots in the spectrum where the nuclear frequencies are suppressed. To detect all nuclear modulations, it is therefore necessary to repeat the experiments with different values of τ.
2. Phase cycling

   The $\pi/2-\tau-\pi/2-T-\pi/2$ pulse sequence does not only result in the formation of the stimulated echo but also gives rise to unwanted echoes created by pairs of the pulses. Thus, HYSCORE generates even more unwanted echoes, which disturb the detected signal. These echoes can be suppressed by phase cycling. With an appropriate phase cycle, only the pathway of the wanted echo is selected. A complete suppression of all echoes is achieved by 4-step phase cycle for 3-pulse ESEEM and by a 16-step phase cycle for HYSCORE. However, since many coherence transfer paths do not lead to a refocusing, a 4- or 8-step phase cycling is usually sufficient.

5.7 MULTIRESONANCE ELECTRON PARAMAGNETIC RESONANCE

It involves application of either two or more MW (at different frequencies) called electron–electron double resonance (ELDOR) one MW and radiofrequency (RF) (ENDOR) to cause resonance between the magnetic nuclear spins and unpaired electrons. Based on that, there are two major types of double resonance techniques [9–12].

1. ENDOR (Electron nuclear double resonance)
2. ELDOR (CW and pulsed) stands for electron–electron double resonance. Pulsed ELDOR is also called double electron–electron resonance (DEER).

5.8 ELECTRON–NUCLEAR DOUBLE RESONANCE (ENDOR)

So far, we have seen in the EPR spectrum, the nuclear HF splitting was always clearly detectable and well resolved. However, more often the HF (due to coupling of unpaired electron with several nuclei resulting in overlapping of signals) splitting is not resolved because it is hidden in the inhomogeneous broadening of EPR peaks in frozen samples. Another drawback is, for example, the solution EPR spectrum of the radical cation of 9,10-dimethylanthracene theoretically produces 175 lines due to the three sets of inequivalent protons (Fig. 5.20). Hence, simulation of the spectrum is required to extract the HF couplings constants. However, the ENDOR spectrum produces only three pairs of lines (one for each set of equivalent nuclei) from which the HF couplings can be determined directly. Thus, ENDOR is able to greatly reduce the number of peaks showing only 2N peaks as opposed to 2^N peaks for N set of equivalent nuclei [19–23].

Hence, the advanced techniques viz., ENDOR spectroscopies, ESEEM, and HYSCORE are used. In complex organic radicals, transition metal

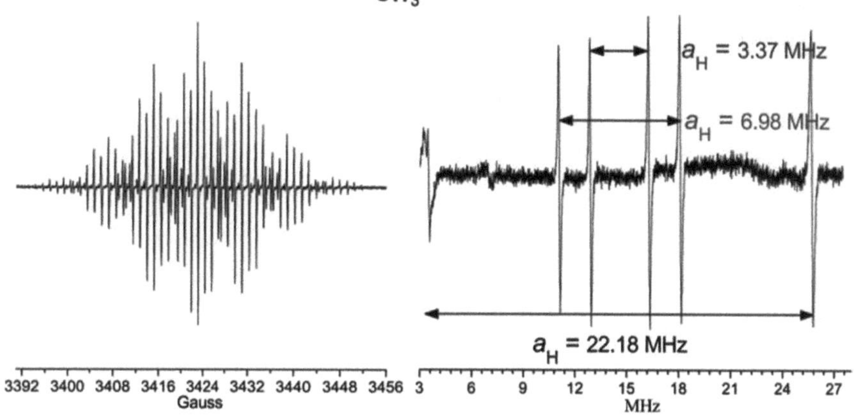

Figure 5.20 Electron paramagnetic resonance (EPR) and electron–nuclear double resonance (ENDOR) spectra of 9,10-dimethylanthracene. *Reproduced with permission from Royal Society of Chemistry.*

complexes, metalloproteins, and magnetic nuclei, such as ^{1}H, ^{2}H, ^{13}C, ^{14}N, ^{15}N, ^{17}O, ^{31}P, and ^{33}S, in the vicinity of the paramagnetic radical/metal ion can be detected by these techniques. These methods thus enable one to identify the presence of a particular ligand nucleus but under favorable circumstances metal–ligand nuclei distances and angles can be obtained as well. Although these techniques give similar types of information, the sensitivity of these techniques can be drastically different.

5.8.1 Theory of Electron–Nuclear Double Resonance

Let us now consider the spin Hamiltonian relevant for these techniques. The spin Hamiltonian equation for the simplest system, with effective electron spin $S = 1/2$ and a single nucleus with $I = 1/2$ is as follows:

$$\widehat{H}_S = \beta_e B g_e S - \beta_n B g_n I + SAI \quad (5.8)$$

Where β_e and β_n are electronic and nuclear magnetons, g and g_n are the electronic and nuclear g-tensors, and A is the anisotropic HF tensor describing the interaction between the electron and nuclear spins. In this Eq. (5.8), the first term is the electronic Zeeman interaction, the second is the NZI, and the third is the HF interaction.

The energies to first order are given by the expression

$$E(m_s m_I) = g_e \beta_e B m_S - g_n \beta_n B m_I + hA m_s m_I \quad (5.9)$$

$$E(m_s m_I)/h = \upsilon_e m_s - \upsilon_n m_I + A m_s m_I \quad (5.10)$$

$$E_1 = -\upsilon_e/2 - \upsilon_A/2 - A/4$$
$$E_2 = -\upsilon_e/2 + \upsilon_A/2 + A/4$$
$$E_3 = \upsilon_e/2 - \upsilon_A/2 + A/4$$
$$E_4 = \upsilon_e/2 + \upsilon_A/2 - A/4$$
$$\upsilon_{12} = E_2 - E_1 = \upsilon_n + A/2$$
$$\upsilon_{34} = \upsilon_n - A/2$$

Two types of transitions are shown in Fig. 5.21. First are the usual EPR transitions occurring at frequencies υ_{13} and υ_{24}. These transitions obey the usual EPR selection rule $|\Delta m_s| = 1$, $|\Delta m_I| = 0$ where only electron spin flips but no nuclear spin flips are allowed. The second transitions, shown by dark lines, are NMR transitions ($|\Delta m_s| = 0$, $|\Delta m_I| = 1$) and occur at the frequencies υ_{12} and υ_{34}. These frequencies are also called as the ENDOR or ESEEM frequencies. The stick spectra on the right hand

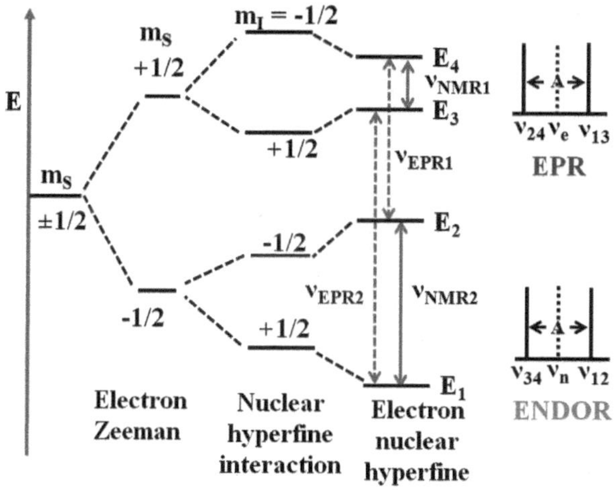

Figure 5.21 Energy level–splitting diagram for $S = 1/2$, $I = 1/2$ spin system. The g-value and hyperfine coupling are considered isotropic showing the allowed nuclear magnetic resonance (NMR) (electron–nuclear double resonance (ENDOR)) and electron paramagnetic resonance (EPR) transitions.

side/top show the resolved EPR spectrum. The bottom stick spectrum shows the resolved ENDOR spectrum. The splitting by the coupling constant A is obtained as shown in Fig. 5.21 (right).

There are two cases depending on the magnitude of HF coupling as discussed in ESEEM (1) Weak coupling and (2) Strong coupling. The energy-level diagrams for ($S = 1/2$, $I = 1/2$) for weak and strong couplings are shown in Figs. 5.22 and 5.23, which are same as in ESEEM. For weak coupling, two ENDOR lines are detected, equally spaced about υ_n and separated by the HF coupling constant. This pattern occurs in the weak-coupling case where $\upsilon_n > |A/2|$. However, for strong coupling ($\upsilon_n < |A/2|$), here again two ENDOR lines are observed centered around A/2 and, however, separated by $2\upsilon_n$ (Fig. 5.23).

5.8.2 Continuous Wave Electron–Nuclear Double Resonance (Electron Paramagnetic Resonance Detected Nuclear Magnetic Resonance)

The first ENDOR technique invented was CW ENDOR. The EPR transitions shown in Fig. 5.21 are first saturated using MW frequency by keeping the applied magnetic field constant. Then an RF field is swept to desaturate these EPR transitions. Furthermore, the changes in the EPR signal intensity as a function of the frequency of the RF field applied to obtain the

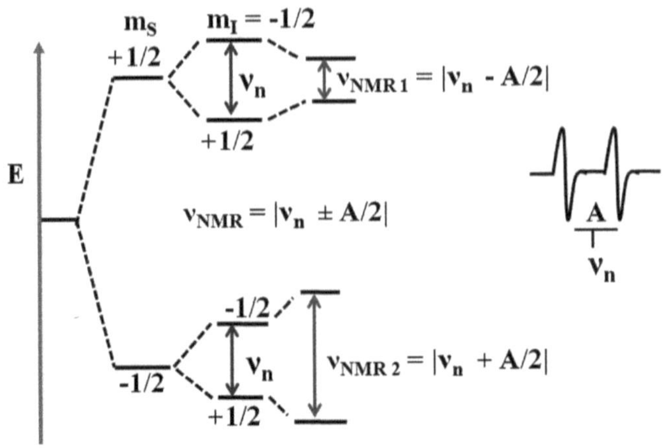

Figure 5.22 Energy-level diagram for weak coupling (I = 1/2, S = 1/2) and the corresponding electron–nuclear double resonance spectra. *NMR*, nuclear magnetic resonance.

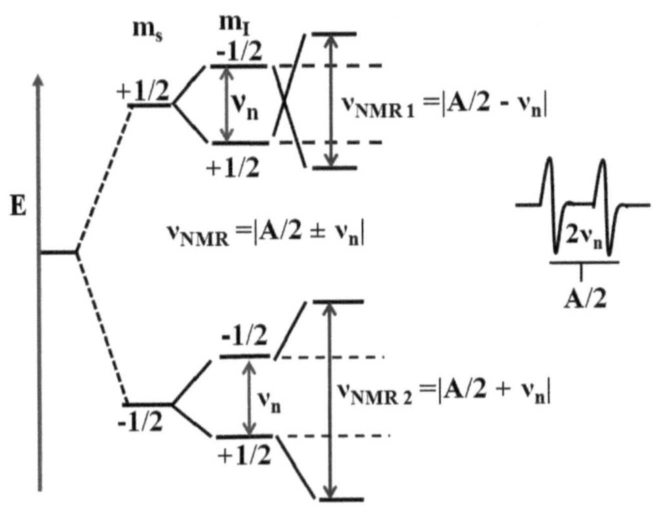

Figure 5.23 Energy-level diagram for strong coupling (I = 1/2, S = 1/2) and the corresponding electron–nuclear double resonance spectra. *NMR*, nuclear magnetic resonance.

CW ENDOR spectrum. As, the RF source is modulated, the first derivative ENDOR spectrum is obtained via this technique. This is the simplest ENDOR experiment to discuss, however, it has become a less commonly used method for acquisition of ENDOR spectra in recent years because of the advent of pulse ENDOR technique.

The ENDOR spectrometer essentially consists of a standard CW EPR spectrometer plus an NMR RF source and an RF coil within the MW cavity. This arrangement enables EPR and ENDOR experiments to be performed simultaneously on the sample. In CW ENDOR, the EPR signal usually most intense (at a given field) is partially saturated by an increase of MW power and followed by sweeping the RF to obtain the ENDOR spectrum. When the RF matches a nuclear frequency, a nuclear transition is induced, the population of the partially saturated EPR transition will be desaturated and hence the recovery of this EPR signal will be realized by detecting net MW absorption. The resonance lines are inherently narrower in ENDOR than in EPR spectroscopy. Therefore, the use of ENDOR in the metalloprotein investigations enables weaker HF interactions to be resolved. In addition, nuclei can be unambiguously identified by their *Larmor frequency*. However, the principle disadvantage of this technique is the greatly reduced sensitivity compared with EPR. This means that samples need to be more than 10-fold concentrated to yield acceptable ENDOR spectra.

5.8.3 Steady State Electron–Nuclear Double Resonance

To observe a continuous ENDOR, there must be complete relaxation pathways available. For S = 1/2, I = 1/2, the energy levels and the various relaxation pathways are represented in Fig. 5.24. The radiation induced

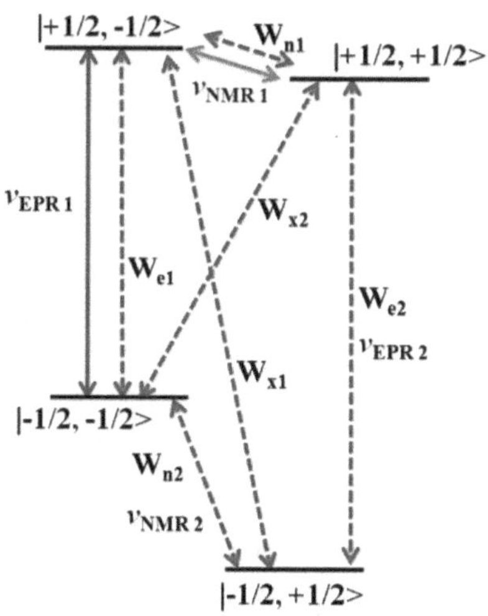

Figure 5.24 Mechanism for steady-state electron–nuclear double resonance. *EPR*, electron paramagnetic resonance; *NMR*, nuclear magnetic resonance.

transitions are marked by solid lines. The dashed line represents the following radiation less relaxation pathways (1) Electron spin lattice (W_e) (2) Nuclear spin lattice (W_n), and (3) Cross relaxation pathways (W_x) [19].

5.8.4 Electron–Nuclear Double Resonance of *p*-Benzoquinone Radical Anion

In the above examples discussed, two lines were only observed both in EPR and ENDOR spectra. To illustrate resolution enhancement observed in ENDOR, more complex energy-level diagrams are required. A good example is the EPR and ENDOR spectra of p-benzoquinone radical anion (Fig. 5.25). In this radical, the four equivalent protons produce a quintet (five) of lines in the EPR spectrum. However, only two lines are observed in the ENDOR spectrum, because all the NMR transitions in the same m_s manifold are degenerate. It is expected one set of all for equivalent protons. The energy level–splitting diagram for an $S = 1/2$ spin system interacting with four equivalent $I = 1/2$ nuclei and the solution EPR and corresponding ENDOR spectra (where $\upsilon_n > |A/2|$) are shown in Fig. 5.25 [19].

5.8.5 Orientation Selective Electron–Nuclear Double Resonance

All of the preceding discussions have only considered the ENDOR spectra for organic radicals in isotropic solution. In frozen solution, or in polycrystalline media, all orientations of the paramagnetic species are observed, and

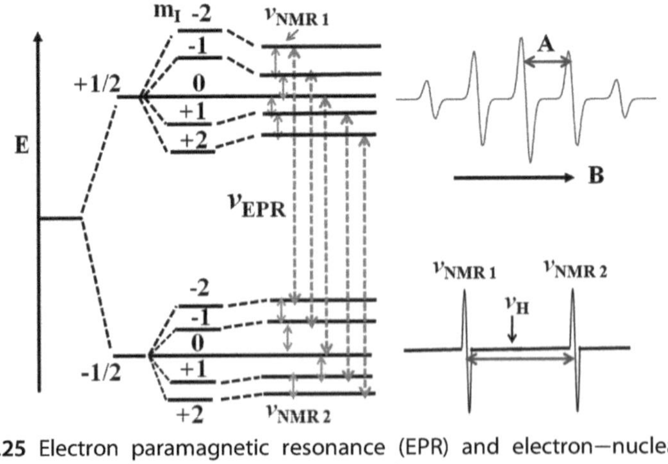

Figure 5.25 Electron paramagnetic resonance (EPR) and electron–nuclear double resonance spectra of *p*-benzoquinone. *NMR*, nuclear magnetic resonance.

one thus obtains an anisotropic EPR spectrum (see Section 4.11), particularly for transition metal species. The ENDOR spectra of such systems are consequently more complex, requiring a detailed understanding of the g- and A-anisotropy of the system. However, orientation selection ENDOR technique has the advantage of its ability to obtain the principal components of the magnetic tensors for each interacting nucleus from the powder ENDOR spectra of metal complexes or paramagnetic proteins in frozen solution [18–20]. The frozen solution EPR spectrum (or powder spectrum) is composed of a superimposition of the individual resonances from the randomly oriented molecules in which the applied magnetic field assumes all possible orientations with respect to the molecular frame. When an ENDOR measurement is made at a selected field position in the EPR spectrum, it comprises only that subset of molecules having orientations that contribute to the EPR intensity at the chosen value of the observing field. If EPR turning points are selected, namely magnetic field values, which correspond to defined molecular orientations, so-called "*single crystal like*" ENDOR spectra can be obtained. Using "*angular selective*" or "*orientation selective*" ENDOR experiment, the principal components of the magnetic tensors for each interacting nucleus can be obtained by simulation of the data, and this information can be used to provide structural information on the distance and spatial orientation of the remote nucleus. This ability to obtain three-dimensional structures is extremely important for paramagnetic systems lacking long range order or in cases where single crystals cannot be prepared.

For example, by fixing the field at $g_{||}$ (or $g_{\perp}$) for Cu(II) and VO(II) complexes (Fig. 5.26), the principal components corresponding to parallel and perpendicular orientations (molecular) were obtained.

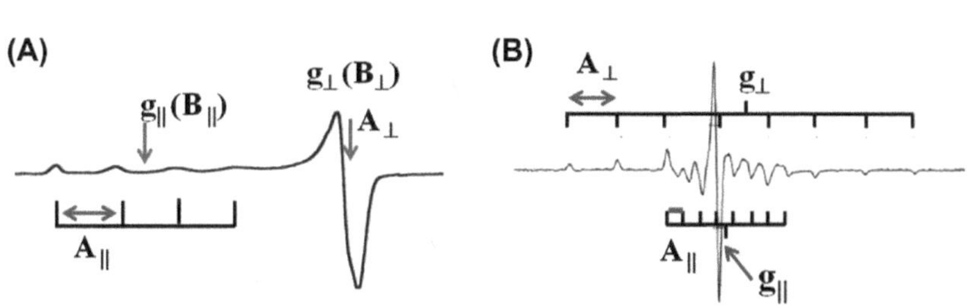

Figure 5.26 Powder electron paramagnetic resonance spectra of (A) Cu(II) and (B) VO(II) complexes and their principal values. *Reproduced with permission from Elsevier.*

5.8.6 Pulsed Electron–Nuclear Double Resonance

Pulsed techniques offer several advantages over CW ENDOR. CW ENDOR is the most sensitive technique; however, pulsed ENDOR provides higher resolution as well as allowing observation of weakly coupled nuclei. It also offers insight into the relaxation times of the system, is less susceptible to artifacts, and does not require balancing of relaxation times with RF power like the CW technique. Therefore, pulsed-ENDOR has been the most common technique in the past several decades, using pulse sequences predominantly developed by Mims and Davies. For an ENDOR experiment, either Mims or Davies, the homogenous applied magnetic field B_0 is held constant. The magnetic field is chosen where the EPR transition occurs by performing an ESE field sweep prior to beginning the ENDOR experiment, thus fixing both the B_0 and MW-B_1 fields. This leaves the RF spectrum to be swept during the mixing period, much like in a standard NMR experiment. Repeat the pulse sequence several (millions) times as shown in Figs. 5.27 and 5.28. The ENDOR spectrum is obtained. ESE, or ESE–EPR, is a technique used for signal detection in most pulsed advanced EPR methods [9–12].

5.8.6.1 Spin Population Polarization

The diagram below shows the changes in populations that occur during an ENDOR experiment. These changes lead to the ENDOR effect and are detected. Before the application of MW frequency photons, (or a preparation pulse) the system exists with the majority of the spin population in the $M_s = -1/2$ spin manifold, as determined by the Boltzmann population distribution and shown in Fig. 5.29A. After the application of MW field, i.e., preparation MW pulse, the population is moved from the $M_s = -1/2$

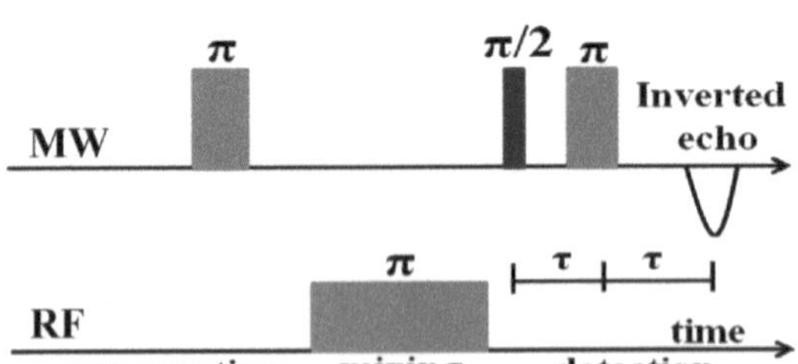

Figure 5.27 Davies electron–nuclear double resonance pulse sequence and its measurement.

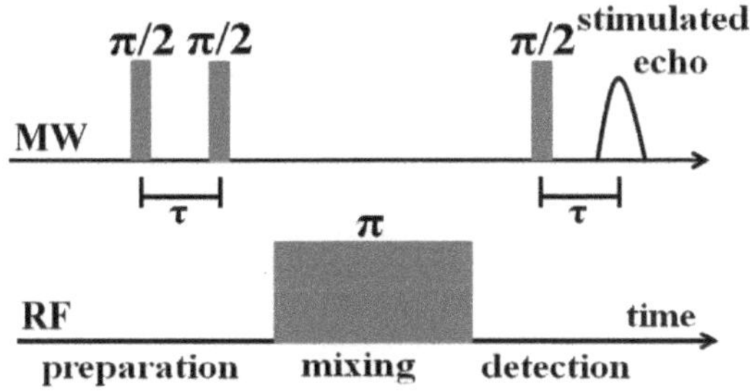

Figure 5.28 Mims electron–nuclear double resonance pulse sequence and its measurement.

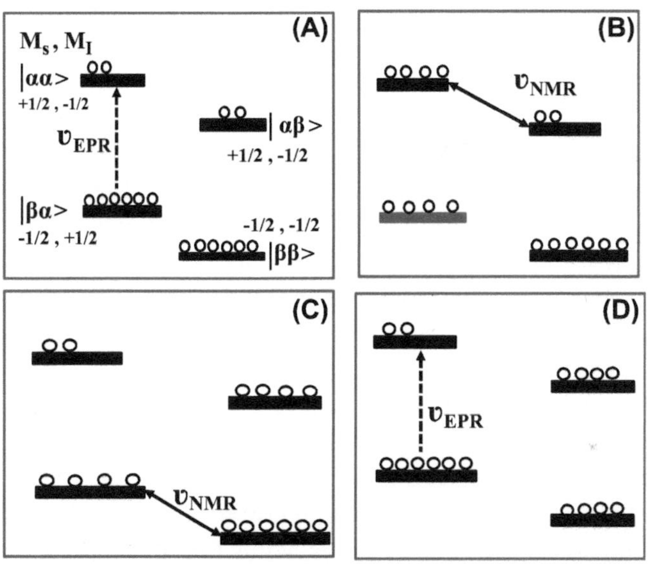

Figure 5.29 Changes in population of lower and upper levels (A–D) during nuclear magnetic resonance (NMR) induced electron paramagnetic resonance (EPR) transition.

to $M_s = +1/2$ manifold as expected by the inversion sequence (π-pulse), seen in Fig. 5.29B. Finally, after the mixing period and induction of on-resonance NMR transitions, population transfer occurs between the upper M_I levels as seen in Fig. 5.29C. This is then detected as a change in the EPR signal with the CW technique, or with an ESE, sequence using a pulsed-ENDOR. On the other hand, if the RF π pulse is off-resonance, then this final population polarization does not occur.

5.8.6.2 Davies Electron–Nuclear Double Resonance

Davies ENDOR involves three steps as shown in Fig. 5.27. (1) In the first step, a preparation π, 180 degrees, MW pulse is applied to invert the magnetization of the spins in the applied static B_0 field. This essentially creates a hole in the EPR spectrum, whose width and depth depend on the length of the pulse MW applied, with long pulse producing narrow holes. (2) Then a π RF pulse is applied during the mixing period. If the RF is resonant with an NMR transition, magnetization will be transferred to the other M_s spin manifold, otherwise no mixing will occur to fill in the hole that the inversion pulse creates. (3) During the detection period, the z-component of the magnetization is measured using a two-pulse echo sequence, and one detects essentially the ESE–EPR signal that is restored during the mixing period. This Davis technique is most suited to nuclei with large hyperfine coupling values [9–12].

5.8.6.3 Mims Electron–Nuclear Double Resonance

The Mims ENDOR technique is based on a stimulated ESE sequence (Fig. 5.28). Here, a two π/2 (90°) preparation MW pulse sequence is used to invert the electron spin population, and a final π/2 MW pulse after the mixing period to obtain the stimulate ESE for signal detection. Between the preparation pulses and the third final pulse Fig. 5.28 a RF pulse is used to invert the nuclear spin population, resulting in polarization transfer between the nuclear and electronic transitions during "mixing period."

These results in the actual ENDOR transitions lead to changes in the intensity of the EPR transition that is detected by use of the ESE MW-pulse sequence. The echo intensity is subsequently measured as a function of the RF frequency to give the characteristic Mims ENDOR spectrum.

The two techniques are very similar to each other in their implementation; however, they are relatively complementary in their results and usefulness. The Mims ENDOR technique is most widely suited to weakly coupled nuclei, with small HF coupling constants (A) generally less than 2 MHz and as low as 0.1 MHz. However, the Davies ENDOR pulse sequence makes most useful for those nuclei with relatively large HF couplings. Therefore, by using both the methods, the assignments of strongly and weakly coupled nuclei are greatly simplified.

5.9 TRIPLE RESONANCE

ENDOR is a double resonance technique because two transitions viz., one MW EPR and one NMR are excited simultaneously. So, it is also possible to make two or more NMR transitions excited simultaneously (instead of one) leading to multiple resonance experiments. TRIPLE, a multiple resonance technique, involves electron—nuclear—nuclear triple resonance [19]. There are two types of TRIPLE resonance experiments

1. Special TRIPLE (or) Double ENDOR.
2. General TRIPLE.

In the *special TRIPLE* (or Double ENDOR), the sample is irradiated simultaneously with two RF fields in addition to the saturating MW radiation. So that both the NMR transitions of the same set of protons are excited at the same time. Whereas in *general TRIPLE*, NMR transitions belonging to different set of protons are excited simultaneously. The experiment is performed as follows: One NMR transition is pumped with the unmodulated RF frequency. While, the second modulated RF field is swept over the range of NMR frequencies.

In special TRIPLE, the two NMR transitions lead to substantial enhancement in sensitivity and resolution. Furthermore, the signal intensities reflect the ratio of the numbers of equivalent nuclei. A single special TRIPLE resonance peak is observed instead of two peaks in ENDOR. However, the general TRIPLE resonance gives information on the relative signs of the HF interactions.

For example, the TRIPLE resonance spectra of phenylenyl radical are shown in Fig 5.30. Two peaks are observed at 2.545 and 8.835 MHz

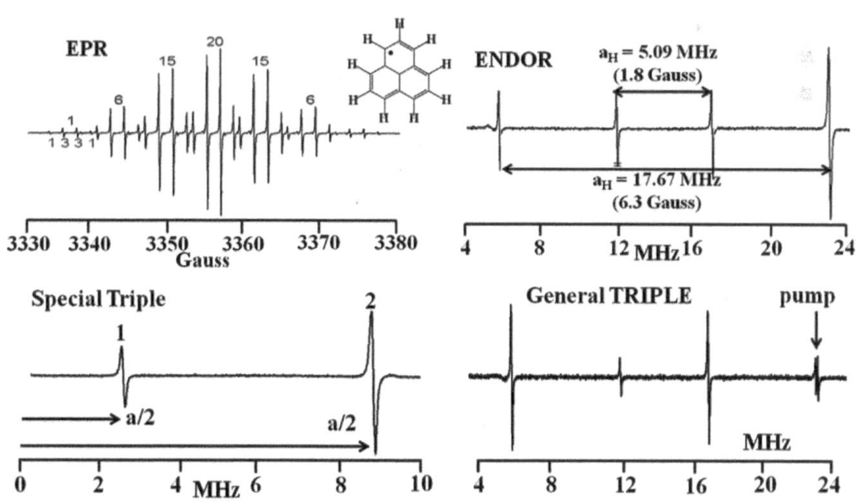

Figure 5.30 (A) Electron paramagnetic resonance (EPR), electron—nuclear double resonance (ENDOR), special TRIPLE, and general TRIPLE resonance of the phenylenyl radical at room temperature in mineral oil. *Reproduced with permission from Royal Society of Chemistry.*

(i.e., A/2 for both HF couplings) with a relative intensity ratio of 1:2. It clearly indicates the presence of two sets of inequivalent nuclei 3:6 ratio in agreement with the structure of the radical. From the general TRIPLE spectrum, one low frequency is enhanced, while the other is diminished when pumping a high frequency line (see Fig. 5.30) transition. This reveals that the two HF couplings have opposite signs.

5.10 CONTINUOUS WAVE ELECTRON–ELECTRON DOUBLE RESONANCE

ELDOR requires two different MW frequencies within one resonator; one is called a "*pump*" MW source and another is the *observing* MW source [21–23]. It is used to monitor saturation transfer mechanisms such as Heisenberg exchange, nitrogen nuclear relaxation, rotational diffusion, etc. In the CW mode, the observing MW frequency is fixed at a non-saturating power level. The "pump" MW source set at a strongly saturating power level irradiates a portion of the EPR spectrum (either the same line or a matching line related by a HF coupling). The change in EPR intensity of a line at the fixed observing source is monitored by sweeping the applied magnetic field. The effect observed is due to a transfer of saturation from the spins irradiated by the pump source to those spins detected by the observing source. "Saturation transfer" means that the z-component of the population difference at the observing frequency is no longer at the Boltzmann population difference. In the CW mode, the ELDOR effect is measured by the ELDOR reduction factor R given by

$$R = \frac{(\text{signal with pump off}) - (\text{signal with pump on})}{(\text{signal with pump off})} \tag{5.11}$$

A plot of R versus pump is linear and the intercept yields the saturation—transfer rate/electron spin—lattice relaxation rate. Thus, the R-factor is a measure of how competitive saturation transfer is with respect to the spin-lattice relaxation time (T_{1e}). Values of T_{1e} are preferably measured by time domain methods (see Section 5.3). In general, the CW ELDOR measurements provide a means to study a variety of dynamic processes that govern the relaxation between the pumped and observed spins.

5.10.1 Pulsed Electron–Electron Double Resonance or Double Electron–Electron Resonance

It measures the dipolar coupling and thereby the distance between the spatially separated unpaired electrons of spins (A or B) 1 and 2 (1.5–8.0 nm) as shown in Fig. 5.31 in chaotically oriented solid-state samples using pulsed-EPR. This method uses the pulse sequences (Fig. 5.32) which eliminates the dipolar broadening of lines at large distances masked by other contributions to the line width viz., g-anisotropy and HF and quadrupole interactions [24]. The spin Hamiltonian equation relating to radical separation is given as follows:

$$\widehat{H}_o = \omega_A S_A + \omega_B S_B + \widehat{H}_{dd} + \widehat{H}_{ex} \quad (5.12)$$

$$\widehat{H}_{dd} = -\left[(\mu_o/4\pi)\left\{g_A g_B \mu_\beta^2\right\}\{(S_A S_B/r^3) - 3(S_A r)(S_B r)/r^5\}\right] \quad (5.13)$$

$$H_{dd} = dS_z I_z - d/2(S_x I_x + S_y I_y)$$

$$d = \left(1 - 3\cos^2\theta\right)\omega_{dd} \quad (5.14)$$

$$\omega_{dd} = \left(1/r^3\right)\left[(\mu_o/4\pi h)\left\{g_A g_B \mu_B^2\right\}\right] \quad (5.15)$$

Here ω_A and ω_B are the frequencies of A and B spins. The distance between the spins A and B is r. θ is the angle between the direction of the external magnetic field and vector r. At large r, the $\widehat{H}_{ex}$ is negligible and not considered.

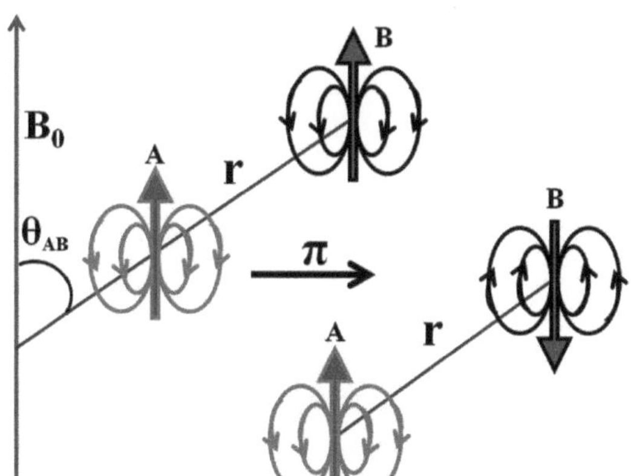

Figure 5.31 Dipolar coupling between spins A and B and effect of π pulse on spin B.

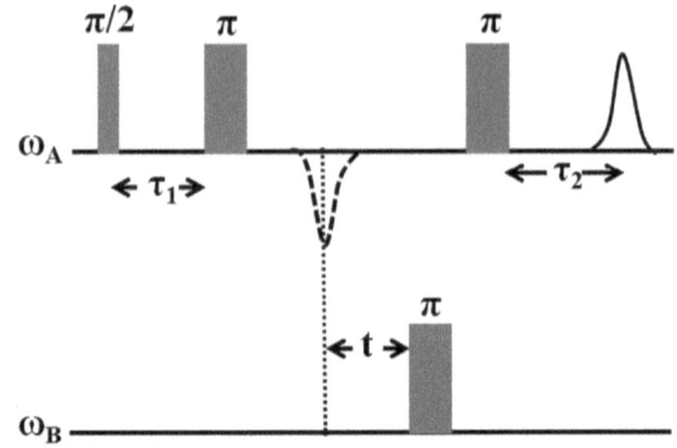

Figure 5.32 Double electron–electron resonance four-pulse sequence used for distance measurements.

In this double-resonance method, spins B are additionally exposed to pumping pulses at frequency ω_2. A pump π-pulse acting on the spins B changes their orientation and thereby changes the local magnetic field on spins A. The value of this change is determined by the magnetic dipole coupling interaction between the two spins A and B, which is proportional to r^{-3}. The pulse sequence of four-pulse DEER technique widely used is shown in Fig. 5.32. This dipolar coupling between these spins causes a modulation of the ESE frequency observed as the pump pulse is incremented as shown in Fig. 5.33. The frequency of the modulation (which is related to interradical separation by r^{-3}) in Eq. (5.16) can be found by Fourier transforming the signal into the frequency domain (Fig. 5.33).

$$V_{DEER}(r,t) = 1 - \int_0^{\pi/2} \sin\theta_{12}\lambda(\theta_{12})[1 - \cos(\omega_{dd}(\theta_{12})t)]d\theta \qquad (5.16)$$

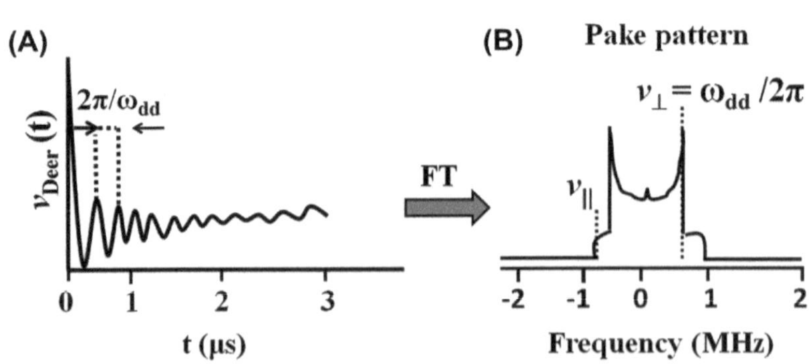

Figure 5.33 The electron spin echo modulation (A) and its Fourier transform (FT) to give Pake pattern (B).

The spectrum in Fig. 5.33 is the dipolar frequency pattern for the case of weak coupling between radicals in powders (averaged over all angles with respect to the strong magnetic field) is obtained by the FT of time-domain signal. The spectrum is often referred to as the Pake pattern or doublet in the frequency domain with characteristic frequency positions $\upsilon_{\perp}$ and $\upsilon_{\parallel}$. From the ω_{dd} $(= \upsilon_{\perp}\ 2\pi)$ the distance between the two spins can be calculated using Eq. (5.17). The simplified expression for distance calculation is

$$r(\mathrm{nm}) = [52.04/\upsilon_{\perp}(\mathrm{MHz})]^{1/3} \tag{5.17}$$

The sample is frozen here or made into a powder to avoid averaging of the dipolar coupling term.

5.11 ELECTRON PARAMAGNETIC RESONANCE IMAGING

So far, we have discussed that EPR spectra are sufficient to provide valuable chemical and/or physical information about the sample. However, in many cases, it can be extremely useful to detect paramagnetic ions/free radicals in animals or cells known as in vivo EPR spectroscopy or to generate images showing the spatial distribution of a free radical/contrast agent throughout the volume of the sample called EPR imaging (EPRI). So, there is a need for EPR-based imaging. A brief account of the fundamentals of EPRI, their types, and commonly used spin probes in imaging applications will be presented here.

5.11.1 Comparison Between Electron Paramagnetic Resonance and Nuclear Magnetic Resonance Imaging

Although EPRI is very closely related to NMR-based Magnetic Resonance Imaging (MRI), there are several fundamental differences between them as follows [6]:

1. EPRI is based on electron spin resonance (ESR) using MW radiation and MRI is based on nuclear spin resonance phenomena using RF radiation.
2. Both use magnetic field gradient to "encode" spatial information about the sample.
3. The major difference between EPRI and MRI are due to the large difference in relaxation times. For example, the transverse relaxation times for water hydrogen nuclei are of the order of tens to hundreds of milliseconds, while those for nitroxides or trityl free radical are hundred of nanoseconds to microseconds.

4. The MRI data are largely acquired with pulsed or time domain techniques using Fourier imaging techniques. However, because of very short EPR relaxation time, the time of application of gradients and pulse is relatively too long. Thus, the EPR signal could have disappeared. Hence, the method of signal detection for EPRI is largely CW-ESR. Therefore, fixed and stepped gradients are used in EPRI and applied during either time domain acquisition or during a CW magnetic field sweep.
5. EPR line widths tend to be large again due to short relaxation time resulting in relatively poor spatial resolution. Hence, the images are blurred in comparison with proton MRI. These are circumvent by using EPR spin probe with the lowest line width (e.g., trityl).
6. Using fluoroanthenyl radical, which has long spin relaxation time, pulsed EPRI similar to MRI was carried out.
7. Using CW-EPRI, slice selective excitation is difficult, so 2D imaging with well defined slices can't be done. However, these problems are overcome by using proton–electron double resonance imaging (PEDRI) or Overhauser Enhanced MRI (OMRI) (see Section 5.11.2.5).
8. EPRI is always carried out at low frequency using X- and L-bands because the higher frequencies used in most EPR spectroscopy can't penetrate far into wet tissue due to huge dielectric loss.

5.11.2 Types of Electron Paramagnetic Resonance Imaging

The following major types of EPRI were developed and studied.

1. Spatial imaging,
2. Spectral–spatial imaging,
3. Pulsed-EPRI, and
4. PEDRI or OMRI.

5.11.2.1 Spatial Imaging

MRI anatomic images from tissue water hydrogen nuclei give very good anatomic resolution because of 90 M concentration of the proton nuclei. However, spin probe concentrations in EPRI will be very low by several orders of magnitude than this huge water proton concentration. Thus, the spatial resolution obtained in MRI images is not likely to be obtained with EPR. Using very high spin–density probes, EPRI can generate high-resolution images of anatomy, approaching the resolution available from water proton MRI. Spatial imaging involves *filtered back projection (FBP) reconstruction* using the highest gradients available consistent with signal-to-noise limits [25]. In spatial imaging, fixed magnitude gradients in different directions are applied to the sample and the magnetic field is swept to produce a projection. Because of this gradient, spectral components at different positions within the sample are observed at different magnetic field positions in a measured spectrum. As the gradient strength increases, the separation between the observed signals increases. For diffuse samples, as the gradient magnitude increases, the signal amplitude diminishes. It is known that the resolution of the images is deteriorated by the finite line width of the EPR substrate. Resolution is improved if a large field gradient is achievable. The combination of a narrow line width and high field gradient yields high spatial resolution [26].

5.11.2.2 Spectral–Spatial Imaging

This imaging method gives the EPR spectrum as a function of position in the sample. Spectra are obtained at a series of magnetic field gradients that corresponds to projections in a spectral–spatial plane. The advantage of EPRI lies in their ability to create spectral–spatial images. The magnetic moment of the electron is 658 times that of the hydrogen nucleus used in MRI. This results in an increase in coupling that increases the sensitivity and specificity of the unpaired electron to environmental magnetic conditions. Spectral–spatial imaging uses gradients of different magnitudes as well as directions. A spectrum, obtained in the presence of a magnetic field gradient with a given magnitude and direction, is a projection of a spectral–spatial object. A spectral–spatial object has both spectral and spatial extent and is the representation of the spectrum at each point in an object. A set of spectra, acquired with a range of gradient magnitudes and directions, may be used to reconstruct an image of the spectral–spatial object through conventional image reconstruction techniques [26].

5.11.2.2.1 Projection-Reconstruction Continuous Wave Electron Paramagnetic Resonance Imaging

In contrast to the pulsed-magnetic field gradients used in Fourier MRI, however, in CW-EPRI the magnetic field gradients must be applied for the duration of each sweep of the magnetic field through resonance. In other words, the usual sweep of the magnetic field is performed, *while the magnetic field gradient is applied* across the sample. Considering a gradient applied in the x-direction, for example, while the magnetic field gradient G_x is switched on, parts of the sample located at positions designated by larger values of x will always experience a higher magnetic field than those

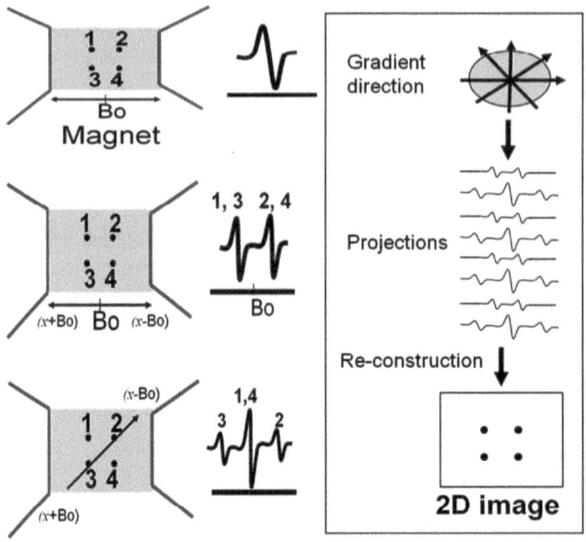

Figure 5.34 Gradient directions, projection-reconstruction to obtain continuous wave electron paramagnetic resonance image.

located at smaller x-values. When the magnetic field sweep is made, unpaired electrons located at large x-values will come into resonance earlier than those at smaller x positions. This is demonstrated in Fig. 5.34, which shows the effect on the EPR spectrum of applying a field gradient across a sample consisting of four free radical-containing tubes. The EPR spectrum is split into two, the separation ΔB being directly proportional to the spatial separation Δx of the tubes along the gradient direction.

In general, an object under study will have a continuous variation of concentration of free radical, rather than being confined to two well-defined tubes. In this case, the integral of the EPR spectrum recorded in the presence of a magnetic field gradient represents a 1D projection of the spin-density of the object along the direction of the applied magnetic field gradient. To generate data from which a 2D image can be formed, a number of such projections at a range of angles through the sample are obtained, by sequentially applying magnetic field sweeps with the magnetic field gradient applied each time at a different angle with respect to the sample. Following data collection, the image can be reconstructed using FBP. This procedure can easily be extended to obtain three-dimensional images of the distribution of paramagnetic species in a sample. By changing the direction of the gradient, as well as its magnitude, this can be done for a three-dimensional spatial object, giving a four-dimensional spectral–spatial image.

The normalized projections of the magnetization obtained in this fashion can be shown to be similar to those obtained using a computed tomography scan. The mathematical description of such projections of the information in an image is the Radon transform of the spectral–spatial object. Thus, the back projection reconstruction techniques commonly used in other forms of imaging can be applied to the calculation of the image from these projections [26].

5.11.2.3 Pulsed Electron Paramagnetic Resonance Imaging

Analogous to MRI, ESE can be obtained using a 90°–t to 180°–t echo pulse sequence. In this pulsed imaging method, a pulsed magnetic gradient is applied between the MW pulses and monitored the dependence of echo amplitude on the duration of the gradient. Using 1 T/m field gradients and solid fluranthenyl salts having very long relaxation times, a resolution of 10 μm in solids was obtained [27].

5.11.2.4 Two-Dimensional Electron–Nuclear Double Resonance Imaging

2D ENDOR images were obtained by rotating the coils in the xy plane by applying 5 G/cm field gradient. The images were obtained based on the dependence of ENDOR signal on oxygen concentration and viscosity. By selecting the HF line to image, individual images of two probes can be constructed [27].

5.11.2.5 Overhauser Effect-Based Imaging

Free radical imaging methods based on the Overhauser effect have been developed to address the above problems of especially CW-EPRI. The resulting technique is known as PEDRI or as OMRI; henceforth, the former acronym will be used here.

5.11.2.5.1 The Overhauser Effect

The Overhauser effect, also known as "dynamic nuclear polarization," was predicted by Overhauser in Ref. [27a]. In any sample containing paramagnetic molecules in solution (such as biological sample or animal containing free radicals) interactions will occur between the unpaired electrons and the NMR-sensitive hydrogen nuclei (protons) in the solvent water molecules. In the Overhauser effect an NMR experiment is carried out to detect an

NMR signal from the water protons, and at the same time the sample is irradiated at the EPR frequency to excite the unpaired electron spins. The result of the EPR irradiation can be a significant enhancement (or amplification) of the measured NMR signal, relative to the signal obtained without EPR irradiation [28].

5.11.2.5.2 Proton Electron Double Resonance Imaging

In PEDRI the sample is irradiated at the EPR frequency of the free radical under study, during the collection of a magnetic resonance (MR) image. NMR signals from regions of the sample containing the free radical are enhanced by the Overhauser effect, and will exhibit correspondingly increased intensity in the MR image, revealing indirectly the distribution of the free radical. An advantage is that PEDRI must be implemented at sufficiently low magnetic field (i.e., low enough EPR frequency) that the ESR irradiation (1) is able to penetrate into the sample and (2) does not overheat the sample through nonresonant absorption. In vivo PEDRI studies have been performed on mice at a magnetic field of 20 mT, with EPR irradiation at 560 MHz, and on rats at 10 mT, with EPR irradiation at 237 MHz. Fig. 5.35 shows a typical PEDRI pulse sequence. The EPR irradiation (shown as RF EPR) lasts for a time TEPR, which should ideally be at least three times the NMR T_1. Frequently PEDRI is applied in an interleaved fashion, EPR-on and EPR-off, to allow simultaneous collection of Overhauser-enhanced and unenhanced images. So, the subtraction of the data sets then produces an image showing only the distribution of free radicals.

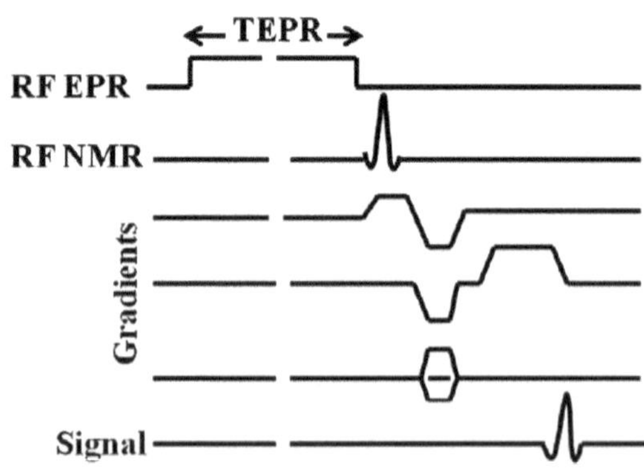

Figure 5.35 Typical proton–electron double-resonance imaging pulse sequence. *EPR*, electron paramagnetic resonance; *NMR*, nuclear magnetic resonance; *RF*, radiofrequency.

This approach gives high resolution, fast image acquisition, and the availability of anatomical information from MRI techniques.

5.11.3 Applications of Electron Paramagnetic Resonance Imaging Using Different Spin Probes

In vivo EPR spectroscopy and imaging have been used in the measurements of redox status using nitroxide imaging, radical generation by using spin trapping, oximetry, i.e., partial pressure of oxygen from the line width of EPR signal, diffusion and reaction kinetics, nitroxide metabolism, and transport phenomena [30—32]. There are several spin probes developed by various groups depending on the type of applications.

5.11.3.1 Redox Status, Radical Generation, and pH Using Nitroxides

The original early in vivo images shown by Berliner and coworkers used nitroxides as their spin probe. The basic nitroxide structure is shown in Fig 5.36. The most common nitroxides are either five- or six-membered cyclic rings of varied saturation. The methyl groups are flanking the N—O and providing the stability by preventing attack on the N—O through steric hindrance and also preventing internal H-transfer to reduce the N—O moiety. Nitroxides have also been used to measure oxygen concentration, self-concentration, pH, and microviscosity of the solvent environment. For example, imidazolinyl or imidazolidinyl nitroxides, or biradical nitroxides were used for pH measurements. Nitroxides have been demonstrated to respond to solvent polarity [29].

5.11.3.2 Electron Paramagnetic Resonance Oximetry

Although molecular oxygen in its ground state is biradical, its EPR signal is weak and hence can not be monitored directly. Therefore, the interaction of stable paramagnetic compounds with molecular oxygen shortens the EPR relaxation times of the compounds, resulting in an increase in the peak-to-peak line width of their EPR spectra. The increase in the line width

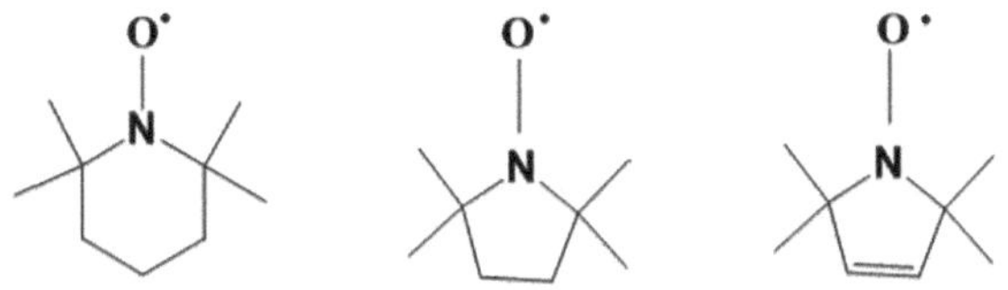

Figure 5.36 Basic chemical structure of piperidine, pyrrolidine, and pyrroline nitroxide spin probes.

correlates well with a concentration of molecular oxygen. Thus this phenomenon can be used for the measurement of partial pressure of oxygen, pO_2 known as EPR oximetry.

There are two types of magnetic probes used for oximetry

1. *Water insoluble*—Lithium phthalocyanin, fusinite, synthetic chars, Indian ink.
2. *Water soluble*—nitroxyl and triaryl methyl radicals (trityl).

Lithium phthalocyanine, LiPc, is an example of high spin−density particulates. The radicals are generated electrochemically during the synthesis of the crystals. The crystals may consist of either single or multiple domains each with a particular line width and each with specific responses to oxygen. The crystals of LiPc appear to consist of stacks of LiPc monomers. The unpaired electrons can hop rapidly between monomers, effectively averaging the magnetic environment encountered. This produces an exchange narrowed EPR line. In the case of LiPc, the line width is often comparable to the spin packet line width of the trityls. The broadening of the LiPc by oxygen is at least five times that of the nitroxides and the trityls at the same oxygen concentration. Moreover, because of the high electron density in these compounds, the EPR signals are quite large for crystals a few tenths of a millimeter in dimension. Some chars and coals have sensitivity to oxygen yet another order of magnitude higher than LiPc.

Trityls: Trityls are variants of the carbon centered radical triphenylmethyl radical. Until recently, trityl radicals useful for imaging were only available from Nycomed Innovations. However, recently a simplified method of synthesis makes at least one of these compounds accessible to groups with organic synthetic ability. Available trityls have three carboxylic acid groups, one on each of the three aryl groups surrounding the radical carbon. Each of these aryl groups has nuclear spin-zero species replacing hydrogen. The first side-chain protons are six bonds from the radical carbon, reducing splittings due to proton HF interactions. This gives very weak unresolved HF splittings. Much of the spin packet line width is due to local vibrational mode effects. The effective lifetime broadening is small giving very narrow spin packet line widths. The spin packet line width of a typical trityl is approximately with an overall peak-to-peak line width of 250 MHz and a spin packet line width of about half that at X-Band. The narrow line gives high oxygen sensitivity, and high spatial resolution for a given gradient strength. Trityls have far less sensitivity to viscosity and temperature than nitroxides. This improves their characteristics as an oxygen sensor by reducing their sensitivity to confounding broadening effects (Fig. 5.37) [30−32].

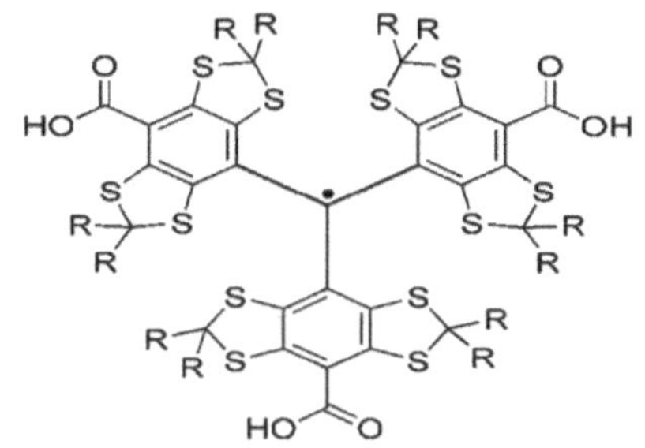

OX063, R = CH_2CH_2OH
CT-03, R = CH_3

Figure 5.37 Chemical formula for the OX063 and CT-03 trityl spin probe developed by Nycomed Innovations.

The line width change is in gauss per torr of oxygen. This makes the trityl quite sensitive. Trityls are highly soluble molecules with molecular weights of between 1 and 2 KDa. They can be infused into animals, and they are resistant to bioreduction. In principle, by changing side groups on the molecule, they potentially could target the extracellular water compartment, or the intracellular water compartment [26].

5.11.3.3 Electron Paramagnetic Resonance Imaging of Nitric Oxide

The measurement of nitric oxide (NO) in vivo and imaging is important due to its key role as a signaling molecule in numerous physiological and pathophysiological processes. However, the methods for NO detection in biological systems are limited because of its short half-life and low concentration in living systems. Fe(II)-dithiocarbamate complexes were widely used for NO detection by EPR spectroscopy. The dithiocarbamate Fe(II) complexes, EPR silent, were used to trap NO to form EPR detectable paramagetic mononitrosyl-iron complex for in vivo spectroscopy and its imaging. These complexes possess an important advantage of relatively high stability of the paramagnetic adduct with NO. The water solubility is increased by using different substituents R1 and R2. By injecting the Fe(II) complex into lipopolysaccharide or nitrite-treated mice, distinct three-line signal of the mononitrosyl complex are observed indicating the unpaired electron localized on nitrogen atom. Both EPRI and in vivo measurements revealed the presence of high concentrations of the Fe—NO complex in the liver.

However, the disadvantages of these complexes include the toxicities of both iron and dithiocarbamate moiety. As an alternative, a low molecular

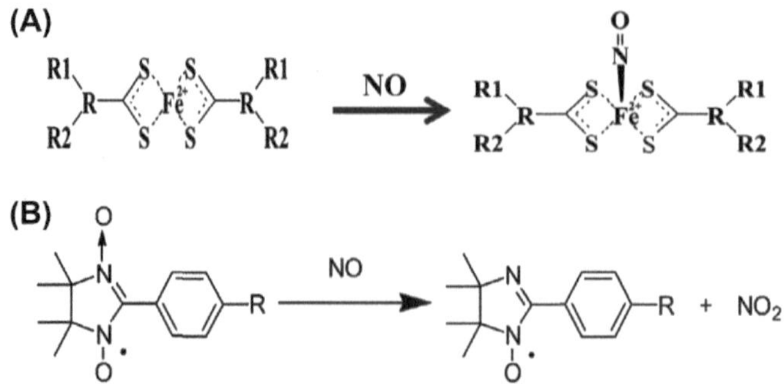

Figure 5.38 (A) Fe(II)-dithiocarbamate complexes as nitric oxide (NO) trap and (B) Nitronyl nitroxides as NO-sensitive probes.

weight nitronyl nitroxides (NNRs), as NO-sensitive probes seems to be very attractive. The scavenging of NO by NNRs has been used for NO detection by EPR spectroscopy. The detection of NO using NNR is based on the specific radical–radical reaction followed by EPR measured spectral changes (Fig. 5.38) [29,33].

REFERENCES

[1] J. Krzystek, A. Ozarowski, J. Telser, Multi-frequency, high-field EPR as a powerful tool to accurately determine zero-field splitting in high-spin transition metal coordination complexes, Coord. Chem. Rev. 250 (2006) 2308–2324.

[2] (a) J. Telser, J. Van Slageren, S. Vongtragool, M. dressel, W.M. Reiff, S.A. Zvyagin, A. Ozarowski, J. Kryzstek, High-frequency/high field EPR spectroscopy of the high-spin ferrous ion in hexaaqua complexes, Magn. Reson. Chem. (2005) S130–S139;
(b) S.K. Misra, Multifrequency Electron Paramagnetic Resonance: Theory and Applications, Wiley-VCH, 2011.

[3] S.K. Misra, Multifrequency Electron Paramagnetic Resonance, Data and Techniques, Wiley-VCH, 2011.

[4] C. Karunakaran, et al., EPR of an exchange-coupled, helical hydrogen-bridged one dimensional Cu(II) complex containing both octahedral and square pyramidal geomentrices in the same unit cell, Mol. Phys. 100 (2002) 287.

[5] J.H. Freed, ESR and molecular dynamics, in: Biomedical EPR, Part B: Methodology, Instrumentation and Dynamics, Biological Magnetic Resonance, vol. 24/B, Springer, 2005, pp. 239–268 (Chapter 9).

[6] S. Subramanian, M.C. Krishna, Time-domain radio frequency EPR imaging, in: S.S. Eaton, G.R. Eaton, L.J. Berliner (Eds.), Biomedical EPR – Part A: Free Radicals, Metals, Medicine and Physiology, Springer, 2005, pp. 321–382 (Chapter 12).

[7] W.E. Antholine, Low frequency EPR of Cu^{2+} in proteins (Chapter 14), in: Biomedical EPR, Part A: Free Radicals, Metals, Medicine, and Physiology, Biological Magnetic Resonance, vol. 23, 2005, pp. 417–454.

[8] D. Hinderberger, S. Ebner, S. Mayr, B. Jaun, M. Reiher, M. Goenrich, R.K. Thauer, J. Harmer, Coordination and binding geometry of methyl-coenzyme M in the red1m state of methyl-coenzyme M reductase, J. Biol. Inorg. Chem. 13 (2008) 1275–1289.

[9] A. Schweiger, G. Jeschke, Principles of Pulse Electron Paramagnetic Resonance, Oxford Universities Press, New York, 2001.

[10] Marina, Electron Paramagnetic Resonance a Practitioner's Toolkit, Brustolon University of Padova Elio Giamello University of Torino, John Wiley & Sons, inc. Hoboken, New Jersey, 2009.

[11] T. Prisner, M. Rohrer, F. MacMillan, Pulsed EPR spectroscopy: biological applications, Annu. Rev. Phys. Chem. 52 (2001) 279–313.

[12] J. Harmer, G. Mitrikas, A. Schweiger, Advanced pulse EPR methods for the characterization of metalloproteins (Chapter 2), in: Biological Magnetic Resonance, vol. 28, 2009, pp. 13–61;
(a) P. Höfer, A. Grupp, H. Nebenführ, M. Mehring, Hyperfine sublevel correlation (hyscore) spectroscopy, a 2D ESR investigation of the squaric acid radical, Chemical Physics Letters 132 (1986) 279–282.

[13] S.S. Eaton, G.R. Eaton, L. Berliner, Saturation recovery EPR, biomedical EPR – part B: methodology, instrumentation, and dynamics (Chapter 1), in: Biological Magnetic Resonance, Biomedical EPR, Part A: Free Radicals, Metals, Medicine, and Physiology. Part B: Methodology, Instrumentation, and Dynamics, vol. 24. Springer, 2005;
(a) N. Bloembergen, On the interaction of nuclear spins in a crystalline lattice, Physica 15 (1949) 386–426.

[14] S.-T. Kim, A. Sancar, C. Essenmacher, G.T. Babcock, Time-resolved EPR studies with DNA photolyase: excited-state FADHo abstracts an electron from Trp-306 to generate FADH-, the catalytically active form of the cofactor, Biophysics 90 (1993) 8023–8027.

[15] M. Weber, N.J. Turro, D. Beckert, A comparative time-resolved CW EPR and FT EPR investigation of the addition of 2-hydroxy-2-propyl radicals to acrylate and methacrylate monomers, Phys. Chem. Chem. Phys. 4 (2002) 168–172.

[16] J. Hernandez-Guzman, L. Sun, A.K. Mehta, J. Dong, D.G. Lynn, Copper(II)-bis-histidine coordination structure in a fibrillar amyloid β-peptide fragment and model complexes revealed by electron spin echo envelope modulation spectroscopy, ChemBioChem 14 (2013) 1762–1771.

[17] K. Chandran, J. McCracken, F.C. Peterson, W.E. Antholine, B.F. Volkman, B. Kalyanaraman, Oxidation of histidine residues in copper-zinc superoxide dismutase by bicarbonate-stimulated peroxidase and thiol oxidase activities: pulse EPR and NMR studies, Biochemistry 49 (2010) 10616–10622.

[18] S. Van Doorslaer, E. Vinck, The strength of EPR and ENDOR techniques in revealing structure-function relationships in metalloproteins, Phys. Chem. Chem. Phys. 9 (2007) 4620–4638.

[19] D.M. Murphy, R.D. Farley, Principles and applications of ENDOR spectroscopy for structure determination in solution and disordered matrices, Chem. Soc. Rev. 35 (2006) 249–268.

[20] C. Karunakaran, H. Zhang, J.P. Crows, W.E. Antholine, B. Kalyanaraman, Direct probing of copper active site and free radical formed during bicarbonate-dependent peroxidase activity of bovine and human copper, zinc superoxide dismutase: low-temperature electron paramagnetic resonance and electron nuclear double resonance studies, J. Biol. Chem. 279 (2004) 32534–32540.

[21] L.D. Kispert, Electron-electron double resonance, in: Biomedical EPR – Part B: Methodology, Instrumentation, and Dynamics, Biological Magnetic Resonance, Biomedical EPR, Part A: Free Radicals, Metals, Medicine, and Physiology. Part B: Methodology, Instrumentation, and Dynamics, vol. 24, 2005, p. 470 (Chapter 6).

[22] A.J. Hoff, Advanced EPR: Applications in Biology and Biochemistry, first ed., Elsevier, 1989.

[23] J. Telser, Electron-Nuclear Double Resonance (ENDOR) Spectroscopy, John Wiley & sons, Ltd., 2011.
[24] O. Schiemann, T.F. Prisner, Long-range distance determinations in biomacromolecules by EPR spectroscopy, Q. Rev. Biophys. 40 (2007) 1–53.
[25] P. Kuppusamy, M. Afeworki, R.A. Shankar, D. Coffin, M.C. Krishna, S.M. Hahn, J.B. Mitchell, J.L. Zweier, *In vivo* electron paramagnetic resonance imaging of tumor heterogeneity, Cancer Res. 58 (1998) 1562–1568.
[26] B.B. Williams, H.J. Halpern, *In vivo* EPR imaging (Chapter 11), in: Biomedical EPR – Part A: Free Radicals, Metals, Medicine and Physiology, 2006.
[27] G.R. Eaton, S.S. Eaton, EPR imaging: progress and prospects, Bull. Magn. Reson. 10 (1988) 22–31;
(a) A.W. Overhauser, Phys. Rev. 92 (1953) 411.
[28] K. Golman, J.S. Petersson, J.-H. Ardenkjær-Larsen, I. Leunbach, L.-G. Wistrand, G. Ehnholm, K. Liu, Dynamic in vivo oxymetry using overhauser enhanced MR imaging, J. Magn. Reson. Imaging 12 (2000) 929–938.
[29] K. Takeshita, T. Ozawa, Recent progress in *In vivo* ESR spectroscopy, J. Radiat. Res. 45 (2004) 373–384.
[30] R. Murugesan, J.A. Cook, N. Devasahayam, M. Afeworki, S. Subramanian, R. Tschudin, J.A. Larsen, J.B. Mitchell, A. Russo, M.C. Krishna, In vivo imaging of a stable free radical probe by pulsed-radio frequency electron paramagnetic spectroscopy, Magn. Reson. Med. 38 (1997) 409–414.
[31] N. Beghein, V. Preat, K. Amighi, J. Marchand-Brynaert, B. Gallez, Nano-emulsions of fluorinated trityl radicals as sensors for EPR oximetry, J. Magn. Reson. 197 (2009) 176–180.
[32] R. Ahmad, P. Kuppusamy, Theory, instrumentation and applications of electron paramagnetic resonance oximetry, Chem. Rev. 110 (2010) 3212–3236.
[33] Iron dithiocarbamate as spin trap for nitric oxide detection: pitfalls and successes, Methods Enzym. 359 (2002) 27–42.

CHAPTER SIX

Applications of Electron Paramagnetic Resonance

Chandran Karunakaran, Murugesan Balamurugan, Murugesan Karthikeyan
VHNSN College (Autonomous), Virudhunagar, India

Contents

Spin Resonance Spectroscopy
ISBN: 978-0-12-813608-9
https://doi.org/10.1016/B978-0-12-813608-9.00006-X

Abstract

This chapter focuses on the applications of electron paramagnetic resonance (EPR) to transition metal complexes of d^1 ions including Ti^{3+}, Mo^{5+}, V^{4+}, d^3 ion Cr^{3+}, EPR of Mn^{2+} and parallel mode EPR of Mn^{3+}, EPR of interconversion of low-spin to high-spin iron complexes and rhombogram for its ground state, iron—sulfur exchange coupled clusters and heme containing proteins, EPR of low spin, high-spin octahedral and tetrahedral $Co^{2+/3+}$ complexes, and EPR of Nickel in +2, +3 and +1 oxidation states. EPR of Cu^{2+} and its proteins and Jahn—Teller coupling and its types viz., static and dynamic Jahn—Teller distortions, temperature dependence of Jahn—Teller effect, structural elucidation of *bis*-salicylaldimine-Cu(II) are discussed. Fast-flow, rapid freeze-quench EPR to direct monitoring of free radicals and indirectly by spin traps, types, and applications of spin traps are presented. Electron spin resonance spin labeling, dosimetry, imaging, spin probes in EPR imaging, in vivo EPR imaging, Overhauser enhanced magnetic resonance imaging, and proton electron double resonance imaging are also described.

Keywords: EPR imaging; EPR of transition metal ions and their proteins; Fast-flow and rapid freeze-quench EPR; Rhombogram; Spin labeling; Spin traps.

6.1 APPLICATIONS OF ELECTRON PARAMAGNETIC RESONANCE TO TRANSITION METAL COMPLEXES

Transition metal ions of d^{1-9} electronic configuration exhibit different patterns of electron paramagnetic resonance (EPR) signals as explained below [1].

6.1.1 Electron Paramagnetic Resonance of d^1 Ions (Ti^{3+}, Mo^{5+}, W^{5+}, V^{4+})

6.1.1.1 Ti^{3+}

Ti^{3+}, $3d^1$ ion, has spin S = 1/2, and hence it is expected to exhibit a single line. The most abundant Ti is $^{46}Ti \approx 87\%$, whose S = 1/2 and I = 0. So, no hyperfine (hf) structure exists. However, the presence of less abundant isotopes ($^{47}Ti \approx 7.7\%$ with I = 5/2 and $^{49}Ti \approx 5.5\%$ with I = 7/2) gives rise to weak hyperfine structure with six and eight components, respectively. The electron and nuclear spin interactions give rise to (2I + 1) hyperfine lines (0, 6, and 8), respectively, and appear as satellite. This weak structure is shown in Fig. 6.1. Because ^{46}Ti abundance is more, the EPR signal contains only one big resonance line as shown in Fig. 6.1. The g-value obtained for this resonance is slightly less than 2.0 [2]. Because of one unpaired electron, it exhibits paramagnetism. The electronic and structural properties could be explored using optical and EPR spectra, which provide information on bonding between ligands and metal ion.

6.1.1.1.1 Relation Between Electron Paramagnetic Resonance and Optical Absorption Spectra

The g-value expression in general for d^1 is given by the relationship

$$g_i = g_e + \frac{n_i \lambda}{\Delta E_i} \quad i = x, y, z \tag{6.1}$$

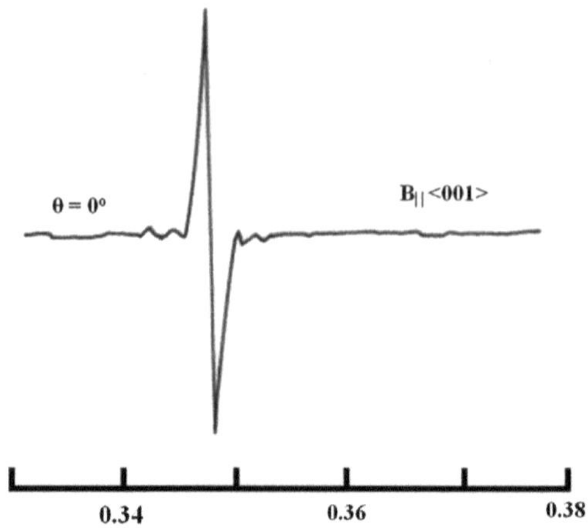

Figure 6.1 Single crystal electron paramagnetic resonance spectrum of Ti(III) showing weak hyperfine structure at θ = 0 degree for plane <001>. *Reproduced with permission from IOP Publishing.*

The first term on the right represents the contribution of μ_S and the second term quantifies μ_L or the shielding term. Δ represents the electronic energy gap between the particular pair of participating orbitals. In general, both n and Δ change when the molecule is observed along x, y, and z; consequently g_i will also vary.

Thus, the EPR data for Ti^{3+} can be correlated with the optical data to obtain the orbital reduction parameter as follows. On rearranging Eq. (6.1), we get

$$\lambda_{complex} = \frac{(g_e - g_{||})\Delta}{n}$$

$$k = \frac{\lambda_{complex}}{\lambda_{ionic}} \tag{6.2}$$

where n is 8 for C_{4V}, Δ is the energy of appropriate transition obtained from the electronic spectra (Fig. 6.2), λ is the spin–orbit coupling (SOC) constant for free ion Ti^{3+}, i.e., 154 cm^{-1}, and k is the orbital reduction parameter.

6.1.1.1.2 Electronic Spectra of Titanium Compounds

Before we discuss the electronic spectra of titanium compounds and transitions, the term symbols and their splitting for various transition metal ions d^{1-10} are listed in Table 6.1–6.3 [3].

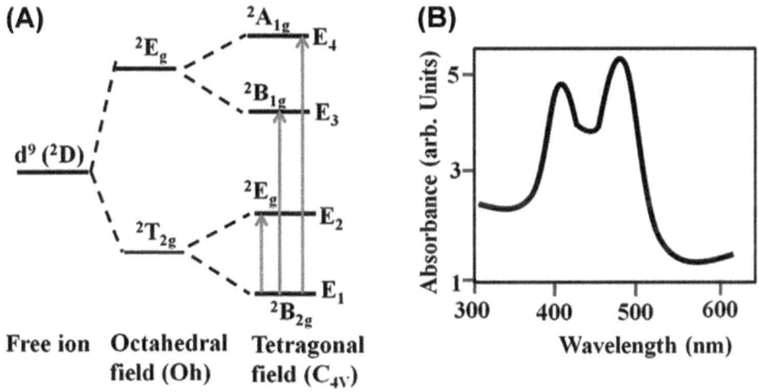

Figure 6.2 (A) The energy level diagram of Ti^{3+} in octahedral and tetragonal fields and (B) its optical spectrum of Ti(III) in yttrium aluminium perovskite. *Reproduced with permission from IOP Publishing.*

Table 6.1 Term symbol of ground and important excited states of $3d^n$ ions[a]

Configuration	Ion	Term symbol
d^1	Ti^{3+}, V^{4+}	^2D
d^2	Ti^{2+}, V^{3+}, Cr^{4+}	^3F, ^3P
d^3	Cr^{3+}, V^{2+}, Mn^{4+}	^4F, ^4P
d^4	Cr^{2+}, Mn^{3+}	^5D
d^5	Mn^{2+}, Fe^{3+}	^6S
d^6	Fe^{2+}, Co^{3+}	^5D
d^7	Ni^{3+}, Co^{2+}	^4F, ^4P
d^8	Ni^{2+}	^3F, ^3P
d^9	Cu^{2+}	^2D
d^{10}	Zn^{2+}	^6S

[a]Ref. 3

Table 6.2 The splitting of energy states of d^n terms[a]

Energy	Substates
S	A_1
P	T_1
D	$E + T_2$
F	$A_2 + T_1 + T_2$
G	$A_1 + E + T_1 + T_2$
H	$E + T_1 + T_1 + T_2$
I	$A_1 + A_2 + E + T_1 + T_2 + T_2$

[a]Ref. 3

Table 6.3 Octahedral field components of the ground and some excited states of d^n configuration[a]

d^n	Ground state	Crystal field states	Important excited states	Crystal field state
d^1, d^9	^2D	$^2T_{2g}$, ^2E_g		
d^2, d^8	^3F	$^3T_{1g}$, $^3T_{2g}$, $^3A_{2g}$	^3P	$^3T_{1g}$
d^3, d^7	^4F	$^4T_{1g}$, $^4T_{2g}$, $^4A_{2g}$	^4P	$^4T_{1g}$
d^4, d^6	^5D	$^5T_{2g}$, ^5E_g		
d^5	^6S	$^6A_{1g}$		

[a]Ref. 3

The electronic configuration of Ti^{3+} is [Ar] $3d^1$. It has fivefold degeneracy and its ground state term symbol is ^2D as shown above. In an octahedral crystal field, the fivefold degeneracy is split into $^2T_{2g}$ and ^2E_g states. Thus only one single-electron transition, $^2T_{2g} \rightarrow {}^2E_g$, is expected in an octahedral crystal field. The separation between these states is 10 Dq, which is crystal field energy. Furthermore, the ground $^2T_{2g}$ state is split due to Jahn–Teller effect and hence lowering of symmetry is expected for Ti(III) ion (see Section 4.21). The ground state splits into $^2B_{2g}$ and ^2E_g states in tetragonal symmetry and the excited term ^2E_g also splits into $^2B_{1g}$ and $^2A_{1g}$ levels. Thus, *three transitions* are expected for *tetragonal* (C_{4v}) symmetry. The energy level diagram in tetragonal environment is shown in Fig. 6.2A. The transitions in the tetragonal field are described by the following equations.

$$\begin{aligned}
&^2B_{2g} \rightarrow {}^2E_g[-4Dq - Ds + 4Dt - (-4Dq + 2Ds - Dt)] = -3Ds + 5Dt \\
&^2B_{2g} \rightarrow {}^2B_{1g}[6Dq + 2Ds - Dt - (-4Dq + 2Ds - Dt)] = 10Dq \\
&^2B_{2g} \rightarrow {}^2A_{1g}[6Dq - 2Ds - 6Dt - (-4Dq + 2Ds - Dt)] = 10Dq - 4Ds + 5Dt
\end{aligned} \tag{6.3}$$

In the above formulae, Dq is octahedral crystal field and Ds and Dt are tetragonal field parameters. The same sign of Dq and Dt indicates an axial elongation and opposite sign indicates an axial compression. However, one broad absorption with shoulder is observed (Fig. 6.2B). It can be explained that the promotion of electron may occur to either of the two closely spaced nondegenerate e_g orbitals, the d_z^2 and $d_{x^2-y^2}$.

6.1.1.2 Mo^{5+} and W^{5+}

Mo(V), also a d^1 ion, is present in the active site of several important enzymes, viz., sulphite oxidase, xanthine oxidase (XO), etc. Analogous to

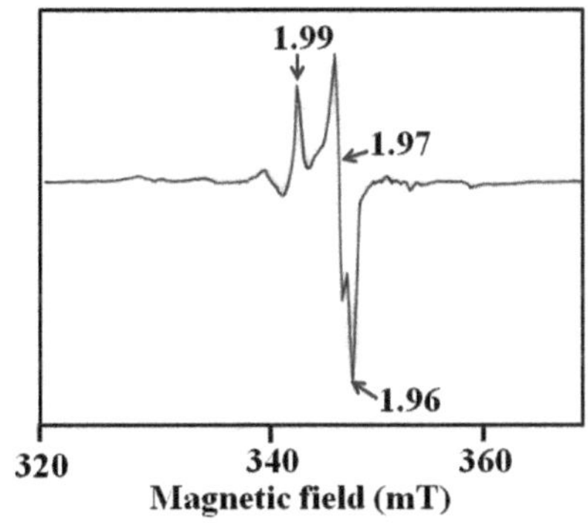

Figure 6.3 Electron paramagnetic resonance spectrum of Mo^{5+} in sulfite oxidase. *Reproduced with permission from American Chemical Society.*

other d^1 ions, the EPR spectrum obtained for the Mo(V) in sulphite oxidase is expected to exhibit a single line as the most abundant isotope (75%) having I = 0. However, it exhibits g-anisotropy, i.e., $g_{xx} = 1.99$, $g_{yy} = 1.97$, and $g_{zz} = 1.96$, in the case of frozen polycrystalline samples as shown in Fig. 6.3. The satellite peaks are obtained due to less abundant isotope (25%) with I = 5/2 [4].

Another enzyme XO is used to convert xanthine to uric acid. This involves the transfer of two electrons from xanthine to the XO Mo(VI) center, which is reduced to Mo(IV), neither of which is paramagnetic. However, when electrons are released from the enzyme during turnover, it occurs in a sequence of single electron transfers from the molybdenum center through a relay of iron—sulfur centers to Flavin Adenine Dinucleotide (FAD). The first-formed Mo(V) species is paramagnetic and therefore EPR is measurable (similar to above) and finally reduced to the Mo(IV). In the absence of any FAD, oxygen occupies the binding site. This fortuitously forms $O_2^{\cdot -}$ superoxide radical anion without FAD. Spin trapping of the superoxide radical anion formed by this method is a benchmark procedure for its detection.

Similar to Mo^{5+}, W^{5+} also contains d^1 electron configuration present in the active site of formylmethanofuran dehdrogenase isolated from cells grown on tungstate. It also exhibits anisotropic spectrum in frozen state with *g*-values 2.048, 2.012, and 1.96 as most tungsten isotopes I = 0 and small isotopes with I = 1/2 giving doublet around each signal [5].

6.1.1.3 VO^{2+}

Similar to Ti(III), VO^{2+}, $3d^1$ ion possesses S = 1/2 and hence it also expected to show a single line whose *g*-value is slightly below 2.0.

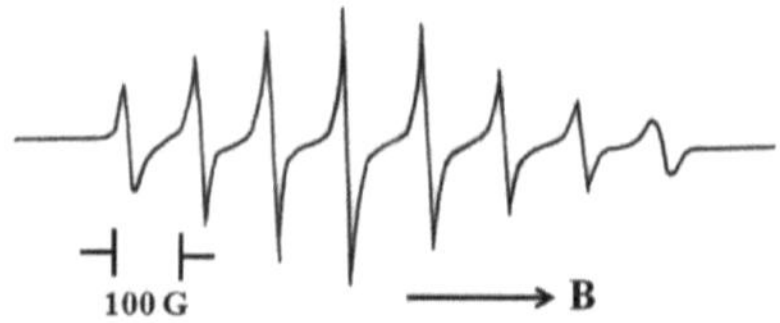

Figure 6.4 Electron paramagnetic resonance spectrum of VO^{2+} complex in solution at room temperature. *Courtesy Dr. C. Karunakaran.*

The most abundant isotope is ^{51}V, $I = 7/2$. So, an eight-line pattern for vanadyl complex $[2(7/2) + 1 = 8]$ is expected. For example, the isotropic EPR spectrum for $[VO(glycolate)]^{2-}$ in water at room temperature is shown in Fig. 6.4 exhibiting hyperfine with eight peaks.

The following EPR signals are obtained for different cases of VO(II) complex:

1. *High concentration of V(IV)*: In general, if the paramagnetic ion vanadium content in the crystal or in solution is high, it gives a single broad resonance line as expected due to exchange interaction between neighboring paramagnetic V^{4+} ions. Therefore, the eight hyperfine lines from ^{51}V cannot be resolved.
2. *Low concentration* or VO^{2+} ion doped in diamagnetic lattice. As expected for V in the $S = 1/2$ and $I = 7/2$ state, the EPR spectrum shows hyperfine structure of eight equidistant lines in octahedral symmetry. For example, the EPR spectrum of VO^{2+}-doped potassium hexaaquazinc zinc(II) sulfate single crystal in the c⋆a plane for the magnetic field at 0 degree shows two sets of eight lines corresponding to the two different crystallographic VO^{2+} sites in the crystal lattice (Fig. 6.5) [6].
3. However, in C_{4v} symmetry, i.e., axial, it exhibits $g_{\parallel}$ and $g_{\perp}$ in the polycrystalline or frozen spectrum. From the hyperfine splitting of eight-line pattern in $g_{\parallel}$ and $g_{\perp}$, respectively, the $A_{\parallel}$ and $A_{\perp}$ are obtained as shown in Fig. 6.7. Furthermore, in tetragonal distortion, $g_{\parallel} < g_{\perp} < g_e$, which

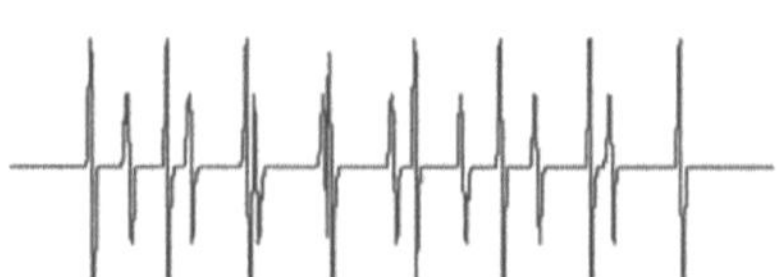

Figure 6.5 Electron paramagnetic resonance spectrum of VO^{2+}-doped potassium hexaaquazinc (II) sulfate single crystal in the c*a plane for the magnetic field at 0 degree from the c*-axis at microwave frequency $v = 9.5$ GHz. *Reproduced with permission from Springer-Verlag.*

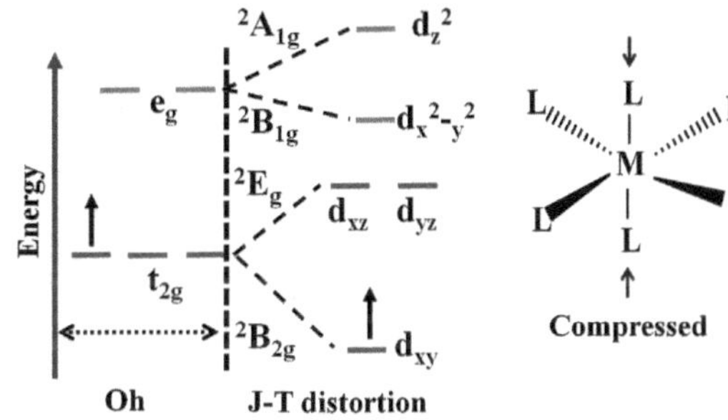

Figure 6.6 d-Orbital splitting and tetragonal compression (d^1 ion).

shows the presence of an unpaired electron in the d_{xy} orbital. This is characteristic of a tetragonally compressed complex due to Jahn-Teller (JT) effect (Fig. 6.6).

For example, V^{4+} *in chloroperoxidase*; in the reduced form of the enzyme, the spectrum V^{4+} (d^1) can be observed (Fig. 6.7). Because vanadium has axial symmetry, its powder spectrum consists of two major peaks ($g_\perp = 1.95$ and $g_{||} = 1.98$). Vanadium possesses one stable isotope ^{51}V with $I = 7/2$. Therefore, each peak will be further split into eight ($2I + 1$) lines [7]. Because of overlap, not all lines are observed. Furthermore, the hyperfine coupling constant $A_\perp$ is very large resulting in the hyperfine lines of $g_\perp$ to pass the $g_{||}$ peak, causing an effect known as overshoot.

4. In C_{2v} *low symmetry*, the EPR spectrum is rhombic with g_{xx}, g_{yy}, and g_{zz} values due to lowering of symmetry. So, three sets of eight hyperfine lines are expected corresponding to the three g-values. The hyperfine constants are designated as A_1, A_2, and A_3, respectively.

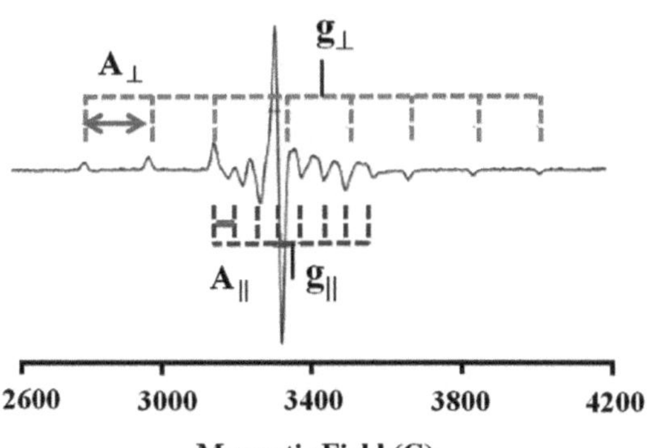

Figure 6.7 Electron paramagnetic resonance spectrum of vanadium-containing chloroperoxidase from the fungus *Curvularia inaqualis. Reproduced with permission from Elsevier.*

6.1.1.3.1 Determination of Ground-State Wave Function, Fermi Contact, and Dipolar-Coupling Terms From Electron Paramagnetic Resonance Data

The ground-state wave function for VO^{2+} ions can be written as

$$\Psi = c_1\left|x^2 - y^2\right\rangle + c_2|xy\rangle + c_3|yz, xz\rangle \tag{6.4}$$

where c_1, c_2, and c_3 are the admixture coefficients, which account for the contributions of d-orbital energy levels, where the unpaired electron of VO^{2+} ions is probable to reside [6]. These coefficients are obtained by using the following relations

$$\begin{aligned} &c_1^2 + c_2^2 + c_3^2 = 1 \\ &g_{\|} = g_z = 2\left(3c_1^2 - c_2^2 - c_3^2\right) \\ &g_{\perp} = \left(g_x + g_y\right)/2 = 4c_1(c_2 - c_3) \end{aligned} \tag{6.5}$$

The dipolar-coupling constant (P) and Fermi contact parameter (K) are evaluated by the expressions below using the hyperfine constants:

$$A_{\|} = A_z = P\left[g_z - \left\{(K + 15/7)\left(1 - 2c_3^2\right)\right\} - \left\{3(1 + 4c_2c_3)\right\}/7\right] \tag{6.6}$$

$$A_{\perp} = A_x + A_y/2 = P\left[(11/4)\left\{\left(g_x + g_y\right)/2\right\} - 2c_1c_2(K + \{9/7\})\right] \tag{6.7}$$

These equations are valid for axial symmetry. Mostly, the rhombic distortion is small, as is evident from the evaluated spin Hamiltonian parameters reported. Therefore, for practical purposes, the symmetry may be considered to be axial and these equations may be applied for calculating the ground-state wave function and the parameters P and K. The values of P and K are evaluated for positive ratio of $A_{\|}/A_{\perp}$ to obtain the best fit values. $A_{\|}$ and $A_{\perp}$ are taken as negative. By substituting the experimental values of the spin Hamiltonian parameters and admixture coefficients (c_1, c_2, and c_3) in the above equations, the values of P and K have been obtained [6].

The dipolar term P is a measure of the radial distribution of the unpaired electron wave function. It accounts for the direct dipole interaction of the electron moment and the nuclear moment, given by

$$P = g_e g_n \beta_e \beta_n e\left\langle r^{-3}\right\rangle \tag{6.8}$$

$\langle r^{-3}\rangle$ gives the average size of the vanadium 3d orbital and the other symbols have their usual meanings. The value of P for free VO^{2+} ions is 0.016 cm^{-1}. In general, the reduced P value obtained for VO^{2+} complex indicates the covalent bonding.

The quantities P and K are also directly related to the isotropic hyperfine coupling constant A_0 as

$$A_0 = -P[K + \{(\Delta g_x + \Delta g_x + \Delta g_z)/3\}]$$
$$= -P\left[K + \left\{g_e - \left(\left(g_{\|} + 2g_{\perp}\right)/3\right)\right\}\right] \tag{6.9}$$

$$A_0 = \left(\frac{A_z + A_y + A_x}{3}\right) = \left(\frac{A_{\|} + 2A_{\perp}}{3}\right)$$

6.1.1.3.2 Molecular Orbital Coefficients

The semiempirical linear combination of atomic orbital-molecular orbital (LCAO-MO) method has been described by Ballhausen and Gray, for the analysis of bonding in the pentaaqua vanadyl ion. The LCAO-MO wave functions for these ligand and metal orbitals are as follows:

$$\begin{aligned} |B_2\rangle &= \beta_2|d_{xy}\rangle - \beta_2'|\varphi_{b2}\rangle \\ |E\rangle &= \gamma|d_{yz}\rangle - \gamma'|\varphi_{ey}\rangle, \gamma|d_{xz}\rangle - \gamma'|\varphi_{ez}\rangle, \\ |B_1\rangle &= \beta_1|d_{x^2-y^2}\rangle - \beta_1'|\varphi_{b1}\rangle \\ |A_1\rangle &= \delta|d_{3z}^2 - r^2\rangle - \delta'|\varphi_{a1}\rangle \end{aligned} \tag{6.10}$$

where the metal orbitals are the usual d functions and the ligand functions, φ, are group orbitals of appropriate symmetry. The information about the bonding in the complexes through the EPR spectra is obtained by shifts of g-values from the free electron value, which are the measure of the spin–orbit mixing of antibonding states into the ground state. The anisotropy of the metal nuclear–hyperfine interaction is also related to the molecular orbital coefficients of the ground state as follows. By correlating the optical and EPR data, the molecular orbital coefficients, i.e., bonding coefficients $\left(\beta_1^2, \beta_1^2, \gamma^2\right)$, are estimated.

$$\begin{aligned} g_{\|} = g_z &= 2.0023 - \left(8\beta_2^2\beta_1^2\lambda/\Delta E_{x^2-y^2}\right) \\ g_y &= 2.0023 - \left(2\beta_2^2\gamma_1^2\lambda/\Delta E_{xz}\right) \\ g_x &= 2.0023 - \left(2\beta_2^2\delta^2\lambda/\Delta E_{yz}\right) \end{aligned} \tag{6.11}$$

where λ (=170 cm^{-1}) is the SOC constant for the free vanadium ion, whereas $\Delta E_{x^2-y^2}$, ΔE_{xz}, and ΔE_{yz} are the electronic transitions $^2B_{2g} \rightarrow {}^2B_{1g}$ ($d_{xy} \rightarrow d_{x^2-y^2}$), $^2B_{2g} \rightarrow {}^2E_g$ ($d_{xy} \rightarrow d_{xz}$), and $^2B_{2g} \rightarrow {}^2E_g$ ($d_{xy} \rightarrow d_{yz}$), respectively. The transitions corresponding to ΔE_{xz} and ΔE_{yz} cannot be differentiated, so the same value is taken for both transitions in the evaluation of the bonding coefficients by using the above equation. Before evaluating the various bonding coefficients, one of the bonding coefficients β_2^2 can be calculated using the expression as follows taking the negative sign of hyperfine values of the spin Hamiltonian. For axial symmetry, $A_x = A_y$, $\Delta g_x = \Delta g_y$

$$\beta_2^2 = 7/12\left[\{(A_x + A_y - 2A_z)/P\} + 2\Delta g_z - \{(5/14)(\Delta g_x + \Delta g_y)\}\right] \tag{6.12}$$

The values of β_1^2 and β_2^2 represent the in-plane σ-bonding and in-plane π-bonding, respectively, whereas bonding coefficients γ_1^2 and δ2 both represent out-of-plane π-bonding [6].

6.1.2 Electron Paramagnetic Resonance of Cr^{3+}

Cr(III), a d^3 ion, has three unpaired electrons with $S = 3/2$ exhibiting $(2S + 1)$ three-line *fine structure* (Fig. 6.8). The most abundant ^{52}Cr has $I = 0$ and does not exhibit hyperfine structure. However, less abundant ^{53}Cr with $I = 3/2$ gives rise to $(2I + 1)$ hyperfine structure with four components. This hyperfine structure is not generally observed because of the low abundance of ^{53}Cr [8].

Cr^{3+} ion splits into $|\pm 1/2\rangle|$ and $|\pm 3/2\rangle$ Kramers' doublets due to zero-field splitting (ZFS) in the absence of magnetic field, separated by 2D. Here, D represents the ZFS parameter. This degeneracy can be lifted only by an external magnetic field. In such a case, three resonances are observed corresponding to the transitions, $|-3/2\rangle \leftrightarrow |-1/2\rangle|$, $|-1/2\rangle \leftrightarrow |1/2\rangle$,

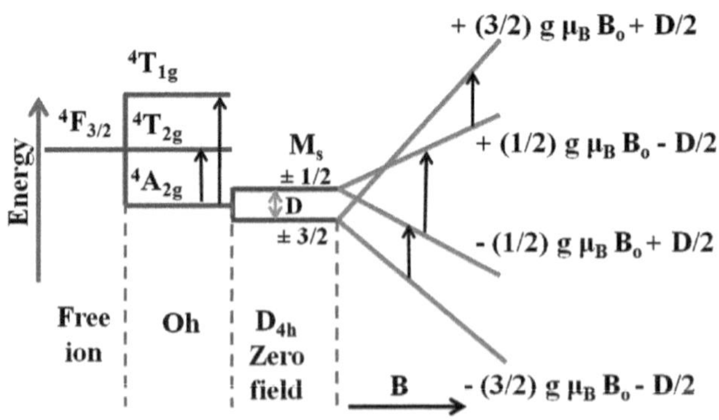

Figure 6.8 Energy level splitting diagram for Cr^{3+} complex showing electron paramagnetic resonance and optical transitions.

and $|1/2\rangle \leftrightarrow |3/2\rangle$ at $g\beta B - 2D$, $g\beta B$, and $g\beta B + 2D$, respectively. In a powder spectrum, the intense perpendicular component $g_\perp$ is mainly visible (and parallel $g_{||}$ component is absent). When all the three transitions are observed, the separation between the extreme sets of lines is 4D $[g\beta B + 2D - (g\beta B - 2D) = 4D]$. However, a single resonance line appears with $g \sim 1.98$ if D is equal to zero. Furthermore, when D is very large compared to microwave frequency, a single line is seen around $g = 4.0$.

The EPR spectra observed and their angular variation for single crystal of Cr^{3+} doped NiMTH [MTH = maleate tetrahydrate] can be described by a spin Hamiltonian with electron spin (S = 3/2) in the following form:

$$H = \beta BgS + D\left[S_z^2 - 1/3S(S+1)\right] + E\left(S_x^2 - S_y^2\right) \tag{6.13}$$

where β is the Bohr magneton, g is the spectroscopic splitting factor (or g-factor), and D and E are the axial and rhombic crystal field terms, respectively.

The g-values were calculated at each interval from the angular variation spectra, i.e., by rotating the single crystal in the three mutually perpendicular planes. The diagonalization procedure was carried out following the standard procedure given by Schonland for the analysis of EPR spectra to obtain the principal g-tensor corresponding to three principal directions (see section 4.14.3). From the observed g-values $g_x = 2.098$, $g_y = 1.975$, and $g_z = 1.968$, it is confirmed that the symmetry around Cr^{3+} ions is triclinic symmetry crystals. The parameters D and E

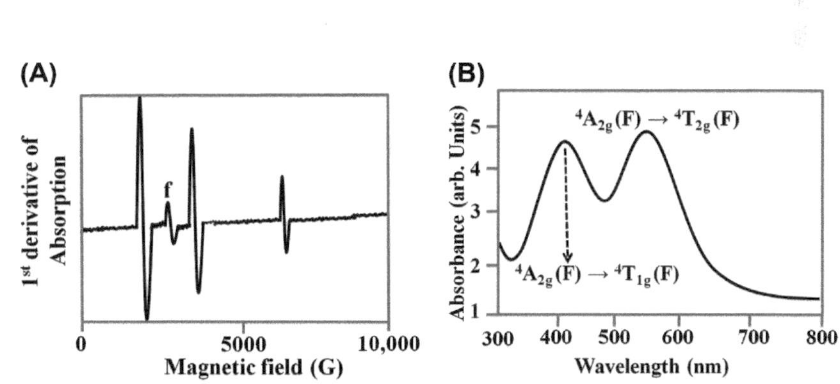

Figure 6.9 (A) Electron paramagnetic resonance spectrum of Cr^{3+}: NiMTH single crystal, the peak marked "f" is weak forbidden transitions when the external magnetic field is along the Z-axis in the ZX′ plane at room temperature and (B) the optical absorption of chromium maleate tetrahydrate crystal recorded at room temperature. *Reproduced with permission from Elsevier.*

have been evaluated by computer simulation using the resonance field position of the above spin Hamiltonian to the third-order perturbation and are found to be −0.0954 and 0.0288 cm^{-1}, respectively. The sign of D value is found to be negative because the value of $[g_z - (g_x + g_y)]/2$ is negative [8].

6.1.2.1 Relation Between Electron Paramagnetic Resonance and Optical Absorption Spectra

A comparison is made between the observed g_{eff} from EPR results and the calculated one from the optical spectrum. For Cr^{3+}, the EPR and optical results are related by,

$$g_{||} = g_e - 8\lambda/\Delta E[^4T_{1g}(F)]$$

$$g_{\perp} = g_e - 2\lambda/\Delta E[^4T_{2g}(F)] \quad (6.14)$$

λ is the SOC constant for Cr^{3+} ion in complex and ΔE values are obtained from optical data. Here $g_{||}$ and $g_{\perp}$ are the spectroscopic splitting factors parallel and perpendicular to the magnetic field direction; the free electron value g_e is 2.0023. These values give,

$$g_{eff} = \left(2g_{||} + g_{\perp}\right)/3 \quad (6.15)$$

6.1.2.2 Optical Absorption Studies

The two broad bands observed at room temperature at 17,725 and 24,505 cm^{-1} have therefore been attributed to the transitions $^4A_{2g}(F) \rightarrow {}^4T_{2g}(F)$ and $^4A_{2g}(F) \rightarrow {}^4T_{1g}(F)$, respectively, as shown in Fig. 6.9. The value of the crystal field splitting parameter Dq is then obtained from the energy of the transition $^4A_{2g}(F) \rightarrow {}^4T_{2g}(F)$:

$$Dq = [E(^4T_{2g}) - E(^4A_{2g})]/10 \quad (6.16)$$

The Racah parameter B can be evaluated using the value of Dq above calculated and the expression as follows:

$$Dq/B = 15(x - 8)/x^2 - 10x \quad (6.17)$$

$$\text{where } x = [E(^4T_{2g} \rightarrow {}^4T_{1g}) - E(^4A_{2g} \rightarrow {}^4T_{2g})]/Dq$$

The value of interelectronic repulsion parameter B for free Cr^{3+} ion is 918 cm^{-1}. A comparison with the observed B value (668 cm^{-1}) for Cr^{3+}

ion in NiMTH complex lattice shows that the B value is decreased by 27%. This decrease is caused by influence of covalent bond [8].

6.1.2.3 Calculation of Reduction Factor (k)

By correlating the optical absorption and EPR data, it is possible to find out k, which is also the measure of covalent nature of the complex. For Cr^{3+} ion in octahedral or distorted trigonal symmetry, g is given by

$$g_z = g_e - [8k\lambda/\Delta] \tag{6.18}$$

where Δ ($\Delta = 10$ Dq) is the crystal field splitting parameter, which is found to be 1773 cm^{-1} for Cr^{3+}: NiMTH and k is the reduction factor given by

$$k = \lambda_{eff}/\lambda \tag{6.19}$$

where λ_{eff} is the value of SOC constant for the Cr^{3+} in the crystal.

By substituting $\Delta = 1773$ cm^{-1} and $g_z = 1.968$ in equation and solving the equation, the values of k and λ_{eff} obtained are 0.82 and 75 cm^{-1}, respectively. The deviation of λ_{eff} value from the free ion value (91 cm^{-1}) for [CrMTH] can be attributed to the covalent bonding effects in the crystal, which produce a reduction in the effective value of SOC and also in orbit moment [8].

6.1.3 Electron Paramagnetic Resonance of Mn^{2+} and Mn^{3+}

6.1.3.1 Mn^{2+}

Mn(II), a d^5 ion, has a total spin of $S = 5/2$. Thus, there are six spin states labeled as ±5/2>, ±3/2>, and ±1/2> and are known as the three Kramers' doublets (see Section 6.1.2). In the absence of external magnetic field, they are separated by 4D and 2D, respectively, due to ZFS parameter (D) (Fig. 6.10). These three doublets further split into six energy levels by the application of an external applied magnetic field. Transitions between these six energy levels give rise to five resonance lines based on $\Delta m_s = \pm 1$ called fine splitting/structure. Using the selection rule $\Delta m_I = 0$, each of these resonance lines, in turn, splits into (+5/2, +3/2, +1/2, −1/2, −3/2, −5/2) to give a sextet due to the interaction of the electron spin with the nuclear spin of ^{55}Mn, ($I = 5/2$). Thus one expects a 30-line pattern as in Fig. 6.10. However, depending on the relative magnitudes of D and A (hyperfine coupling constant of manganese), these 30 lines appear as a separate bunch of 30 lines or 6 lines (if $D = 0$). The separation between the extreme set of resonance lines is approximately equal to 8D (first order). If D is very small compared with hyperfine coupling constant (A), the 30

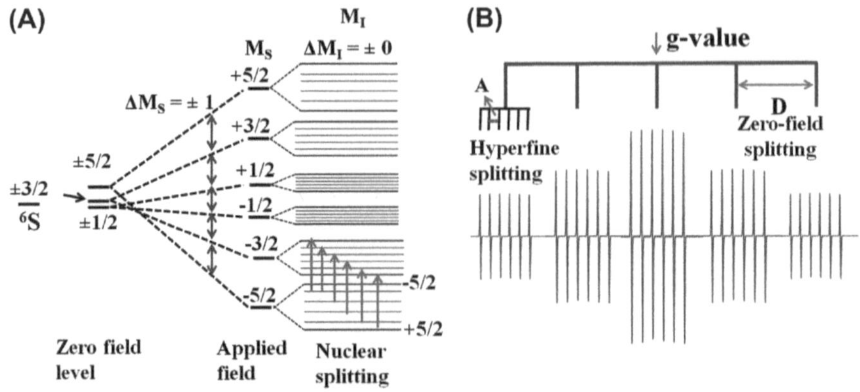

Figure 6.10 (A) Energy level splitting due to zero field (*five lines*) and further hyperfine splitting (*six lines*) and (B) the electron paramagnetic resonance spectrum of Mn^{2+}.

lines are so closely packed and sometime not resolved. If $D = 0$, the system is perfectly octahedral resulting in a single-line spectrum. Any deviation from axial symmetry leads to a term known as E in the spin Hamiltonian. The value of E can be easily calculated from single-crystal measurements. A nonzero value of E results in making the spectrum unsymmetrical about the central sextet.

A typical EPR spectrum recorded, when an applied magnetic field B is parallel to the axis for a crystal with one magnetic site in the lattice, is shown in Fig. 6.10. The spectrum can be stimulated based on the total Hamiltonian equation below to estimate the spin Hamiltonian parameters, g, A, B, D, and E. The total spin Hamiltonian expression for a spin multiplet due to second-order effects and other zero-field terms is given as follows:

$$\begin{aligned}\hat{H} &= g\mu_B BS + D\left[S_z^2 - (1/3)S(S+1)\right] + E\left(S_x^2 - S_y^2\right) \\ &\quad + (a/6)\left[S_x^4 + S_y^4 + S_z^4 - (1/5)S(S+1)\left(3S^2 + 3S - 1\right)\right] \\ &\quad + AS_zI_z + B(S_xI_x + S_yI_y)\end{aligned} \tag{6.20}$$

where μ_B or β is the Bohr magneton, B is the external magnetic field, g is the spectroscopic splitting factor, and S is the effective spin vector. The parameters a, D, and E are the cubic, axial, and rhombic ZFS parameters, respectively [9].

The first term represents the electronic Zeeman interaction, the second and third terms represent the axial and rhombic components of the ZFS, the fourth term represents the fourth-rank cubic ZFS term, and the fifth and sixth terms are the hyperfine interaction terms ($I = 5/2$).

6.1.3.1.1 Determination of Covalency of Mn-Ligand

The percentage of covalency (C) of Mn-ligand bond can be calculated in two ways using (1) Matumura's plot and (2) electronegativities, χ_p and χ_q using the below equation:

$$C = 1/n\left[1 - 0.16(\chi_p - \chi_q) - 0.035(\chi_p - \chi_q)^2\right] \tag{6.21}$$

here n is the number of ligands around Mn(II) ion; $\chi_p = \chi_{Mn} = 1.6$ for Mn(II) and $\chi_q = \chi_{ligand}$.

Also the hyperfine constant is related to covalency by,

$$A_{iso} = (2.04C - 104.5) \times 10^{-4}\ cm^{-1} \tag{6.22}$$

Furthermore, the g-value for the hyperfine splitting is indicative of the nature of bonding. If the g-value shows a negative shift with respect to the free electron g-value (2.0023), the bonding is ionic and conversely if the shift is positive, then the bonding is said to be more covalent in nature. Because all the excited states of Mn(II) ion will either be quartets or doublets compared with the ground state ($^6A_{1g}$), the optical absorption spectra of Mn(II) ions will have only spin-forbidden transitions. Therefore, the intensity of optical transitions is weak.

6.1.3.2 Electron Paramagnetic Resonance of Mn^{3+}

The energy levels of Mn^3 (d^{4+}) ion in high-spin state in Mn superoxide dismutase (SOD), $S = 2$ integer spin system with a positive axial ZFS value, D, are shown in Fig. 6.11. An EPR signal from the Mn^{3+} in MnSOD enzyme ion has not previously been detected with conventional X-band EPR methods. Because of the large ZFS values ($D > 2\ cm^{-1}$), the conventional perpendicular polarization EPR allowed only $\Delta M_s = \pm 1$ transitions, which are not possible at normal X-band microwave frequencies ($0.3\ cm^{-1}$) (See Fig. 6.11 left). However, EPR signals from integer spin systems can be readily detected when the oscillating magnetic field applied to induce a spin-state transition is oriented parallel to the static magnetic field. This method is called *parallel psolarization EPR*. Here, the parallel magnetic field orientation allows transitions between the closely spaced $M_s = \pm 2$ energy levels (Fig. 6.11 inset) to give six lines in the EPR spectrum. The signal intensities are observed of many orders of magnitude larger than the corresponding perpendicular magnetic field orientation [10].

The parallel polarization EPR spectrum of the Mn^{3+} ion of native MnSOD is shown in Fig. 6.11, right. This signal is centered at an effective g-value of 8.17 and consists of six hyperfine lines separated by ~100 G.

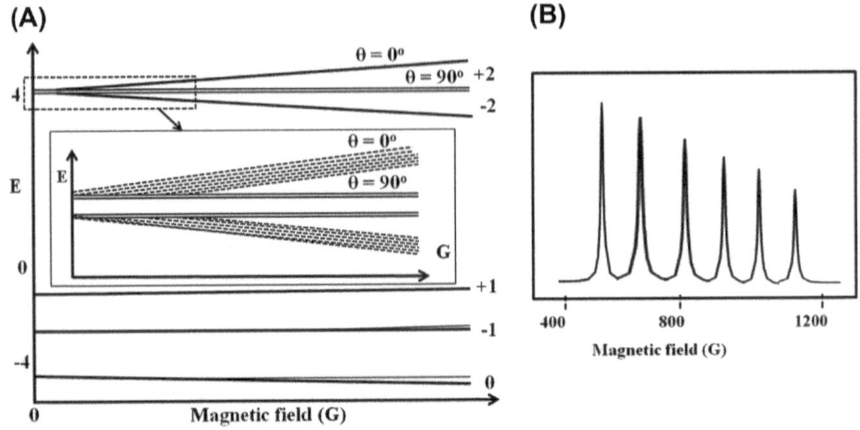

Figure 6.11 (A) Energy level diagram of an $S = 2$ spin system, obtained with $D = 2.10\ cm^{-1}$, $E = 0.243\ cm^{-1}$, $A_x = A_y = 101$ G, $A_z = 100$ G, $g_x = 2.00$, $g_y = 1.99$, $g_z = 1.98$, and $\theta = 0°$ and 90° with $\phi = 0$ degree. The inset figure is an expanded view of the $M_s = \pm 2$ energy levels in the region of the observed parallel mode electron paramagnetic resonance (EPR) transitions (indicated by *arrows*). (B) Parallel mode continuous wave EPR spectrum of Mn^{3+} superoxide dismutase at microwave frequency, 9.41 GHz and microwave power, 32 mW and 10K. *Reproduced with permission from American Chemical Society.*

This EPR signal is absent in the conventional perpendicular mode spectrum. The spin Hamiltonian describing this system is:

$$\hat{H} = \beta_e \boldsymbol{B}_0 g_e \hat{S} + D\left[\hat{S}_z^2 - \{1/3\}S(S+1)\right] + (E/2)\left(\hat{S}_+^2 + \hat{S}_-^2\right) + \hat{S}A\hat{I} \tag{6.23}$$

where β_e is the electron Bohr magneton, $\boldsymbol{B}_0$ is the static magnetic field vector, g_e is the electron **g** tensor, $\hat{S}$ is the electron spin operator (total spin $S = 2$), D is the axial ZFS parameter, E is the rhombic ZFS parameter, A is the nuclear hyperfine tensor, and $\hat{I}$ is the nuclear spin operator for the 100% abundant ^{55}Mn nucleus ($I = 5/2$). The ZFS values $D = 2.10\ cm^{-1}$ and $E = 0.24\ cm^{-1}$ were obtained from the simulation of observed spectrum.

6.1.4 Electron Paramagnetic Resonance of Iron Complexes and Proteins

Iron exists widely in +3 oxidation state, d^5 ion most stable configuration with two spin states: (1) low-spin $\left(t_{2g}^5 e_g^o\right)$ and (2) high-spin $\left(t_{2g}^3 e_g^2\right)$. Their respective spin values of 1/2 and 5/2 are both detectable by EPR, exhibiting

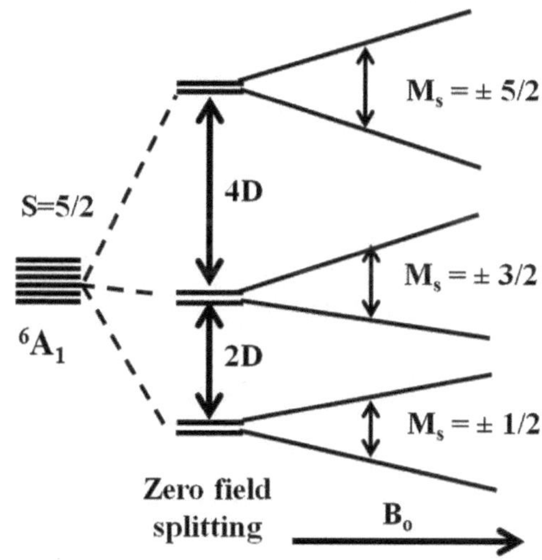

Figure 6.12 Energy level splitting diagram of Fe(III) in high-spin state (S = 5/2) with D value greater than the microwave frequency.

many spectral characteristics depending on ligands, molecular environment, and solvent medium [11–13].

For *high-spin complexes* with S = 5/2, there are three Kramers' doublets of $|S = 5/2>$ and are split into $|S = \pm5/2\rangle$, $|S = \pm3/2\rangle$, and $|S = \pm1/2\rangle$ separated by 4D and 2D, respectively, where D is the ZFS parameter (Fig. 6.12). If the $|S = \pm1/2\rangle$ doublet is low lying then the value of D is positive. However, D is negative when the $|S = \pm1/2\rangle|$ doublet is having the highest energy. Depending on the relative populations of these doublets, the g-values ranging from 0 to 9.0 are observed. The line widths are larger in low magnetic field when compared with high magnetic field EPR.

Because the transitions from the ground-state doublet to higher doublet states are forbidden (because the spin multiplicity changes by two and hence they are too weak), there are only three transitions between the respective doublets (shown in Fig. 6.12) as follows:

1. When the *lowest doublet*, $|S = \pm1/2\rangle$, is populated, it gives a g-value of 2–6. For example, heme and its proteins, viz., cytochrome *c*, hemoglobin, etc.
2. If the *middle Kramers'* doublet $|S = \pm3/2\rangle$ is populated, a g-value 4.30 is observed for transferrin, FeSOD proteins.
3. When the *third doublet* $|S = \pm5/2\rangle$ is populated, it exhibits a g-value of 2/7–30/7.

The EPR spectra of Fe^{3+} compounds may be described by the spin Hamiltonian,

$$H = gB\beta S + D\left[\left(S_z^2 - S(S+1)/3\right) + E\left(S_x^2 - S_y^2\right)\right] \quad (6.24)$$

The second and third terms in Eq. (6.23) represent the effects of axial and rhombic components of the crystal field, respectively. When D=E = 0, it corresponds to a free ion in the magnetic field, B. If E = 0, it implies an axial symmetry field. If λ (=E/D) increases, it results in the variation of rhombic character.

1. When $\lambda = 1/3$ (0.33), a maximum rhombic character is observed and further increase in λ from 1/3 to 1 results in the decrease of rhombic character. When $\lambda = 1/3$, the g-value is 4.27 and when λ is less than 1/3, g-value is around 4.
2. When $\lambda = 1$, it exhibits the axial field symmetry, which has signals at g = 6.0 and 2.0 for ferric high-spin states.

6.1.4.1 Rhombogram

For half-integer spin, viz. S = 3/2, 5/2, 7/2, etc., systems exhibiting Kramer's degeneracy, the effective g-values are a function of rhombicity (λ = E/D). This relation between the effective g-values versus the rhombicity can be plotted in two-dimensional graphs, so-called rhombograms (Fig. 6.13 right). In other words, the effective g-values are plotted as

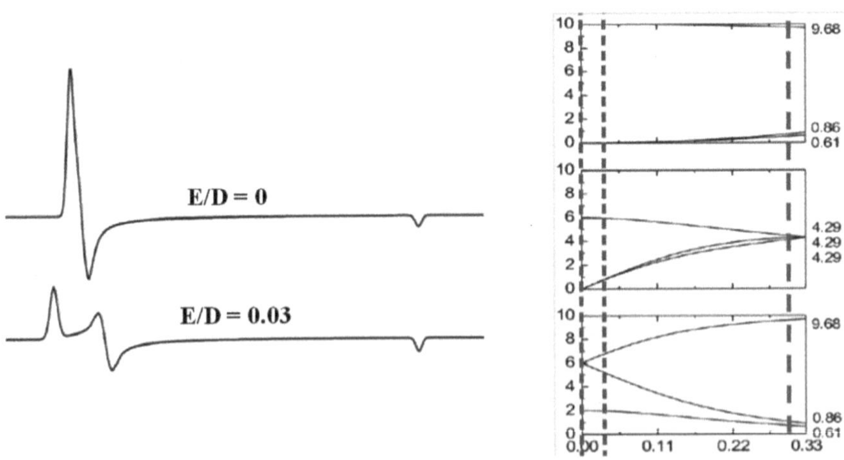

Figure 6.13 Rhombogram for spin system S = 5/2 and simulated electron paramagnetic resonance spectra for E/D = 0 and 0.03.

function of the rhombicity ranging from 0 to 0.33. It will help with the identification of the spin state, determination of which spin doublets are detectable, and determination of the E/D value. For example, Fe(III) complex with S = 5/2, the EPR spectra obtained for (A) E/D = 0 (axial symmetry) and (B) increase rhombic symmetry E/D = 0.03 are shown in Fig. 6.13.

6.1.4.2 Electron Paramagnetic Resonance Spectra of Transferrin and FeSOD

Human serum transferrin, a bilobal ~75 kD glycoprotein that has the ability to reversibly bind Fe^{3+} with nanomolar affinity, represents one of the major vehicles for iron delivery to cells both in vivo and in vitro. It contains a high-spin Fe^{3+} at its active site (Fig. 6.14) exhibiting g-values at 4.36, 4.25, and 4.12 as shown in Fig. 6.14. The ZFS parameters are $D = 0.32\ cm^{-1}$, $E = 0.10\ cm^{-1}$, and E/D = 0.315. For transferrin, the EPR transitions at g-values 9.7, 4.3, and <2 are possible based on the rhombogram at E/D = 0.315 (Fig. 6.13 right). However, the resonance in the region, g = 4.3, arises only from the relatively highly populated middle doublet. Thus, the three transitions are explained. The increase in E/D value also indicates the rhombic low-symmetry system as evident from the active site structure [14—16].

Fig. 6.15 shows the X-band EPR spectrum of the FeSOD recorded at 10K. The signals corresponded to a high-spin ground state S = 5/2. The g-values of the major features are 4.78, 4.05, and 3.73. This spectrum is thus characteristic of a rhombic species corresponding to a high-spin ferric iron in low-symmetry environment (Fig. 6.15) [17].

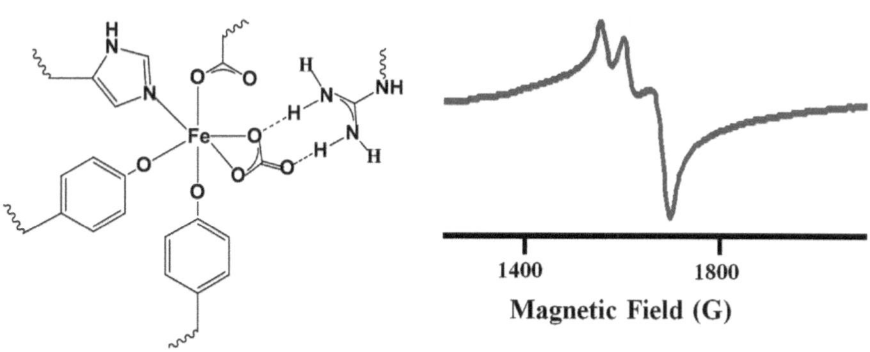

Figure 6.14 Active site structure and electron paramagnetic resonance spectrum of transferrin. *Courtesy Dr. C. Karunakaran.*

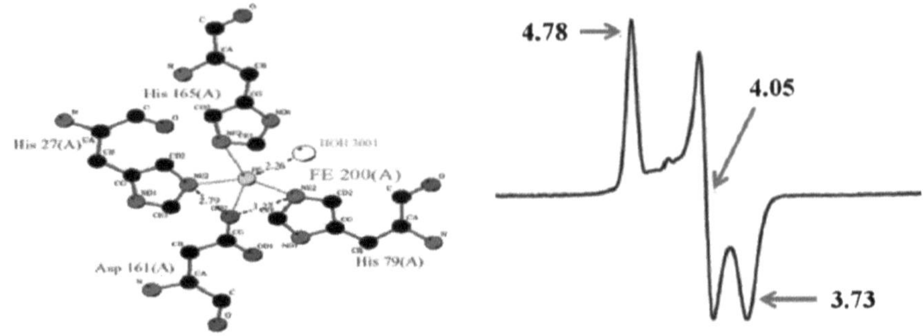

Figure 6.15 Active site geometry and X-band electron paramagnetic resonance of FeSOD recorded at 10K; microwave frequency 9.65 GHz; and microwave power 1 mW. *Reproduced with permission from Elsevier.*

6.1.4.3 Electron Paramagnetic Resonance Spectra of Fe(III) High-Spin to Low-Spin Conversion

EPR spectra of the ferric mono *p*-nitrophenolate complexes 1 and heme d_1 exhibit signals at $g = 6.0$ and 2.0 and indicate a ferric high-spin states (Fig. 6.16). With titration or addition of 1-methylimidazole, the axial EPR spectrum changes to a mixture of a rhombic EPR spectrum with $g = 6.5$, 5.4, and 1.98, typical for ferric high-spin species, and a new EPR spectrum with $g = 2.51$, 2.25, and 1.84, typical signals for ferric low spin complex. Furthermore, addition of 1-methylimidazole gave the EPR spectrum predominantly of ferric bis-1-methylimidazole complex at $g = 2.61$,

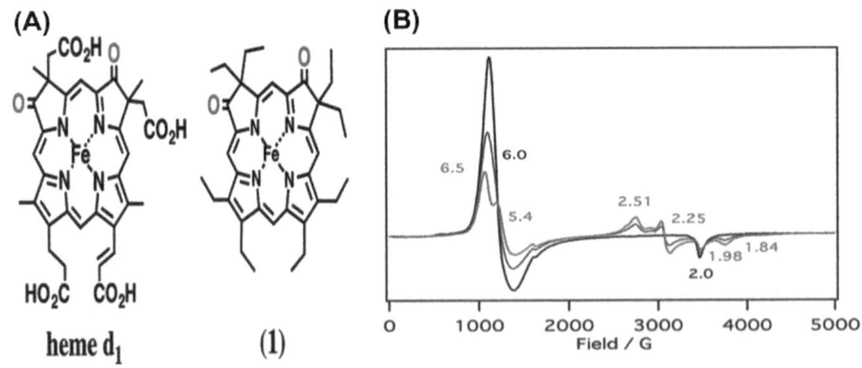

Figure 6.16 (A) Structure of heme d_1 and ferric porphyrin complex (1) and (B) electron paramagnetic resonance spectral change (4K) for the titration of the *p*-nitrophenolate complex of **1** with 1-methylimidazole in acetonitrile. *Black line* (1): 0 equiv., *gray line* (2): 0.5 equiv., and red (light gray in print versions) line (3): 1 equiv. *Reproduced with permission from Royal Society of Chemistry.*

2.34, and 1.66 indicating a mixture of small amount of ferric high-spin and largely low-spin states [18]. A similar EPR spectrum was also reported for a ferric porphyrin complex, but ferric low-spin EPR signals are quite of less intensity because of weaker binding of 1-methylimidazole. These EPR parameters are similar to those of the ferric resting state of the heme d_1 site of cytochrome cd_1 from *Paracoccus pantotrophus* ($g = 6.86$, 4.99, and 2.00 for the high-spin state and 2.52, 2.19, and 1.84 for the low-spin state).

Similar EPR spectra were observed for Fe^{3+} derivatives of hemoglobin in high- and low-spin spectroscopic states. The high-spin state Fe^{3+} shows strong features at $g = 6$ and a complimentary weak absorption at close to free spin, $g = 2$. For the low-spin form, three lines can be observed at g-values of 2.82, 2.20, and 1.67.

6.1.4.4 Electron Paramagnetic Resonance of Fe^{2+}, $Fe^{IV}{=}O$ and Iron Storage Proteins

Divalent Fe^{2+} is a d^6 configuration with two possibilities of spin states; a low spin $t_{2g}^6 e_g^0$ and a high-spin $t_{2g}^4 e_g^2$ depending on crystal field splitting energy of coordinated ligands. Their respective spin values are hence $S = 0$ and $S = 2$, both of which are not observable by normal X-band EPR. However, nitric oxide can be detected by EPR by binding to ferrohemoproteins to form Fe-N=O complex in which nitrogen p-orbital overlaps with the Fe d-orbital such that the unpaired electron is mainly localized on to the NO ligand.

The *tetravalent state* Fe^{IV} d^4, $t_{2g}^4 e_g^0$ has a spin value $S = 1$, which is again not detectable by EPR. This is the form of Fe in the oxocomplex of hemoglobin or other heme proteins after their reaction with hydrogen peroxide known as ferryl species. Proof of formation of the ferryl [$Fe^{IV}{=}O$] species by EPR spectroscopy was obtained from its one-electron reduction to the easily detectable Fe^{III} derivative.

Iron storage proteins, viz., ferritin and hemosiderin, have capacity for storage of iron (up to 500) in the form of Fe^{3+} with gated channels for release as per demand. The stored species is in a special form of Fe_2O_3, which in nature is the hard mineral oxide of iron. Because of a high concentration of Fe^{3+} in the storage "cluster," the EPR spectra of ferritin and hemosiderin are broad lines unlike those observed for dilute solutions of hemoproteins such as hemoglobin and myoglobin.

6.1.4.5 Electron Paramagnetic Resonance of Iron—Sulfur Clusters

Iron—sulfur [Fe-S] clusters are ubiquitous and evolutionary ancient prosthetic groups that are required to sustain fundamental life processes. Owing to their remarkable structural plasticity and versatile chemical/electronic features, Fe-S clusters participate in electron transfer, substrate binding/activation, iron—sulfur storage, regulation of gene expression, enzyme activity, and serve as sulfur donors during the synthesis of lipoic acid and biotin. The most common and simplest forms of Fe-S clusters are of the [2Fe-2S] and [4Fe-4S] type, but also [3Fe-4S] forms or more complex clusters containing additional heavy metal ions are known. The Fe ion of the cluster is typically coordinated by the sulfur of protein-bound cysteine residues or the nitrogen of histidine residues. The different types of iron—sulfur clusters and their characteristic EPR spectra are shown in Fig. 6.17. The basic active site structure of the cluster types is shown on the left column. The redox states and their respective spin states are indicated in the center column. The EPR spectra of the iron—sulfur clusters in ferredoxins from *Mastigocladus laminosus* [2Fe-2S] (a), *Desulfovibrio gigas* [3Fe-4S] (b), *Bacillus stearothermophilus* [4Fe-4S] (c), *Chromatium vinosum* high-potential iron—sulfur protein (HiPIP)[4Fe-4S] (d), and *Clostridium pasteurianum* 8Fe ferredoxin (e) are shown as follows [19].

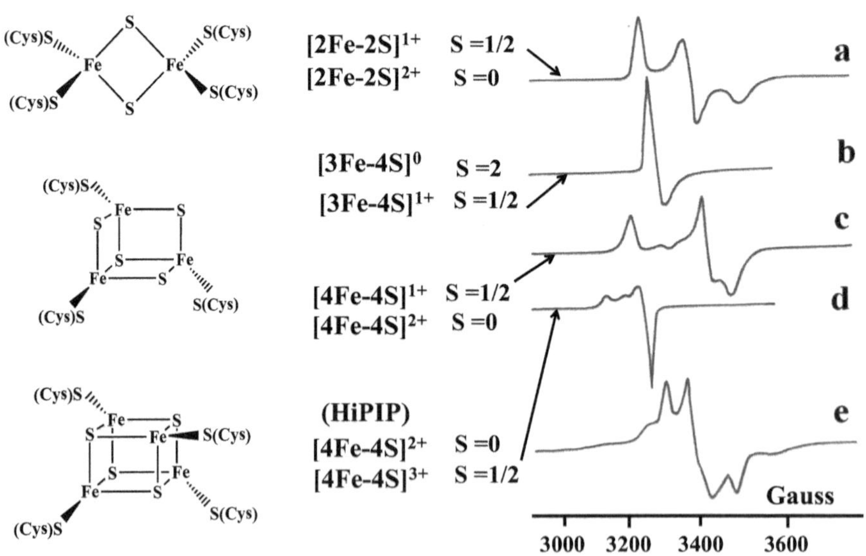

Figure 6.17 Structure and electron paramagnetic resonance of iron—sulfur clusters. *Reproduced with permission from Portland Press.*

6.1.5 Electron Paramagnetic Resonance of $Co^{2+/3+}$

The $3d^7$ electronic configuration of Co^{2+} exists either in low-spin or high-spin states as shown in Fig. 6.18. The existence of high-spin or low-spin ground state of Co(II) complex can be resolved with the aid of the EPR spectroscopy. EPR is also very sensitive technique to study the coordination of high-spin Co^{2+}, but fast spin-lattice relaxation often precludes the observation of EPR signals, especially those from octahedral Co^{2+}. Consequently, all our spectra were taken at very low temperatures (~10K) where the relaxation rates are sufficiently slow to allow a saturation of electron and nuclear spin transitions.

6.1.5.1 High-Spin Co^{2+} Electron Paramagnetic Resonance

For high spin, ^4F is the ground state for the free ion. A simplified energy level diagram of the high-spin cobalt(II) based on crystal field and symmetry considerations showing influence of a rhombic distortion (C_{2v}) and SOC is presented in Fig. 6.19. The cubic ligand field yields an orbitally degenerate $^4T_1$ state (derived from octahedral coordination) and the SOC effects further give rise to an orbitally nondegenerate $^4A_2$ depending on the actual ligand arrangement. The combined action of the symmetry distortion and the SOC lead to the ZFS of two Kramers doublets ($m_s = \pm 1/2$ and $m_s = \pm 3/2$). This effect is known as the ZFS $|\Delta|$ and is given by $|\Delta| = 2(D^2 + 3E^2)^{1/2}$ value, where D and E are the axial and rhombic ZFS parameters, respectively. The $^4A_2$ ground state with the quenched orbital momentum is typical for tetra- or pentacoordinate cobalt(II). Neglecting the rhombic deformation, the sign of Δ value determines whether or not the $m_s = \pm 1/2$ or $m_s = \pm 3/2$ state is the lowest. In the presence of magnetic field B, two degenerate spin doublets are split, and

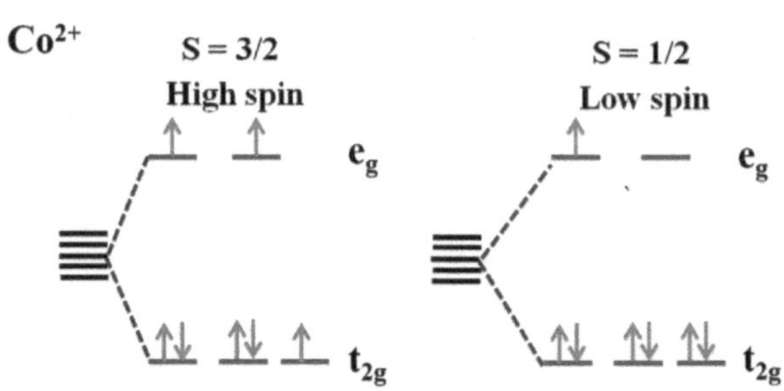

Figure 6.18 High- and low-spin electronic configurations of Co^{2+}.

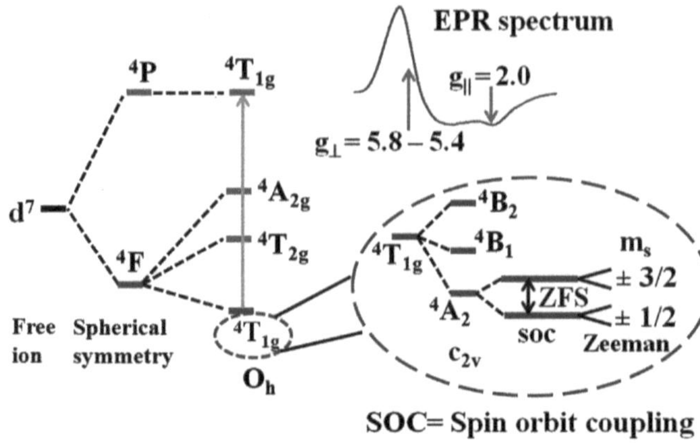

Figure 6.19 Energy level splitting of Co^{2+} high spin (left) and its electron paramagnetic resonance (EPR) spectrum (top right) at 4K. *Reproduced with permission from American Chemical Society.*

the incident microwave radiation will induce transitions according to the selection rule $\Delta m_s = \pm 1$. For positive values of Δ, when $|\Delta| \gg h\nu$, the $m_s = \pm 1/2$ or $m_s = \pm 3/2$ states become well separated, and the Co(II) paramagnet can be described by an effective spin $S_{eff} = 1/2$. However, the low-lying excited level $m_s = \pm 3/2$ leads to fast relaxation; therefore, such spectra can be observed exclusively at liquid helium temperatures, allowing for simple distinction between the high- and low-spin states of the investigated complex [20].

6.1.5.1.1 Electron Paramagnetic Resonance Spectrum of $Co(acac)_2$ Complex

The X-band EPR spectrum of the $Co(acac)_2$ complex recorded at 5.8K in a toluene/ethanol glass is shown in Fig. 6.19. The spectrum exhibits two broad features with apparent $g_{\perp} \approx 5.8-5.4$ and $g_{||} \approx 2.0$, due to the resonances occurring when the magnetic field is coincident with the perpendicular (stronger feature) and parallel (weaker feature) directions of the complex. However, because of the large width of the EPR signal spanning 350 mT, accurate spin Hamiltonian parameters cannot be extracted from such a powder spectrum recorded at X-band only. The effective g-values are in agreement with those of other molecular complexes of cobalt(II) with oxygen-donor ligands. There is no evidence of even poorly resolved hyperfine structure because of the ^{59}Co ($I = 7/2$, 100%) nucleus, in contrast to low-spin complexes, where a hyperfine pattern is often observed. The EPR signal was detectable only below 40K, implying a clear high-spin

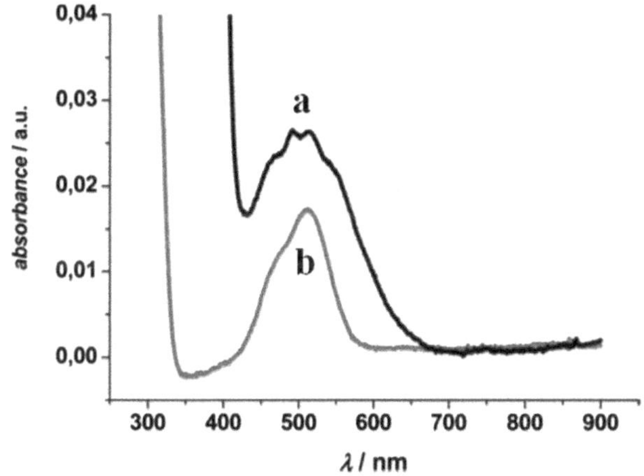

Figure 6.20 Optical spectra of the (A) $Co(acac)_2$ complex and (B) $[Co(H_2O)_6]^{2+}$. *Reproduced with permission from American Chemical Society.*

quartet state, i.e., S = 3/2, of the $Co(acac)_2$ complex. The observed spectrum can be rationalized in terms of an axial spin Hamiltonian:

$$\widehat{H}_s = \mu_B B\widehat{g}S + D\left[S_z^2 - S(S+1)/3\right] \quad \text{with } S = 3/2 \tag{6.25}$$

6.1.5.1.2 Optical Spectra of $Co(acac)_2$ Complex

The electronic UV-vis absorption spectrum of the complex in solution compared with that containing cobalt(II) nitrate in water is shown in Fig. 6.20. The UV-vis signal of the complex is characteristic of a hexacoordinated $Co(H_2O)_6{}^{2+}$ cation. The spectrum is centered at 515 nm because of the $^4T_{1g}(F) \rightarrow {}^4T_{1g}$ (P) transition (Fig. 6.19) with a shoulder at 470 nm resulting from the SOC in the excited $^4T_{1g}$(P) state that lifts its degeneracy [20]. Additional slight splitting of the bands is associated with lowering of the local OH symmetry in the $Co(acac)_2(EtOH)_2$ complex to C_{2v} (Fig. 6.20).

6.1.5.2 Low-Spin Co^{2+} Electron Paramagnetic Resonance

The EPR of low-spin Co(II) complexes with S = 1/2 is illustrated with examples as follows

6.1.5.2.1 Electron Paramagnetic Resonance of Co(II)OEP

The EPR spectra (X-band) of Co(II)OEP (OEP = octaethyl porphyrinato) with (A) ^{14}N pyridine and (B) ^{15}N pyridine as axial ligands, respectively, in chloroform at T = 140K (frozen solution) as shown in Fig. 6.21. The cobalt

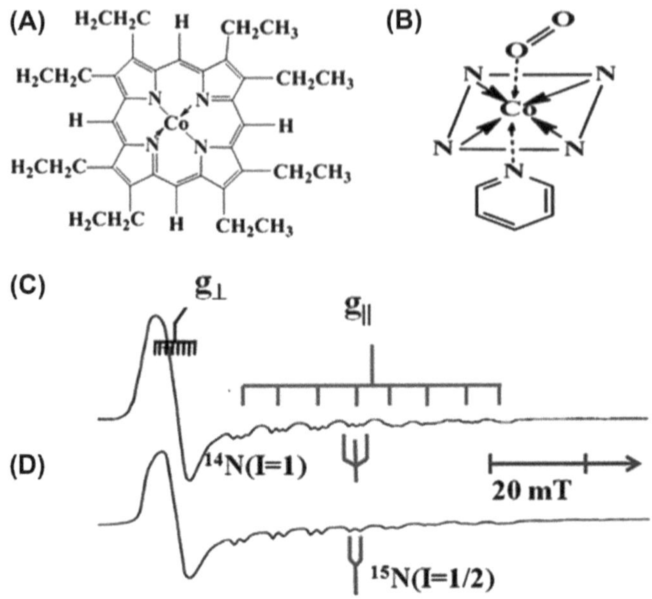

Figure 6.21 Structures (A & B) and Electron paramagnetic resonance spectra of Co(II) OEP (C) $I(^{14}N) = 1$ and (D) $I(^{15}N) = 1/2$. *Reproduced with permission from Springer-Verlag.*

hf structure, $I(^{59}Co = 7/2)$, is only well resolved as eight-line hf pattern for $g_{||}$. The superhyperfine (shf) splitting in this region arises from the coupling of the unpaired electron at the Co(II) with one nitrogen atom, (A) $I(^{14}N) = 1$ and (B) $I(^{15}N) = 1/2$. Coordination of only one pyridine molecule to the Co(II) ion is evident from the three-line ^{14}N (^{14}N $I = 1$) shf pattern centered at each of the $A_{||}$ (^{59}Co) components. Consequently, the use of ^{15}N pyridine (^{15}N $I = 1/2$) leads to an shf splitting, resulting in two components with equal intensity [21].

6.1.5.2.2 Electron Paramagnetic Resonance of Vitamin B_{12}

A low-spin $S = 1/2$ state is also likely to occur when Co^{2+} ions are in a square planar or square-pyramidal environment such as $\{CoN_4\}$ or $\{CoN_2O_2\}$. Two possible ground states, $^2A_2$ or $^2A_1$, associated with d_{yz} and d_z^2 semioccupied molecular orbitals, respectively, can be distinguished based on the analysis of the g and A tensor anisotropies. The values of $g_z \sim 2.0 < g_{x,y}$ and $A_z > A_{x,y}$ are indicative for the $^2A_1$ state, whereas the opposite ordering is valid for the $^2A_2$ ground state. Owing to the usually well-separated energy levels, the EPR spectra of the low-spin Co(II) can be readily observed at liquid nitrogen temperature and are characterized

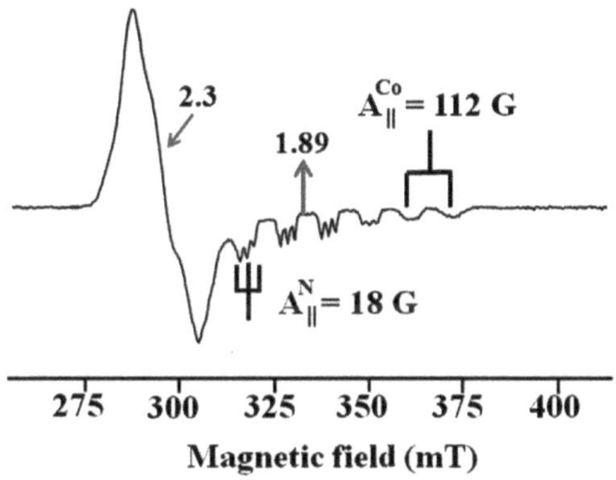

Figure 6.22 X-band electron paramagnetic resonance spectrum of vitamin B_{12}. *Reproduced with permission from Royal Society of Chemistry.*

by an anisotropy smaller than that characteristic of the S = 3/2 species. The cobalt hf structure, $I(^{59}Co) = 7/2$, is only well resolved as eight-line hf pattern for $g_{||}$. Vit B_{12} also exhibits shf nitrogen splitting from axial adenosyl (Fig. 6.22) [22].

6.1.5.3 Electron Paramagnetic Resonance of Tetrahedral Co^{2+}

A comparison of energy level splitting of Co^{2+} in tetrahedral and octahedral ligand fields is shown in Fig. 6.23. Due to ZFS and distortion in tetrahedron, the $^4A_2$ ground state is split into two Kramer's doublets ($m_s = \pm 1/2$ and $\pm 3/2$), which are degenerated in the absence of a magnetic field B (Fig. 6.24). The corresponding energy splitting between the two doublets

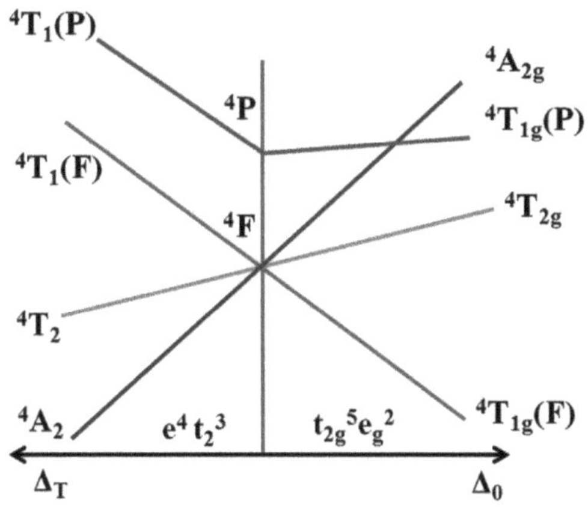

Figure 6.23 Energy level diagram of Co^{2+} in a tetrahedral and octahedral ligand field.

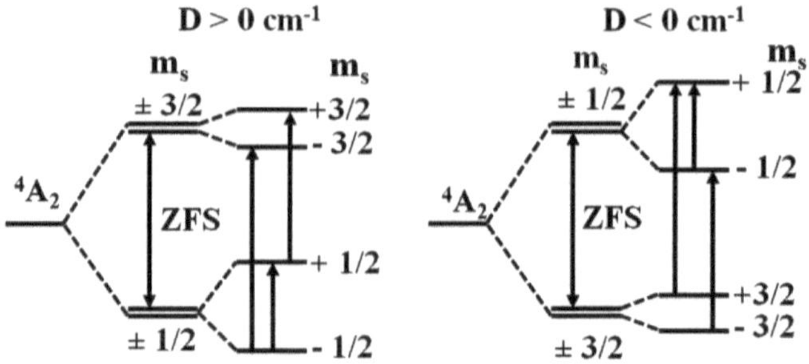

Figure 6.24 The possible electron paramagnetic resonance transitions for tetrahedral Co^{2+}. *ZFS*, zero-field splitting.

is given by $2(D^2+3E^2)^{1/2}$, with D and E the ZFS parameters. In the case of axial distortions ($E = 0$), the energy difference between the two doublets equals 2D. The sign of the ZFS, i.e., 2D, is determined by whether the $m_s = \pm 1/2$ or $m_s = \pm 3/2$ doublet lies lowest. By convention, the value of D is positive when $m_s = \pm 1/2$ is lowest and negative when $m_s = \pm 3/2$ is lowest. Tetrahedral D values are usually between -10 and $10\ cm^{-1}$. In the presence of a magnetic field B, the two degenerate doublets $m_s = \pm 1/2$ and $\pm 3/2$ are split up and an incident microwave will induce transitions according to the selection rule $\Delta m_s = \pm 1$. The three possible transitions, giving rise to an EPR signal, are included in Fig. 6.24. In the case of (distorted) tetrahedral Co^{2+} high spin ($S = 3/2$), typical g-values around 1.77, 2.62, and 5.82 are observed below 77K [23].

6.1.6 Electron Paramagnetic Resonance of $Ni^{2+/3+/1+}$

In square planar complexes, $Ni^{2+}(d^8)$ has no unpaired electron (square planer) in its orbitals in Fig. 6.25. Therefore, it does not exhibit EPR signal at room temperature. However, in Ni(II) octahedral or elongated complexes, there are two unpaired electrons ($S = 1$), which have large ZFS and hence shows no EPR spectra (Fig. 6.25A). However, under favorable conditions, it shows one EPR signal (Fig. 6.25A) [24,25]. The EPR data could be related with the optical data by the following equation $g = 2.0023 - 8\lambda/\Delta$, where Δ is the energy of the transition of the perfect octahedral site. λ is $324\ cm^{-1}$ for free Ni^{2+} ion.

For example, the anisotropic EPR spectrum of the Ni(II) complex in a glassy matrix at 115K in CH_2Cl_2 shows g_1, g_2, and g_3 along the three principal axes orientations in Fig. 6.26.

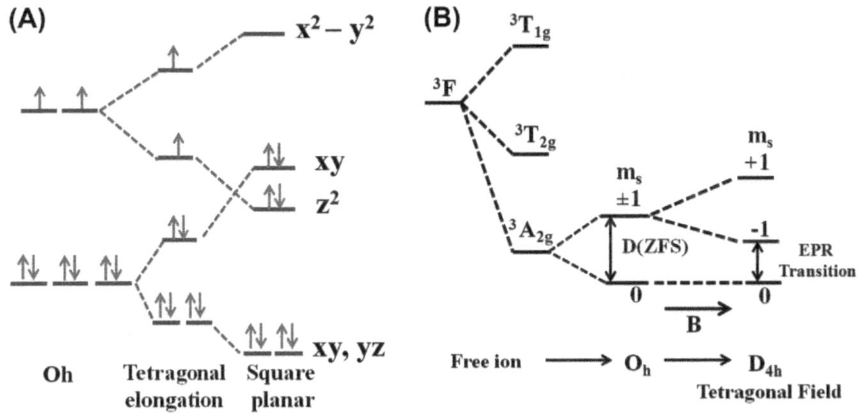

Figure 6.25 (A) Energy level splitting of Ni(II) and (B) electron paramagnetic resonance transition.

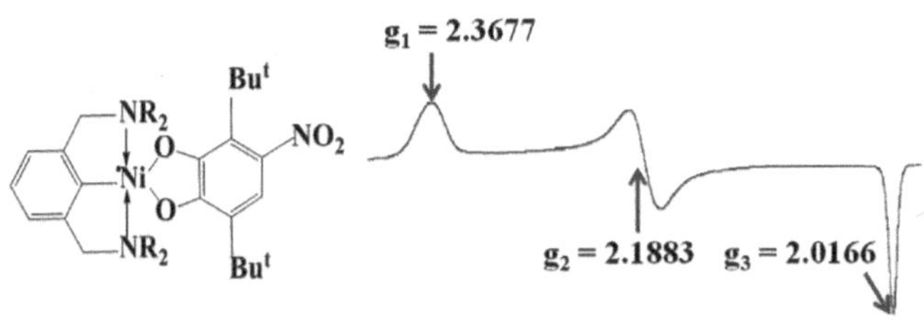

Figure 6.26 Ni(II) complex structure and its anisotropic electron paramagnetic resonance spectrum at 115K in toluene.

6.1.6.1 *Electron Paramagnetic Resonance of $Ni^{3+/1+}$*

Ni^{3+} complex (low spin, $S = 1/2$) with d^7 configuration exhibits a single line similar to Co^{2+} for isotropic case. However, $g_{||}$ and $g_{\perp}$ features are observed with axially compressed geometry due to JT effect in the powder/frozen state as shown in Fig. 6.27. For example, the EPR spectrum of $[Ni(cyclam)]^{3+}$ is shown in Fig. 6.27 [25]. On the other hand, Ni(I) with d^9 contains one unpaired electron similar to Cu(II). For example, the EPR spectrum of Ni(I) complex characteristic of $g_{||} > g_{\perp}$ is shown in Fig. 6.28.

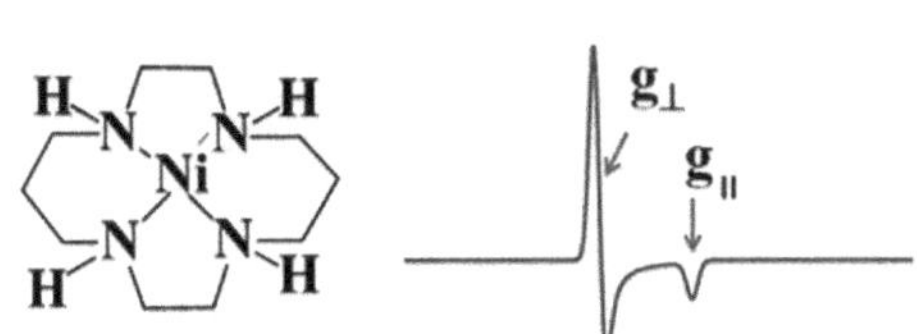

Figure 6.27 Electron paramagnetic resonance spectrum of $[Ni(cyclam)]^{3+}$.

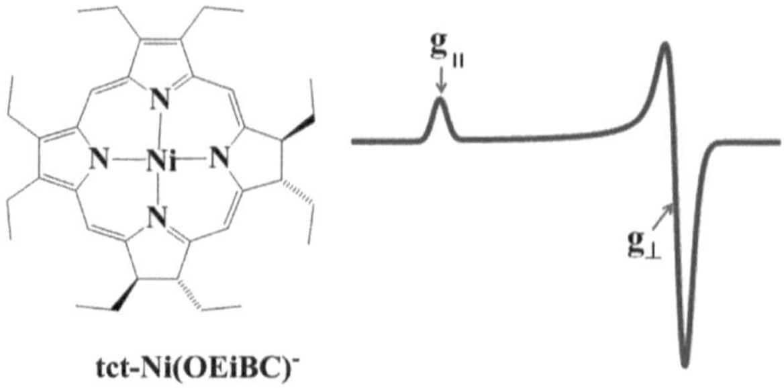

Figure 6.28 The structure of Ni(I) complex and its electron paramagnetic resonance spectrum.

6.1.6.2 Electron Paramagnetic Resonance of NiSOD (Ni^{3+})

NiSOD in oxidized state contain Ni^{3+} in its active site with square pyramidal geometry (Fig. 6.29). Addition of an axial histidine ligand to the oxidized active site tunes the reduction potential for superoxide dismutation. The axial His1 (ND1) ligand exhibits a much longer bond length to Ni^{3+} than typical Ni-imidazole molecules or metalloproteins. This axial His ligand explains the EPR three-line hyperfine splitting pattern (Fig. 6.29) characteristic of ^{14}N (nuclear spin I = 1) interactions with an unpaired electron in the d_z^2 orbital. EPR spectrum observed as shown in Fig. 6.29 suggests that *Streptomyces seoulensis* NiSOD contains oxidized Ni^{3+} with an axial nitrogen ligand. The oxidized samples exhibit g-values of 2.30, 2.22, and 2.01. EPR spectra were collected at 55K using a microwave frequency of 9.678 GHz, microwave power of 10 mW, and modulation amplitude of 5 G. The low-spin (S = 1/2) Ni(III) center in NiSOD (ox) displays a rhombic EPR spectrum with principal g-values of 2.30, 2.24, and 2.01 with a distinct triplet shf splitting pattern (A_{zz} = 24.9 G) in the g_z component consistent with axial N-ligation (I = 1) [26].

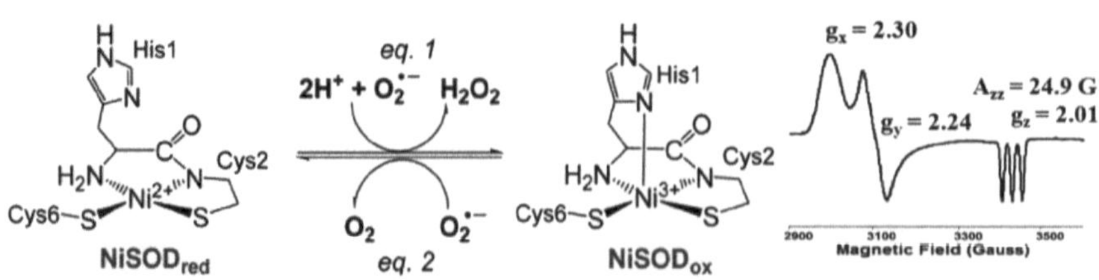

Figure 6.29 Mechanism of NiSOD and its X-band electron paramagnetic resonance spectrum. *Reproduced with permission from American Chemical Society.*

6.1.7 Electron Paramagnetic Resonance of Cu^{2+} and Proteins

Copper exists in two ionic states viz., Cu(I) and Cu(II). The electronic configuration of Cu(I) is [Ar] $3d^{10}$ and has no unpaired electron in its outermost orbit. Hence, it exhibits diamagnetism and is EPR silent. The Cu(II), d^9 species contain an odd number of electron to give a spin $S = 1/2$ and therefore only one allowed EPR transition. Due to its nuclear magnetic moment $I = 3/2$ for ^{63}Cu(II) (abundance 69%), the complex gives a spectrum comprising a quartet of lines as shown in Fig. 6.30. There are different forms of EPR spectra of Cu(II):

1. Axial (elongated/compressed) spectra
   Fig. 6.30A shows the axial elongated EPR spectrum ($g_{||} > g_{\perp}$) of Cu(II) complex for square planar (or) elongated octahedron coordination geometry. The quartet of lines arising from $S = 1/2$, $M_I = 3/2$, is generally resolved only for the ($g_{||}$) parallel features (Fig. 6.30A), which come at low-magnetic fields (high g-values). The perpendicular feature ($g_{\perp}$) unresolved with an intense peak representative of a combination of the x and y components (Fig. 6.30). For axial compressed system ($g_{\perp} > g_{||}$), the Cu(II) hyperfine is resolved in $g_{||}$ and not resolved in $g_{\perp}$ similar to the elongated.
2. Orthorhombic
   For low symmetry, the Cu(II) exhibits rhombic EPR spectrum with $g_1 \neq g_2 \neq g_3$ as shown in Fig. 6.30C.

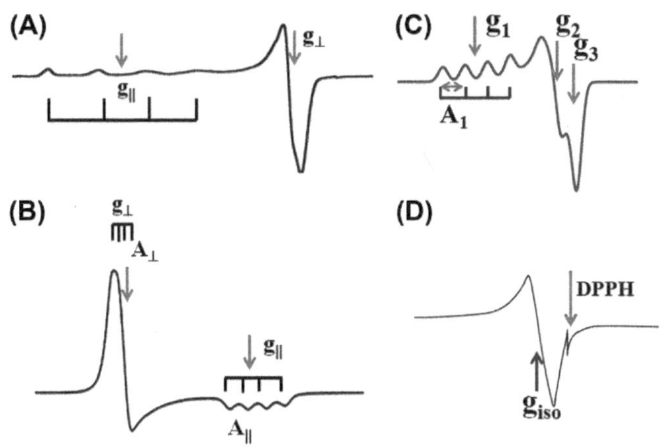

Figure 6.30 Electron paramagnetic resonance spectra of Cu(II) complex (A) axial elongated, (B) compressed symmetry, (C) orthorhombic, and (D) exchange narrowed (no hyperfine splitting). *Courtesy Dr. C. Karunakaran.*

3. Exchange-narrowed spectra
In the case of concentrated samples where exchange interaction occurs between the neighboring Cu(II) spins, no hyperfine and shf features from the Cu(II) and ligand nitrogen were resolved in the undiluted powder sample. For example, the X band spectra of a Cu(II) complex at both 300 and 77K exhibit a single exchange-narrowed signal centered at $g = 2.076$ as shown in Fig. 6.30D. However, there is line broadening caused by the anisotropy of the g factors and hyperfine interactions and by the interaction between the neighboring Cu(II) spins in powder samples [27,28].

6.1.7.1 Estimation of the Ground State Wave Function From Electron Paramagnetic Resonance Data

The ground state wave function for Cu^{2+} complex is given by

$$\alpha|x^2 - y^2\rangle + \beta|3z^2 - r^2\rangle \tag{6.26}$$

where α and β are mixture coefficients, α is very near to unity and β is less than unity. The expressions used for calculation of various bonding parameters, viz. α, β, κ, and P, are as follows in terms of spin Hamiltonian parameters. κ is the Fermi contact term or spin exchange polarization parameter and P is the hyperfine interaction parameter. From the below equation, the β value may be obtained.

$$\delta g = g_y - g_x = -2\sqrt{3}\beta(\Delta g_x + \Delta g_y) \tag{6.27}$$

The hyperfine interaction parameter P is given by

$$P = 2\gamma_{Cu}\mu_B\mu_N\langle r^{-3}\rangle \tag{6.28}$$

where γ_{Cu} is the magnetogyric ratio of copper nucleus, μ_B is the Bohr magneton, μ_N is the nuclear magneton, and r is the distance of the electron from the central nucleus. The measured value of P_{exp} can be obtained by multiplying P_{fi} (fi = free ion) by the probability of the Cu^{2+} electrons actually being in the ^2D ground state. The parameter P_{exp} can be used to determine the delocalization of the d-electron of the magnetic Cu(II) ion in the host lattice.

$$P_{exp} = \alpha^2 P_{fi} \tag{6.29}$$

The Fermi contact term κ is calculated using the following expression

$$\kappa = [(A_x + A_y + A_z)/3P] + [\{(g_x + g_y + g_z)/3\} - g_e] \tag{6.30}$$

g_e is the free electron g-value (2.0023). In many cases, the ground state is not purely $|x^2 - y^2|$. As the value of α is less than unity the Cu^{2+} ion is said to have undergone a distortion of rhombic or still lower symmetry. For example, in Cu^{2+}: DHMS, the α, β, and κ values were calculated to be 0.998, −0.061, and 0.327, respectively, using the spin Hamiltonian parameters g- and A values. The evaluated value of P_{exp} is 291×10^{-4} cm^{-1}, whereas its free ion value is 360×10^{-4} cm^{-1}. The percentage of the unpaired spin density on the copper ion can be measured by the ratio of P_{exp} to P_{fi} [=P (complex)/P (free ion)], which is evaluated as 81%. Therefore, the unpaired spin density distributed on the ligands in the complex is 19%.

Furthermore, if the ratio $R = (g_y - g_x)/(g_z - g_y)$ is less than unity, the unpaired electron exists predominantly in the $d_x^2 - y^2$ state, and if R is greater than unity, the unpaired electron exists in the $d_{3z}^2 - r^2$ state. The calculated value of R in Cu^{2+}: DHMS is 0.071, i.e., R <1; therefore, the ground state of the unpaired electron in the complex is $d_x^2 - y^2$. In fact, when the site symmetry is rhombic or lower, the ground state will be neither $d_x^2 - y^2$ nor $d_{3z}^2 - r^2$ but an admixture of both. The value of α^2 in the complex tells us whether the bonding is partly covalent or ionic in nature. If $\alpha^2 = 0.5$, the bond would be completely covalent. The smaller the value of α^2, the greater is the covalent nature of the bond [29].

6.1.7.2 Relation Between Electron Paramagnetic Resonance and Optical Absorption Spectra

The optical absorption and EPR data are related as follows. In tetragonal symmetry (Fig. 6.31), EPR studies are correlated with optical data to obtain the orbital reduction parameter [29].

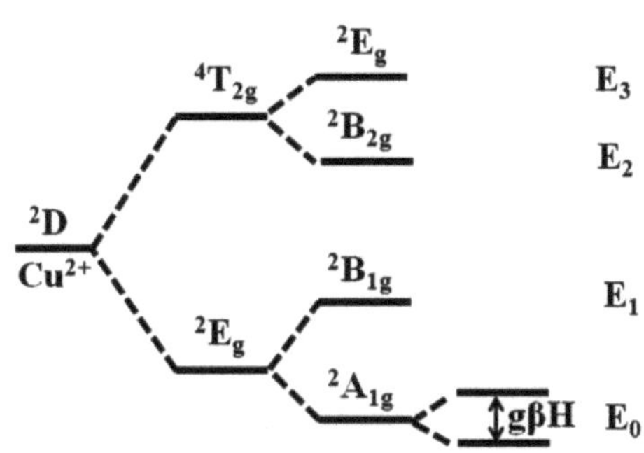

Figure 6.31 The energy level splitting based on consideration of molecular orbital theory.

For a *tetragonal compression*, the ground state is $^2A_{1g}$ (d_z^2).

$$\begin{aligned} g_{\|} &= 2.0023 = g_e \\ g_{\perp} &= g_e - (6\lambda/\Delta E_{\perp}[^2B_1 \rightarrow {}^2E]) \end{aligned} \tag{6.31}$$

λ is the SOC constant for free Cu(II) ion $= -830\ cm^{-1}$.

For a *tetragonal elongation*, the ground state is $^2B_{1g}$ ($d_x^2 - y^2$).

$$\begin{aligned} g_{\|} &= g_e - \left(8\lambda/\Delta E_{\|}[^2B_1 \rightarrow {}^2B_2]\right) \\ g_{\perp} &= g_e - (2\lambda/\Delta E_{\perp}[^2B_1 \rightarrow {}^2E]) \end{aligned} \tag{6.32}$$

Furthermore, if $A_{\|}$, $g_{\|}$, and $g_{\perp}$ values are known, α^2 can be estimated using the equation:

$$\alpha^2 = -\left[(A_{\|}/0.036) - (g_{\|} - g_e) + \{3/7\}(g_{\perp} - g_e) + 0.04\right] \tag{6.33}$$

6.1.7.3 Electron Paramagnetic Resonance of Copper Proteins

There are different copper centers identified in biological system, as *type* 1, *type* 2 (Fig. 6.32), and *type* 3 by virtue of respective ligand environments from their unique optical and EPR parameters. The *type* 1 has an intense blue coloration due to a high extinction coefficient in its electronic absorption spectrum and lower hyperfine splitting constants (A-values). However, the *type* 2 has a weaker absorption spectrum and higher hyperfine splitting constants (A-values) in its EPR spectrum than type 1 (Table 6.4). *Type* 3 consists of pairs of copper so close that there is antiferromagnetic coupling

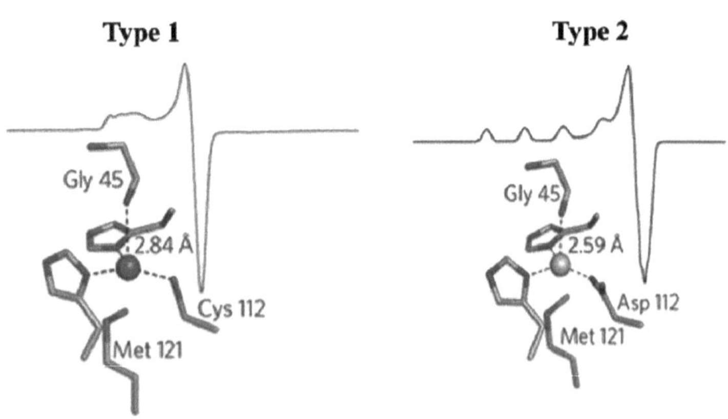

Figure 6.32 Active site geometries and electron paramagnetic resonance spectra of Type 1 and 2 Cu(II) proteins. *Reproduced with permission from Elsevier.*

Table 6.4 Electron paramagnetic resonance (EPR) and optical behavior of Cu sites

Copper center	Examples	Biological role	EPR characteristics	UV-vis spectra
Type 1 (mononuclear)	Blue copper proteins (e.g., azurin, amicyanin) ceruloplasmin lactase ascorbate oxidase, nitrite reductase Fe(III)	Electron transport	Small hyperfine splitting (<100 cm^{-1})	Intense absorption ($\varepsilon = \sim 3-6\ mM^{-1}\ cm^{-1}$) at ~600 nm. LMCT (S → Cu II)
Type 2 (mononuclear)	Amine oxidase galactose oxidase, superoxide dismutase, cytochrome *c* oxidase (Cu_B) nitrite reductase	Catalysis (i.e., substrate oxidation)	Normal hyperfine splitting (100–200 cm^{-1})	Forbidden d–d transition at ~600 nm (low intensity)
Type 3 (dinuclear)	Hemocyanin tyrosinase	O_2 transport and activation	No EPR signal	Intense absorption ($\varepsilon = \sim 20\ mM^{-1}\ cm^{-1}$) at ~ 330 nm. LMCT ($O_2^{2-} \rightarrow Cu$)
Trinuclear Cu site (trinuclear)	Ascorbate oxidase lactase ceruloplasmin Fet3	Catalysis (O_2 reduction to H_2O)	"Normal" hyperfine splitting (100–200 cm^{-1}) from only one of three copper ions	Intense absorption ($\varepsilon = \sim 20\ mM^{-1}\ cm^{-1}$) at ~330 nm. LMCT ($O_2^{2-} \rightarrow Cu$)
Cu_A (Dinuclear)	Nitrous oxide reductase, cytochrome *c* oxidase	Electron transport	Seven-line pattern with a very small hyperfine splitting due to a mixed valence copper pair	Three absorption bands between 360 and 535 nm and an absorption band around 790 nm
Metallothionein-like (mono to multinuclear)	Metallothionein CUP2, Cu transport ATPase	Regulation, storage, and transport of Cu	No oxidized form, thus no EPR	No oxidized form, thus no UV-vis

LMCT, Ligand-to-Metal Charge Transfer

and therefore not observable by EPR spectroscopy. This is best exemplified by the respiratory protein hemocyanin in molluscs and arthropods consisting two coppers per oxygen binding site. It undergoes reversible redox reactions during which the two coppers cycle between Cu(I) and Cu(II) states simultaneously. In the oxygenated state both coppers are in Cu(II) and there is considerable overlap of d-orbitals resulting in strong antiferromagnetic coupling between the unpaired electrons in those orbitals. The effective spin is thus zero and no EPR transition observed. On selective reduction of one of the coppers a mixed valence state is obtained allowing detection of the Cu(II) in the presence of Cu(I) [30—32].

6.1.7.4 Jahn—Teller Coupling

As discussed in Chapter 4, complexes with partial occupancy of the e_g orbitals can often exhibit significant strong Jahn—Teller distortions, especially octahedral copper(II) complexes in d^9 electronic configuration. In these complexes, the Jahn—Teller theorem predicts that in a perfect octahedral environment, the degeneracy of the ^2e_g ground state will be split, reflecting a concomitant elongation or compression of the Cu-ligand bonds parallel to one of the three molecular axes. The JT distortion arises due to the interaction of doublet electronic states with one or more normal modes of vibration called *electron—photon coupling*. The energy level diagram based on crystal field theory in which the bonding considered as electrostatic (ionic) is shown in Fig. 6.33.

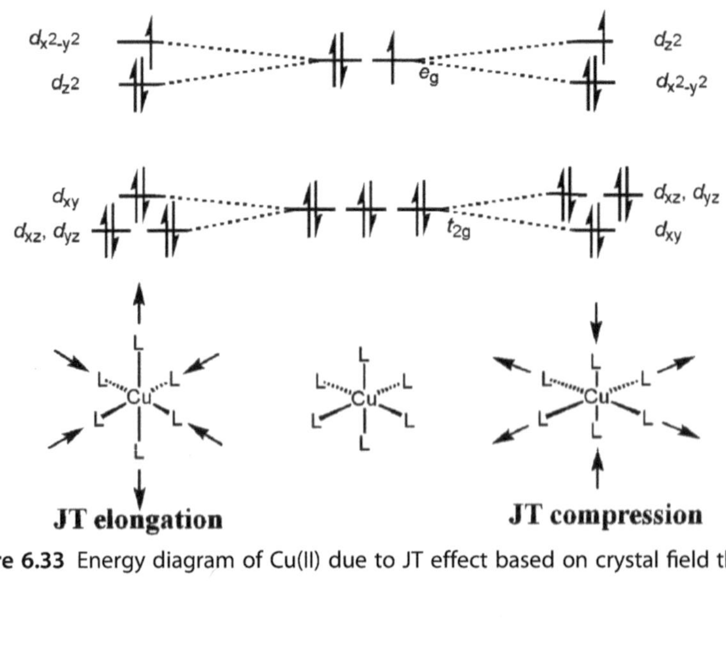

Figure 6.33 Energy diagram of Cu(II) due to JT effect based on crystal field theory.

In principle, the detection of Jahn–Teller distortions in octahedral Cu^{II} complexes may simply be achieved by determining the metal–ligand distances by X-ray crystallography. In general, the long and short Cu-ligand bonds usually differ in length by >20%. However, these bond lengths are ambiguous, owing to the static and/or dynamic disorder. Hence, EPR spectroscopy is an appropriate tool to determine the nature of the Jahn–Teller distortion because the distortion (elongation or compression) dictates the electronic ground state, with the unpaired electron occupying either $3d_x^2 - y^2$ or $3d_z^2$ orbital, respectively. These two configurations yield very different EPR spectra (Fig. 6.30A and B), whose lineshapes are diagnostic of the identity of the singly occupied d-orbital in the molecule [31–33].

Case (*i*) The unpaired electron in $\{3d_{x^2-y^2}\}^1$ complex, a $g_1 > g_2 \geq g_3 > 2.00$ or ($g_{||} > g_{\perp}$), pattern is expected. Examples are as follows:

1. In $[Cu(hfacac)_2(bipy)]$, there are two short CuO bonds and two long Cu–O bonds (where hfacac = hexafluoroacetylacetone anion; bipy = 2,2′bipyridine). It is an example for Z-out distortion.
2. Cu(II) cannot form $[Cu(en)_3]^{2+}$ because JT distortion brings strain into the ethylene diamine molecule that is added along z-axis. Hence only $[Cu(en)_2(H_2O)_2]^{2+}$ is formed.

Case (*ii*) The unpaired electron in $\{3d_z^2\}^1$ complex yields $g_1 \geq g_2 > g_3 \approx 2.00$ or ($g_{\perp} > g_{||}$). For example, in the crystalline $KCuF_3$, the two CuF distances are at 1.96 Å and the remaining four Cu–F distances are at 2.07 Å. It exhibits Z- in compression.

6.1.7.5 Static and Dynamic Jahn–Teller Distortions

6.1.7.5.1 Static Jahn–Teller Distortion

Certain molecules exhibit tetragonal geometry under all conditions, i.e., in solid and in solution states, at low and high temperatures, i.e., temperature independent. This phenomenon is known as static Jahn–Teller distortion. It is observed when the degeneracy occurs in e_g orbitals. Here the distortion is strong and permanent, for example, $[Cu(en)_2(H_2O)_2]^{2+}$.

6.1.7.5.2 Dynamic Jahn–Teller Distortion

For some molecules, the distortion is not seen either because of random movements of bonds, which does not allow the measurement within a time frame, or because the distortion is so weak as to be negligible. However, the distortion can be observed by freezing the molecule at lower temperatures, i.e., temperature dependent. This condition is known as dynamic Jahn—Teller distortion. For example,

1. The complexes of the type $M_2PbCu(NO_2)_6$ show dynamic Jahn—Teller distortion (M = K, Rb, Cs, Tl). They exhibit tetragonal symmetry at lower temperatures due to static Jahn—Teller distortion. But at higher temperatures, these complexes appear octahedral geometry due to the dynamic Jahn—Teller effect.
2. $[Fe(H_2O)_6]^{2+}$ shows dynamic Jahn—Teller distortion and appears to be octahedral. In this complex, Fe^{2+} is in high-spin d^6 system with $t_{2g}^4e_g^2$ configuration. Here, the distortion is small because the degeneracy occurs only in t_{2g} orbitals.

6.1.7.6 Temperature Dependence of Jahn—Teller Effect

The dynamic JT effect can be explained by the three minima in the warped hat potential corresponding to three tetragonal distortions of the cube along its three axes (either elongation or contraction).

At low temperature, the barriers between the wells are large resulting in three different distortions. When a spin resonance experiment is done at this temperature, spectra from all the three types of distortions along x, y, and z axes are seen simultaneously (as if there were three crystallographically equivalent sites). When the temperature is raised with three equivalent wells, thermal averaging of the three spectra leads to dynamic JT giving a single EPR resonance. It is explained because of the barriers between the wells are small enough causing quantum mechanical tunneling between them.

For example, the dynamic JT effect can be studied from the EPR spectra of Cu^{2+}-doped DHMS at different temperatures as shown in Fig. 6.34. At low temperature, it exhibits four hyperfine lines characteristic of Cu(II). As the temperature is raised above 295K, the hyperfine lines of Cu^{2+} become broader through relaxation effects because spin relaxation time (T_1) decreases with the temperature increase. At temperatures of 300 and 320K, the hyperfine lines of Cu^{2+} are so broad that they are no longer resolved and only one broad line is observed. This is characteristic of a dynamic JT effect. So, this dynamic JT occurs when the rate of tunneling through the barrier from one distorted configuration of the Cu^{2+} complex (ML_6 type) to the other exceeds the frequency difference between the corresponding EPR resonance lines for the different distorted configurations. Because of the time-averaging effect, the equations used by Abragam and Bleaney

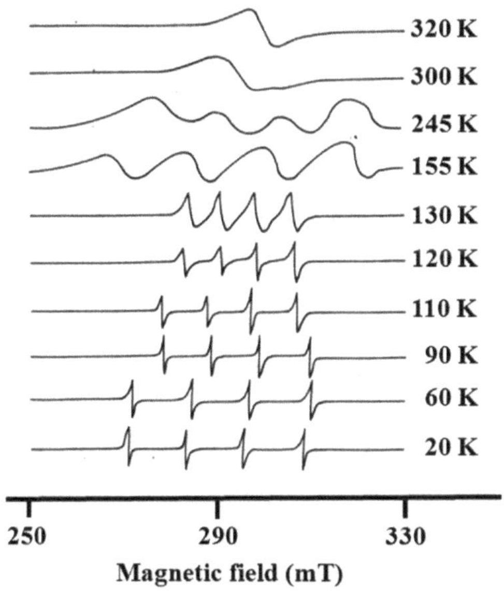

Figure 6.34 Electron paramagnetic resonance spectrum of Cu^{2+}: DHMS (diammonium hexaaqua magnesium sulfate) single crystal recorded in the c*a plane for B at 0 degree from the c*-axis at different temperatures. *Reproduced with permission from Springer.*

become $g_x = g_y = g_z = g_e - 4\lambda/\Delta$. Using cubic crystal field parameter (Dq) at room temperature (=1527 cm^{-1}) found from the optical study of Cu^{2+}: DHMS, the average value of g ($=g_e - 4\lambda/\Delta$) is evaluated as 2.219, which is close to the observed value (=2.204) (obtained from the recorded spectra) [29].

6.1.7.7 Bis-Salicylaldimine-Copper(II)

The EPR spectrum of bis-salicylaldimine-copper(II) is an interesting example summarizing both the hyperfine and shf features. It shows four main groups of lines resulting from coupling of the Cu^{63} nucleus (I = 3/2) with the unpaired electron. The hyperfine structure in each of the four groups further splits into ***11*** peaks of intensity ratio 1:2:3:4:5:6:5:4:3:2:1 in Fig. 6.35. These peaks result from splitting by the two equivalent nitrogens and two hydrogens, H′ in the inset of Fig. 6.35. Theoretically, the total number of peaks expected is 15, $(2\,n_N\,I_N + 1)\,(2\,n_H\,I_H + 1) = 5 \times 3 = 15$. However, only 11 peaks observed for each subgroup in the actual spectrum perhaps result from overlap of some of the 15 peaks as shown in the stick diagram in Fig. 6.35.

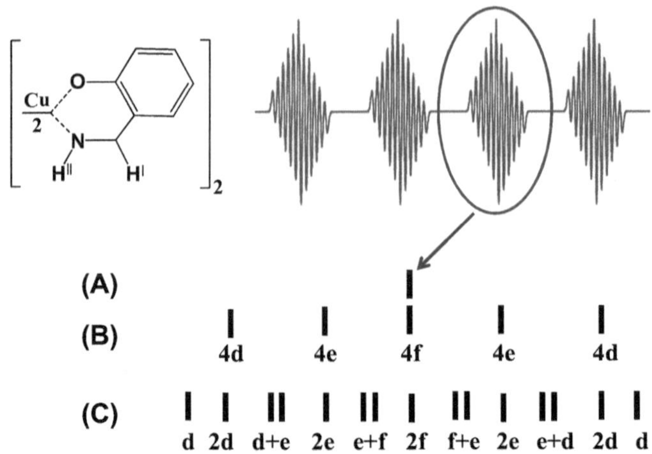

Figure 6.35 Electron paramagnetic resonance spectrum (top right) of bis-salicylaldimine-copper(II) and its stick diagram (A, B, and C).

The splitting by the two equivalent nitrogens is indicated in Fig. 6.35B and the further splitting by two equivalent protons is represented in Fig. 6.35C. Initially, the two nitrogen split the resonance into five peaks $[(2\ n_N\ I_N + 1) = 5]$ of relative intensity 1:2:3:2:1. These values are given in Fig. 6.35B by d, e, and f where the intensities correspond to d = 1, e = 2, and f = 3. Subsequently, the splitting by two equivalent protons will give rise to three lines for each line in Fig. 6.35B, with an intensity ratio of 1:2:1. The intensities shown by letters underneath the lines in Fig. 6.35C result from the summation of the expected intensities; thus the experimental spectrum agrees with this interpretation. This structure is further established by the following two experiments:

1. Deuteration of the N—H″ groups produced a compound to give an identical spectrum.
2. When the H′ hydrogens were replaced by methyl groups, the EPR spectrum for this compound consisted of four main groups, each of which exhibited only five lines due to only nitrogen splitting. The hyperfine splitting due to the N—H″ proton and that by the protons on the methyl groups are either too small to be observed or nonexistent.

The observed spectrum with hyper and shf features provides conclusive proof for the delocalization of the odd electron in Cu(II) onto the ligand. This can be further interpreted by considering covalent bonding in the metal—ligand interaction [34].

6.2 ELECTRON PARAMAGNETIC RESONANCE OF EXCHANGE COUPLED SYSTEMS/CLUSTERS

The different EPR pattern obtained for exchange coupled homonuclear and heteronuclear clusters/systems are illustrated here with the following examples.

6.2.1 Cu(II) Dimer

For Cu(II) dimer, each unpaired electron may couple equally with the two Cu(II) nuclear spin $I = 3/2$ due to covalent bonding to give rise to seven lines, $(2nI + 1 = 2 \times 2 \times 3/2 + 1 = 7)$, in the spectrum. For example, the EPR of frozen $[Cu_2(dimp)\ (H_2O)_2](ClO_4)_4$ dissolved in methanol is shown in Fig. 6.36 exhibiting the behavior [35].

6.2.2 Mn Dimer in Catalase

The EPR spectrum of Mn dimeric catalase is shown in Fig. 6.37. This spectrum can be explained as follows. It is a dimer consisting of Mn^{3+} $(S = 2)$ and Mn^{4+} $(S = 3/2)$ of both high spins. These ions are antiferromagnetically coupled to give the total spin, $S_T = 1/2$ as follows [36].

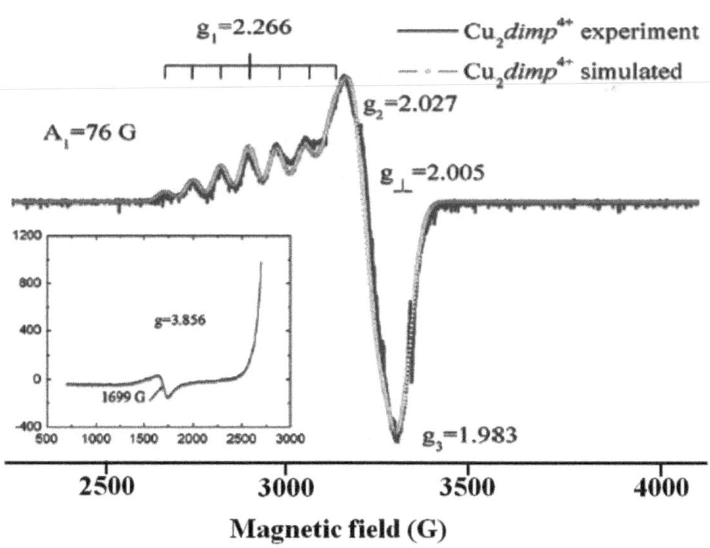

Figure 6.36 Experimental and simulated electron paramagnetic resonance of $[Cu_2(dimp)(H_2O)_2]^{4+}$. Inset shows the half-field transition signal. *Reproduced with permission from Elsevier.*

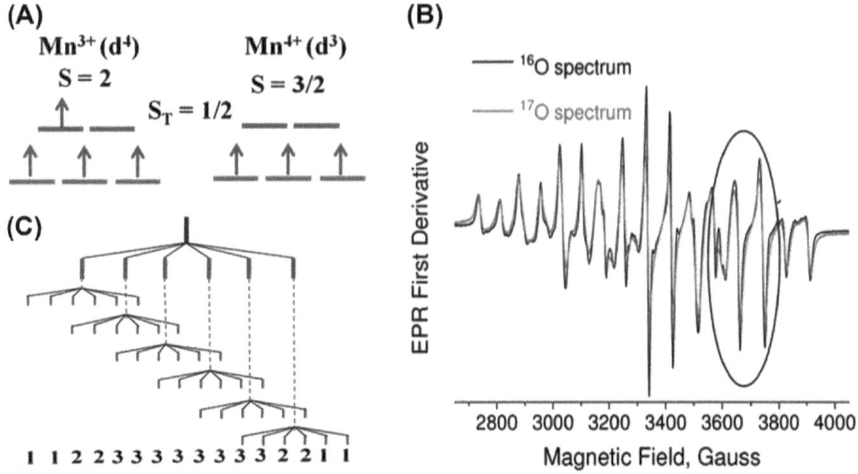

Figure 6.37 (A) Electronic arrangement, (B) X-band electron paramagnetic resonance spectra of 2Mn catalase in $H_2^{16}O$ and $H_2^{17}O$ and (C) Stick diagram. *Reproduced with permission from American Chemical Society.*

Because Mn^{4+} is highly charged (I=5/2), the strong electron−nuclear spin interaction gives rise to six lines with greater coupling constant (80 G) than Mn^{3+} (40 G). Using the stick diagram, the 16 lines are expected and also experimentally observed as shown in Fig. 6.37.

6.2.3 Electron Paramagnetic Resonance of $[4Fe-4S]^+$

The X-band EPR spectrum of iron−sulfur cluster protein is shown in Fig. 6.38 with E/D value 0.24. Using the rhombogram for S = 3/2 as shown in Fig. 6.38, the g-values, 5.6, 5.0, 2.10, 1.87, 1.80, are obtained and are in agreement with the experimental g-values (Fig. 6.38) [37].

6.2.4 Electron Paramagnetic Resonance of $[2Fe-2S]^{2+}$

The iron−sulfur cluster $[2Fe-2S]^{2+}$ is EPR inactive. In this cluster, the two high-spin Fe(III) ions are antiferromagnetically coupled to give total spin, S = 0, diamagnetic ground state. Thus, it does not give EPR spectrum.

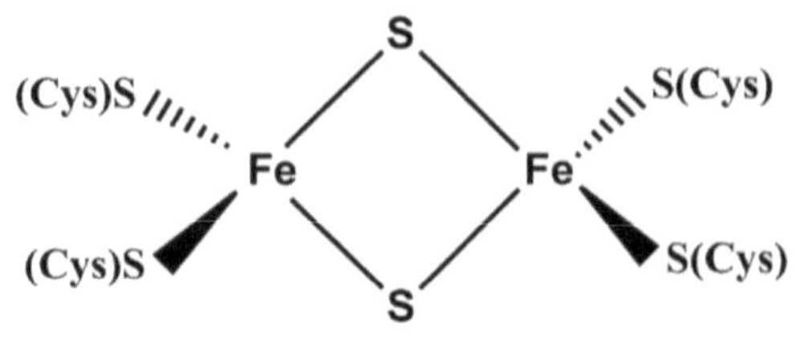

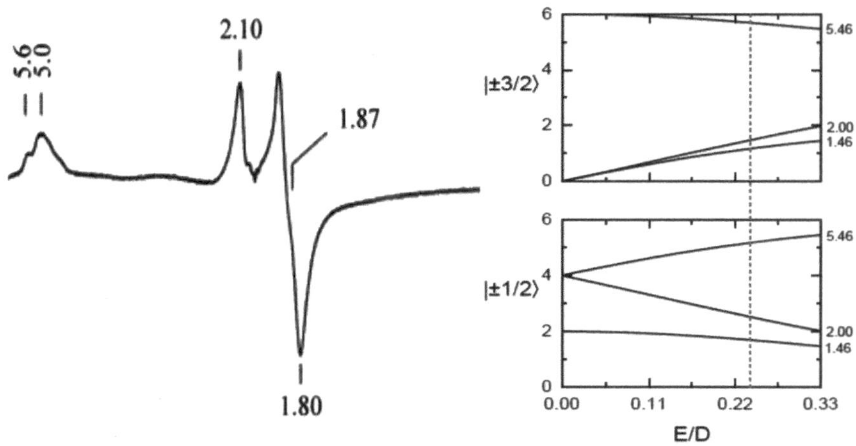

Figure 6.38 Electron paramagnetic resonance spectrum of dithionite reduced wild-type of *Pyrococcus furiosus* [4Fe—4S] at 4.2K and rhombogram of S = 3/2 (right). *Reproduced with permission from Wiley.*

6.2.5 Electron Paramagnetic Resonance of Fe(III) Ni(II)—Coupled Complex

The EPR spectrum of the heteronuclear dimeric complex formed between high-spin Fe(III) and Ni(II) at 10K is shown in Fig. 6.39 with E/D = 0.078. Based on the rhombogram for S = 3/2, the observed g-values are accounted for indicating the ground state of spin system as 3/2 [38].

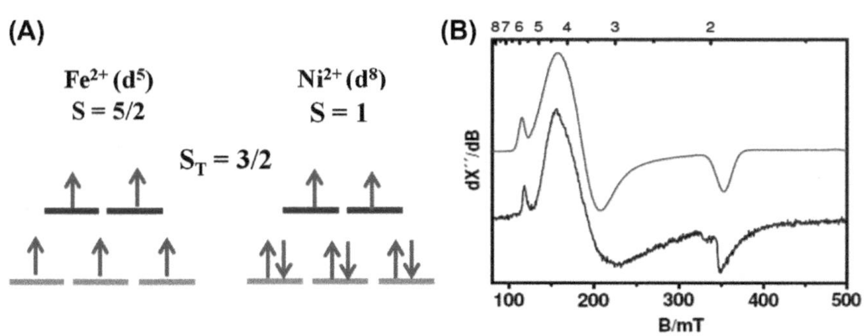

Figure 6.39 (A) Electronic configurations and (B) electron paramagnetic resonance spectrum of Fe(III) Ni(II) complex (S = 3/2) in CH_3CN at 10K together with simulated spectrum. *Reproduced with permission from American Chemical Society.*

6.3 ELECTRON PARAMAGNETIC RESONANCE OF FREE RADICALS AND SPIN TRAPPING

Free radicals are involved in many areas of chemistry, from oxidation processes, combustion, polymerization reactions, organic synthesis, biology, medicine, pollution, and radiation chemistry to biochemistry. These are formed as the intended products or intermediates as well as due to numerous side reactions. These are involved in many physiological process, viz. signaling, protein electron transport, apoptotic, and immune process. Furthermore, many defects in physiological processes have also been attributed to free radical formation damage; however, free radicals have a very short lifetime and will react immediately when formed. For example, tyrosyl (Tyr), tryptophan (Trp), superoxide anion, NO and hydroxyl radicals are produced during the reactions. These are only investigated and characterized by electron spin resonance (ESR) and their parameters. Using fast-flow and rapid freeze-quench (RFQ) EPR techniques, these can be directly monitored.

6.3.1 Fast-Flow Electron Paramagnetic Resonance

The instability and very short lifetime of free radical necessitated the use of fast-flow method where mixing of the reagent solutions in the ESR cavity to generate adequate ESR spectra. Fast-flow techniques are widely used to study reactions occurring on time scales of seconds to milliseconds. There are two major types (1) fast-flow method and (2) stopped-flow method. A general schematic of the fast-flow/stopped-flow methods is shown in Fig. 6.40. For example, tryptophan radical is investigated as follows. The experiment consists of the following:

Solution A: Ce^{4+} (0.5 mM, 4.0 L) + H_2SO_4 (0.225 M);

Solution B: L-tryptophan (1.0 mM, 4.0 L) in H_2SO_4 (0.225 M).

These were placed in solution reservoirs elevated 1.8 m relative to the ESR cavity. These solutions are deoxygenated by bubbling nitrogen through them, equilibrated to ambient temperature, and allowed to flow through plastic hoses to quartz, fast-flow mixing chamber flat cell at rates controlled by Gilmont compact flow meters. The temperature of the system was monitored with a Fisher digithermo temperature probe.

The ESR fast-flow spectrum of L-tryptophan radical cation produced during the mixing of these solutions is shown in Fig. 6.40. Equal volumes of aqueous solutions of L-tryptophan/H_2SO_4 and Ce^{4+}/H_2SO_4 were mixed in milliseconds before entering the ESR flat cell at a total flow rate

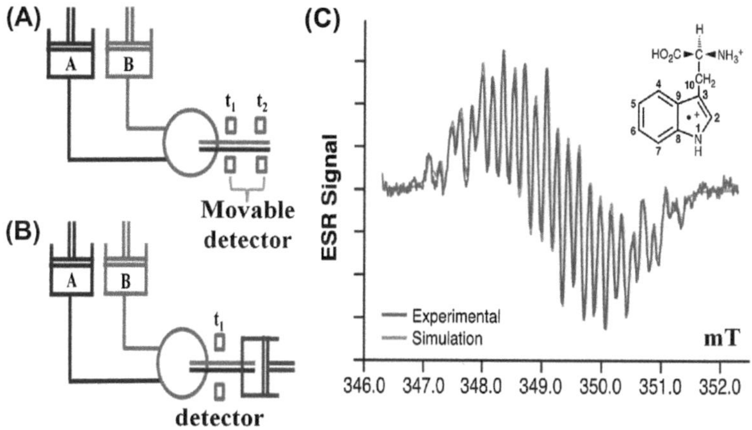

Figure 6.40 (A) Schematic of fast-flow, (B) stopped-flow methods, and (C) L-tryptophan radical cation electron spin resonance (ESR) spectrum. *Reproduced with permission from American Chemical Society.*

of 60 mL min^{-1}. ESR spectra were recorded at 20 mW microwave power, 0.1 mT field modulation, 6.00 mT field sweep width, 163 ms time constant, 81 ms conversion time, and 86 scans of 1024 data points. Also shown is the L-tryptophan radical cation simulation spectrum with coupling constants, viz., 0.195 mT, for the indole ring nitrogen and large unequal couplings for 0.71 and 1.07 mT due to β-methylene hydrogens. Although the two forms of L-tryptophan's radicals viz. neutral (deprotonated) or as protonated cation type were reported, the radical formed here is protonated cation radical [39].

6.3.2 Rapid Freeze-Quench Electron Paramagnetic Resonance

EPR in combination with the RFQ technique is a well-established method to trap the unstable free radicals by physical means and characterize the intermediates in chemical or enzymatic reactions at the millisecond or even shorter time scales.

The standard method to study the pre- and steady state kinetics of enzymatic reaction is stopped-flow technique as shown in Fig. 6.40B. The most basic setup consists of two syringes. One is filled with a solution containing the enzyme and the other is filled with a solution containing the substrate. The two syringes are mounted as in Fig. 6.40B in a setup that the two syringes are to be emptied simultaneously through a mixing device into a cuvette. The absorption spectra are measured as soon as the liquid is injected into the cuvette. This can either be the rapid collection of spectra over a wavelength range or the absorption change at a specific wavelength.

However, most biological samples have to be frozen to be studied in EPR spectroscopy. So, freezing of the two solutions (enzyme—substrate) mixture will quench the reaction. Therefore, this technique is called freeze quench. The quenching is normally achieved by having the aging tube end just above a cold solution of isopentane (125K). The aging of the sample is a function of both the speed which with the syringes are emptied and the length of the aging tube (Fig. 6.41). By changing each of these, samples can be quenched within a couple of milliseconds or longer time intervals. The enzyme—substrate solutions are shot directly into the solution, freeze instantaneously, and form what is called a "snow." The snow can now be collected and packed into an EPR tube.

For example, using RFQ-EPR the tyrosyl radical was detected. The ferric (in resting state) enzyme is mixed with a threefold molar excess of alkyl peroxide (hydrogen peroxide-free), and sample mixtures are frozen in liquid isopentane at −130°C in precision bore EPR tubes. Spectra were then recorded at 77K. A doublet signal similar to Fig. 6.43 was detected (~17 G line width) at the earliest time points (for example, 6.4 ms) and after 50 ms reaction time was only barely detectable. Furthermore, high-field EPR spectroscopy was used to confirm that the majority radical species is a tyrosyl radical [40].

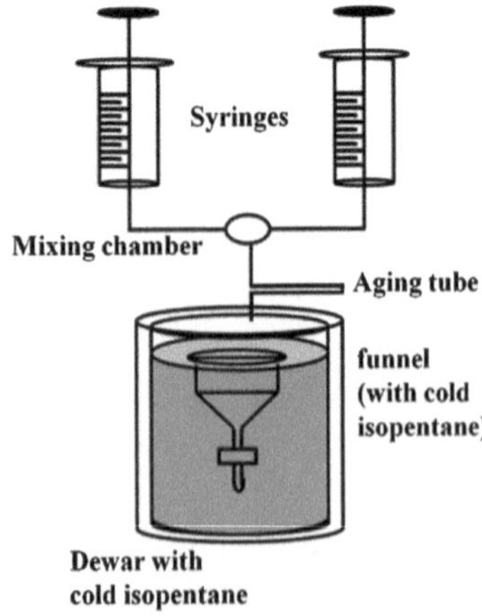

Figure 6.41 Schematic of the freeze-quench experiment.

6.3.3 Free-Radical Enzyme Intermediates

Amino acid side chains and cofactors are known to form radical or paramagnetic intermediates in enzymes. For example, ribonucleotide reductases are involved in electron transfer process to carry out reduction of ribonucleotides to deoxyribonucleotides. The formation of free-radical intermediates on side chains of the amino acids, viz., tyrosine (Tyr), tryptophan, cysteine, and even glycine, has been demonstrated by ESR. Fig. 6.42 shows the frozen solution low-temperature ESR spectra of tyrosine radicals of ribonucleotide reductases from different organisms [41].

6.3.4 Effect on UV Light on Free Radical Formation

In another example, a tryptophan radical was observed in the presence of H_2O_2 and HCO_3^- with hSOD1 (human Cu,Zn superoxide dismutase). A new EPR signal at low temperature appeared in the free radical region, which grew in intensity until 30 min (reaction time) and then decreased. The microwave power saturation $P_{1/2}$ values suggest that the tryptophan-derived radical saturates at a relatively low-microwave power. However, no free radical signal was detected from (tryptophan (W) replaced by phenylalanine (F)) W32F SOD1 under similar conditions. This reveals that the free radical detected from incubations containing wild-type (WT) hSOD1, H_2O_2, and HCO_3^- derived, most likely, from Trp-32. Furthermore, irradiation of WT hSOD1 and bSOD1 with UV light induced tryptophanyl and tyrosyl radicals, respectively, as shown in Fig. 6.43.

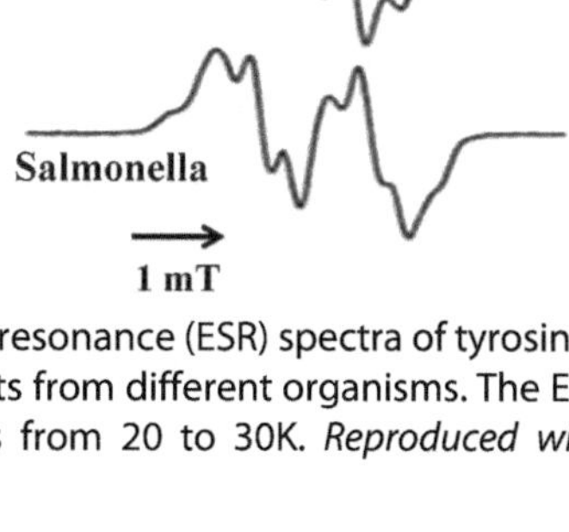

Figure 6.42 Electron spin resonance (ESR) spectra of tyrosine-free radicals of ribonucleotide reductase R2 subunits from different organisms. The ESR spectra at ~9 GHz were recorded at temperatures from 20 to 30K. *Reproduced with permission from Annual Reviews.*

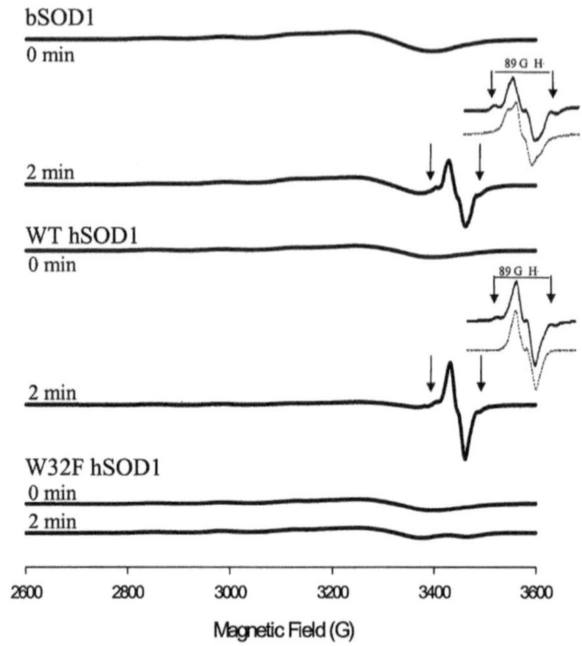

Figure 6.43 Effect of UV light on free radical formation from bovine superoxide dismutase (SOD) 1, WT human SOD1, and W32F human SOD1. *Reproduced with permission from American Society for Biochemistry and Molecular Biology.*

The UV irradiation of W32 F hSOD1, under similar conditions, did not induce the free radical signal (Fig. 6.43, bottom panel). This confirms that the free radical is most likely derived from Trp. On expanding the free radical region, hyperfine features corresponding to Trp and Tyr were resolved (Fig. 6.43). To identify the nature and structure of the free radicals, computer simulations (shown as dotted lines) were performed using the EPR parameters reported for Trp and Tyr radicals obtained through electron nuclear double resonance and high-frequency EPR studies. The experimental spectra could thus be successfully fitted using the literature values. Thus the doublet signal is due to Trp-32-derived radical in WT hSOD1. Unlike hSOD1, bSOD1 has a tyrosine residue instead of Trp-32. So, the corresponding signal (Fig. 6.43) is thus attributed to the tyrosyl radical [42].

6.4 ELECTRON PARAMAGNETIC RESONANCE SPIN TRAPPING—INDIRECT METHOD

Biologists and chemists have recently devised indirect ways to study free radicals. One method consists of using molecules that are capable of reacting with radicals such that the resulting product is stable and can be detected by some means, usually EPR spectroscopy. The molecule which itself EPR inactive reacts with the free radical to form a stable EPR active free radical adduct is known as ESR probe. If the stable adduct formed is also a radical, it may be characterized using EPR spectroscopy. The probe is called as a "*spin trap*" because it has trapped and preserved the unpaired spin for a longer lifetime. The adduct is now monitored at room temperature.

Spin traps were developed for the detection and characterization of free radicals. Until 1945 not much was known about radicals until the first ESR experiment by Zavoiski in Russia. Later on in 1954, Fraenkel, Hirshon, and Walling first reported an EPR spectrum recorded during a polymerization process using a number of cross-linked vinyl polymers.

Thereafter, detection and determination of radicals has witnessed a continuous improvement due to the advent of more sophisticated electronic equipment and, above all, the synthesis and use of spin traps.

The characteristic properties of a good spin trap are as follows:

1. It should possess high stability
2. The spin adduct formed with free radicals must be relatively stable
3. Different spin adducts with different kinds of free radicals must have distinctly different ESR spectra
4. Nontoxic

6.4.1 Classification of Spin Traps

There are two major classes of spin traps:

1. **PBN** [phenyl-*N*-ter-butylnitrone]**-type** and
2. **DMPO** [dimethyl-1-pyrroline *N*-oxide]**-type**.

These two classes are both nitrones. The DMPO-type spin traps are highly characterized, both theoretically and experimentally. However, not much is known about the hyperfine properties of the PBN-type spin traps. Fig. 6.44 shows the structures of the two main classes of spin traps.

Several PBN-type nitrones have been synthesized among which are α-substituted methoxy, amino, cyano, and mercapto nitrones as well

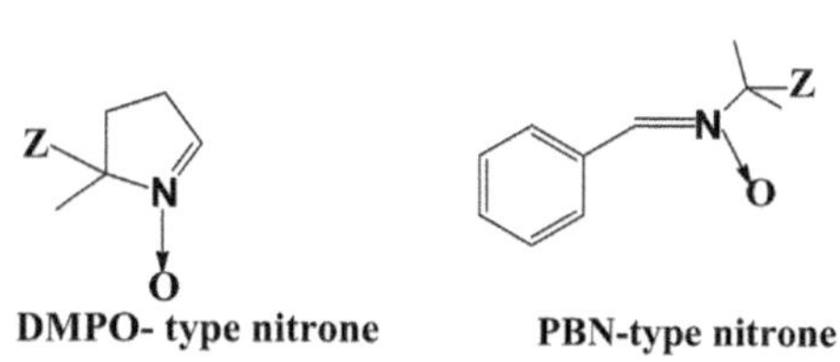

Figure 6.44 Structures of two major types of spin traps. Z represents substituent.

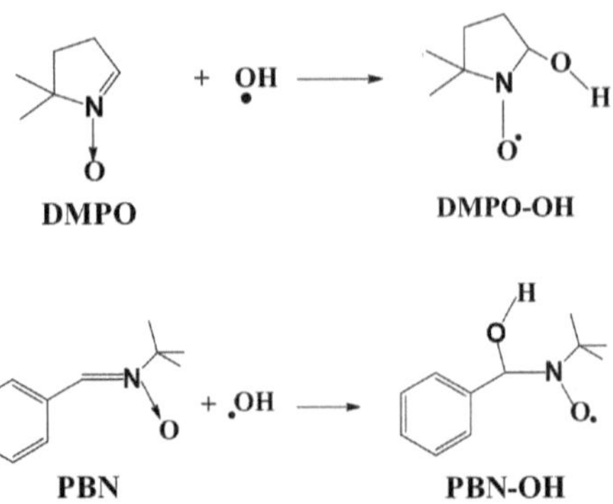

Figure 6.45 Formation of electron paramagnetic resonance detectable DMPO-OH and PBN-OH adducts. *DMPO*, dimethyl-1-pyrroline *N*-oxide; *PBN*, phenyl-*N*-ter-butylnitrone.

as the carboxyl derivatives and the alkoxy phosphoryl derivative. The known DMPO-type spin traps include the alkoxyphosphoryllated nitrones, 5-diethoxyphosphoryl-5-methyl-1-pyrroline *N*-oxide, alkoxycarbonyl nitrones 5-ethoxycarbonyl-5-methyl-1-pyrroline *N*-oxide, and 5-butoxycarbonyl-5-methyl-1-pyrroline *N*-oxide. Based on both theoretical and experimental studies, PBN-type and DMPO-type nitrones have been reported to show spin trapping abilities. Nevertheless, they have certain limitations. PBN-type nitrones are limited by their ability to distinguish the different radicals trapped, whereas DMPO-type nitrones are widely used to characterize the different radicals from their hyperfine splitting. However, DMPO-types are limited by the stability of the adducts formed and the overall spin trapping efficiency.

The spin trapping technique involves the addition of reactive unstable free radicals across the double bond of a diamagnetic spin trap to form a much more stable free radical adduct (Fig. 6.45), which can then be indirectly monitored by EPR spectroscopy. This is the least ambiguous method for detection and characterizing free radicals in chemistry, biology, and medicine [43—46].

6.4.2 Enzymatic Generation of Free Radicals and Their Spin Trapping

For example, xanthine and hypoxanthine in the presence of XO generate unstable superoxide anion radical and hydrogen peroxide, respectively

(Fig. 6.46). In the case of DMPO, two stable radical spin adducts with unstable superoxide and hydroxyl radicals were generated by the XO system (Fig. 6.46). Hydrogen peroxide on disintegration or further reduction gives hydroxyl radical, which further generates methyl radical in the presence of dimethyl sulfoxide as shown below.

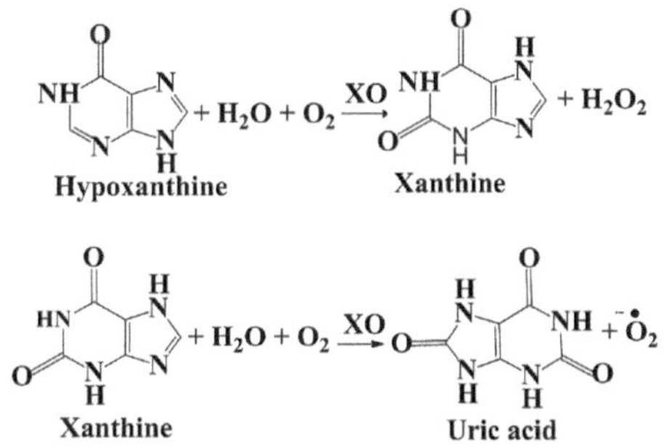

$$Fe^{2+} + H_2O_2 \longrightarrow Fe^{3+} + HO^{\bullet} + HO^-$$

$$\text{DMPO} \xrightarrow[CH_3SOCH_3]{^{\bullet}OH} \text{DMPO} - CH_3$$

Figure 6.46 Formation of superoxide anion radical, hydrogen peroxide, ($^{\bullet}$OH), and methyl radical.

The formation and involvement of superoxide anion, hydroxyl radicals, and methyl radical in the XO system from the ESR hyperfine splitting of DMPO/$^{\bullet}$OOH, DMPO/$^{\bullet}$OH, and DMPO/$^{\bullet}CH_3$ can be established. The ESR spectrum recorded in Fig. 6.47 immediately after adding XO consisting of both the adducts, viz., DMPO/$^{\bullet}$OOH and DMPO/$^{\bullet}$OH. However, in the presence of SOD, the superoxide radical was completely scavenged as evident from only DMPO/$^{\bullet}$OH signals detected (Fig. 6.48). It suggests that the $^{\bullet}$OH radical generation is not mediated by superoxide as shown in mechanism. On the other hand, in the presence of catalase, the hydrogen peroxide were scavenged and gave only EPR spectrum characteristics of superoxide anion adduct as evident in Fig. 6.48. The observed DMPO/$^{\bullet}$OH is probably due to the further reduction of H_2O_2 by XO or due to Fenton reaction.

EPR spin trapping is thus a valuable tool in the detection and study of transient free radicals in biological/chemical systems.

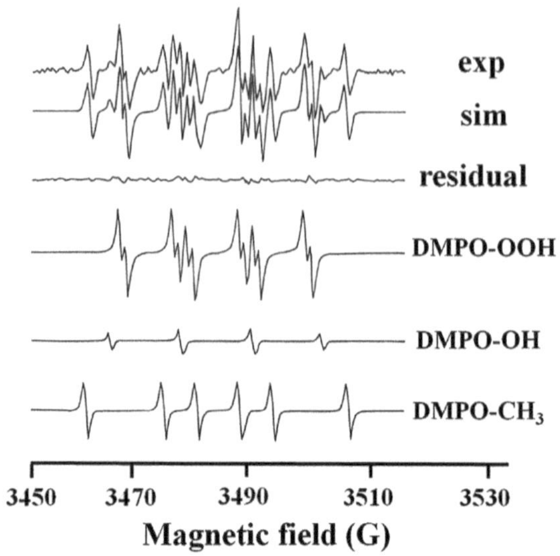

Figure 6.47 Electron spin resonance spectra of composite dimethyl-1-pyrroline N-oxide (DMPO) adducts with superoxide anion radical, hydrogen peroxide (•OH), and methyl radical (top) and simulated spectra for the individual radicals.

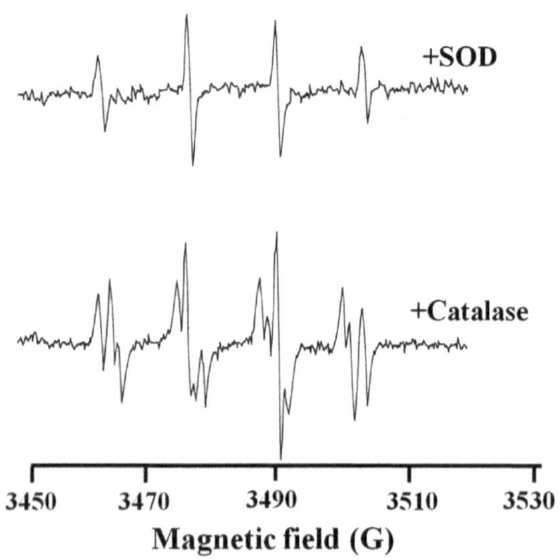

Figure 6.48 Electron spin resonance spectra obtained from solution containing xanthine, xanthine oxidase (XO), and dimethyl-1-pyrroline *N*-oxide in the presence of superoxide dismutase (SOD) (top) and catalase enzymes (bottom). *Courtesy Dr. C. Karunakaran.*

6.4.3 Applications of Spin Trapping

6.4.3.1 Biology

In recent years, spin trapping techniques have been employed extensively to provide insight into the mechanism of free radical biology of phagocyte cells. The ESR measurements demonstrate that spin trapping is of adequate sensitivity to detect superoxide and/or hydroxyl radical generated by these cells. The spin trapping method is capable of measuring phagosomal radicals as well. Spin trapping has thus proven to be a powerful and unambiguous tool for the identification of biologically generated transient organic free radicals.

Spin traps have also been successfully used in direct detection of drug-derived free radicals. For example, vitamin E and the spin trap PBN were shown to prevent and scavenge the radical formed by the drug anthralin. Spin trapping of sulfur trioxide anion radical and phenyl hydrazine-induced formation of sulfur radical in vivo have been used to detect endogenous or drug-derived nitric oxide in a molecule participating in numerous physiological and pathological processes. For example, the fate of toxic metal ion formed from carcinogenic chromium (V) as a result of intravenous and transdermal administrations has also been followed with spin traps. PBN is specifically used for evaluation of beer quality in industry.

6.4.3.2 Photodynamic Therapy

Photodynamic therapy (PDT) is a treatment option in which activation of photosensitizer (PS) drugs with specific wavelengths of light leads to energy transfer to oxygen molecules or other substrates in the surrounding areas, generating cytotoxic reactive oxygen species (ROS), which can trigger apoptotic and necrotic cell death. In the absence of light, the PS drugs are minimally toxic. Therefore, PDT provides a safe and effective way to selectively eradicate target cells/tissues such as cancerous cells while avoiding systemic toxicity and side effects on healthy tissues. Compared with traditional chemotherapy and radiotherapy, PDT-based cancer treatment significantly reduces side effects and improves target specificity because only the lesion under light irradiation is treated [47].

6.4.3.2.1 Mechanism of Photodynamic Therapy Using Photosensitizer

Mechanism of PDT using PS has been elucidated in several studies. Briefly, PS in the ground state absorbs a photon and is activated to an excited singlet state on irradiation with suitable light (Fig. 6.49). The excited singlet state can convert into the triplet state via intersystem crossing caused by a change

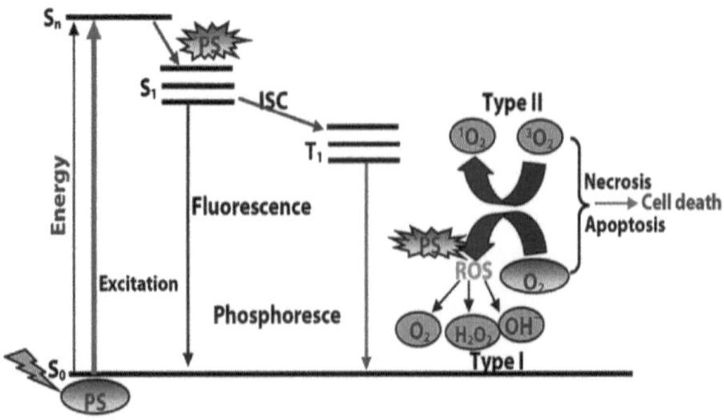

Figure 6.49 Mechanism of photodynamic therapy. *Reproduced with permission from Elsevier.*

in the spin of electrons. The PS in the triplet state interacts with surrounding molecules and thus produces ROS through type I and type II reactions.

1. *Type* I (*electron reaction*): It involves the transfer of either hydrogen atom or an electron between the excited PS and the substrates, leading to the generation of free radicals. These radicals then react with oxygen, resulting in the production of ROS such as superoxide and hydroxyl radicals.
2. *Type* II (*energy transfer reaction*): It involves the energy transfer between the excited PS and the molecular oxygen in the ground state ($^3O_2$), resulting in the formation of highly reactive state of oxygen known as *singlet oxygen* ($^1O_2$). The resulting ROS can cause irreversible damage to target tissues/cells.

$$
\begin{aligned}
&{*PS} + {}^3O_2 \xrightarrow[\text{transfer}]{\text{Energy}} PS + {}^1O_2 \quad \text{(Type II)} \\
&{*PS} + {}^3O_2 \xrightarrow[\text{transfer}]{\text{Electron}} PS + O_2^{-\bullet} \\
&{*PS} + H_2O_2 \longrightarrow PS + HO^{\bullet}
\end{aligned}
\left.\vphantom{\begin{aligned}a\\b\end{aligned}}\right\} \text{(Type I)} \qquad (6.34)
$$

6.4.3.3 Pollution

As discussed above, spin trapping has been widely used in biochemistry and medicine for diagnosis of various diseases such as cancer, stress, leukemia, brain damage, and many radical-initiated diseases. In gas phase, cigarette smoke consisting of high concentrations of both oxygen- and carbon-centered free radicals was detected using ESR and spin trapping technique, including the use of spin traps in the solid state, to show that the radicals are trapped directly from the gas phase. Furthermore, the radicals detected in aged cigarette smoke and other postcombustion gases are not remnants of the actual combustion process but rather are formed within smoke plumes via dynamic processes involving reactions of species such as NO, NO_2, O_2, and incompletely oxidized organic compounds. Because radicals are formed within the aging smoke, they can be biologically important when smoke is inhaled and can play an important role in destroying and forming other organic compounds in the smoke.

Methyl radicals have been detected during the photolysis of acetone and ethyl radicals from the photolysis of 3-pentanone, tetraethyl lead, and diethyl mercury. ESR spin trapping also helped to identify free radicals accompanying the photodecomposition of acetaldehyde. The spin trap PBN has been successfully used to detect radical intermediates in phenol/TiO_2 photoredox system.

Air-borne particulate matter, such as total suspended particulates and fresh wood soot, has recently been studied. It was observed using EPR that the quantities of persistent free radicals are characteristic of a mixture of radicals with different structures and a carbonaceous core of carbon-centered radicals.

6.5 SPIN LABELING VERSUS SPIN PROBES

One of the common drawbacks of EPR spectroscopy is related to the small number of paramagnetic systems. However, the environment, dynamic and structural details of these systems can be obtained by using EPR introduction of paramagnetic tracer molecules called spin probes or spin labels [48—50]. Nitroxides, stable free radicals with the structural group R_2NO, are by far the most important class of these paramagnetic probe molecules. The chemical structures of common nitroxide derivatives are displayed in Fig. 6.50.

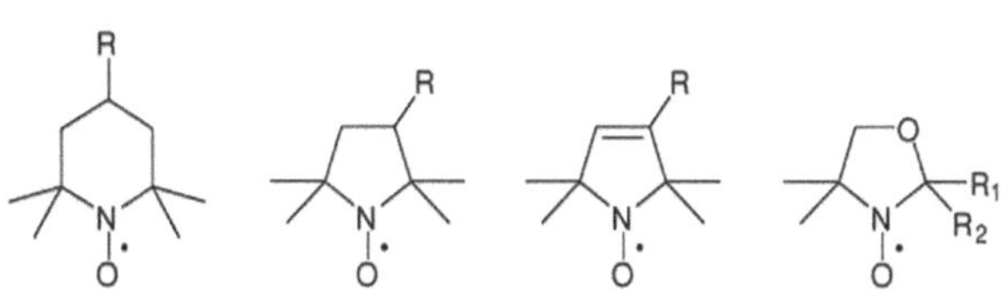

Figure 6.50 Chemical structures of common nitroxide spin probes. (1) 2,2,6,6-Tetramethylpiperidine-1-oxyl, (2) 2,2,5,5-tetramethylpyrrolidine-1-oxyl, (3) 2,2,5,5-tetramethylpyrroline-1-oxyl (dehydro-PROXYL), and (4) 4,4-dimethyloxazolidine-1-oxyl.

The paramagnetic metal ions viz., Cu^{2+}, Mn^{2+}, Fe^{3+}, and Co^{2+} also widely used spin probes to replace substitutionally the diamagnetic ions especially Zn(II), Cd(II), and Ni(II) in their crystal sites because of their comparable size and coordination property to obtain the structural and bonding information of the site.

Spin labels are chemically attached to the material of interest especially proteins while spin probes interact noncovalently with the system. Mostly spin labels are nitroxide molecules. A major advantage of spin label ESR studies is that they can be performed at physiologically relevant temperatures. Therefore, they are sensitive to molecular motion. Nitroxide-labeled nucleic acids, lipids, and enzymes are among the variations of spin label analogs that have been made.

6.5.1 Site-Directed Electron Paramagnetic Resonance Spin Labeling

EPR studies were initially limited to metalloproteins possessing its own paramagnetic centers or to proteins with naturally occurring cysteine residues required for capturing spin labels. Because the application of molecular biology (which allowed engineering of cysteine residues at targeted sites in proteins) combined with EPR spectroscopy, the site-directed spin labeling (SDSL) method has become a powerful new tool to investigate the molecular function and dynamical aspects of biomolecules under the condition close to physiological condition. In the cysteine scanning method, the chosen amino acid residues within the amino acid sequence are substituted with cysteine residues. Now-a-days, these cysteine residues in the protein of interest can be subsequently modified with a variety of sulfhydryl-specific nitroxide spin labels. The different types of nitroxide-based spin labels are shown in Fig. 6.51.

Among them, the methanethiosulfonate spin probe is currently most utilized. The resulting side chain designated as R1 is comparable in molecular volume to phenylalanine or tryptophan side chains and hence exerts a negligible influence on the biological activity and stability as well as it does not affect the secondary structure of the protein (Fig. 6.52). Furthermore, the properties of R1 related to its enhanced sensitivity to protein backbone motion and secondary structure facilitate determination of subtle structural details from the EPR spectra lineshape analysis.

Most recently, an interesting alternative for cysteine spin labeling has been developed. This is based on incorporation of unnatural amino acids

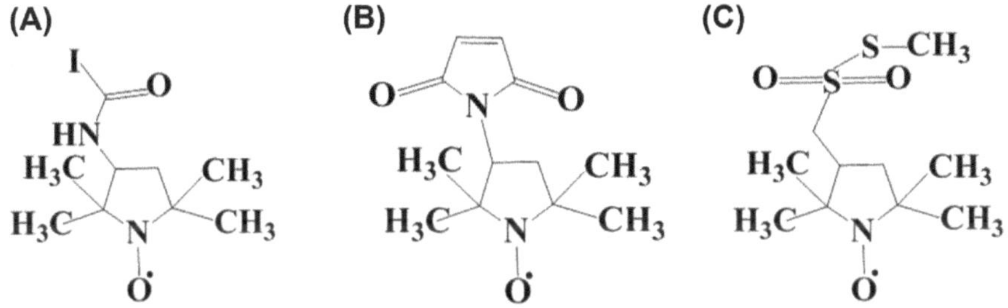

Figure 6.51 Structures of nitroxide-based spin labels commonly used in site-directed spin labeling experiments. (A) proxyliodoacetamide; (B) proxylmaleimide; (C) proxylmethanethiosulfonate.

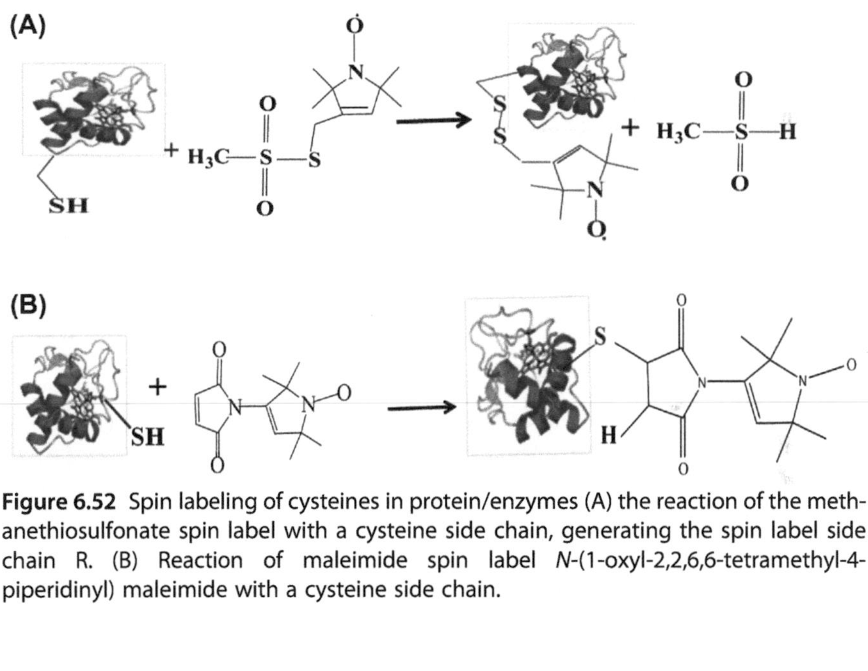

Figure 6.52 Spin labeling of cysteines in protein/enzymes (A) the reaction of the methanethiosulfonate spin label with a cysteine side chain, generating the spin label side chain R. (B) Reaction of maleimide spin label *N*-(1-oxyl-2,2,6,6-tetramethyl-4-piperidinyl) maleimide with a cysteine side chain.

into proteins. Tailor-made amino acids carrying a spin label, such as 2,2,6,6-tetramethyl-N-oxyl-4-amino-4-carboxylic acid, can be incorporated into the protein by solid phase peptide synthesis or alternatively together with recombinant synthesis of proteins or by using synthesized amino acylated tRNA for spin-labeled protein expression in living cells. By those means, individually designed spin-labeled side chains can be incorporated into specific sites of a protein.

6.5.2 Applications of Spin Labeling

By spin labeling proteins/DNA/membranes, ESR spectroscopy can be used to investigate the following:

1. Molecular dynamics
2. Solvent accessibilities
3. Polarity of the medium
4. Interspin label interactions and their distance measurements

6.5.2.1 *Molecular Dynamics*

The lineshape of room temperature continuous wave (CW) EPR spectra of spin-labeled protein/DNA reflects the reorientational motion of the spin label side chain in the macromolecules. The influence of spin label dynamics on the spectral shape has been investigated in detail. The relationship between the dynamics of the spin label side chain and protein structure has been extensively studied for T4 lysozyme.

In general, the EPR spectral features are used to characterize the mobility, i.e., motional rate, amplitude, and anisotropy of the overall reorientational motion of the spin label. For example, spin-labeled sites exposed to the bulk water exhibit weak interaction with the rest of the protein as found for helix surface sites or loop regions and consequently display a high degree of mobility, which is characterized by a small apparent hyperfine splitting and results in narrow line widths (Fig. 6.53 top). If the mobility of the spin label side chain is restricted by interaction with neighboring side chains or backbone atoms of the protein itself or an interaction partner, the line widths and the apparent hyperfine splitting are much larger (Fig. 6.53 middle). If the spin label is strongly immobilized, it corresponds to the slow motion limit (>100 ns) (bottom). The spectrum is very similar to the powder or rigid limit spectrum that is obtained for any nitroxide in the absence of rotational motion and for a dilute powder or frozen solution.

The motion of a nitroxide spin label side chain is essentially characterized by the following three correlation times, (1) the rotational correlation time for the entire protein, (2) the effective correlation time due to the rotational isomerization spin label linker, and (3) the effective correlation time for the segmental motion of the protein backbone. These correlation times can significantly differ in the time scales they occur on. For this case, correlation times from μs (for the overall protein motion) down to ps (for the rotational isomerization) have to be covered by the experiment. EPR spectra at different microwave frequencies are sensitive to motions on different time

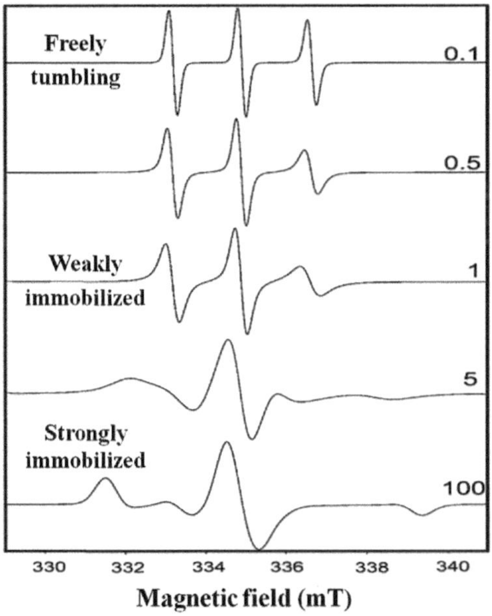

Figure 6.53 Dynamics of spin-labeled side chain and protein structure from electron paramagnetic resonance spectra. *Reproduced with permission from Portland Press.*

scales. For instance, EPR at lower frequencies is sensitive to slower motions, whereas faster motions are completely averaged out. Moreover, high-frequency EPR can be used to resolve fast motions, but slower motions are "frozen" at the high-frequency time scale. Consequently, combining experiments at different microwave frequencies (multifrequency EPR) allow distinction and separation of various motional modes in a spin-labeled protein according to their different time scales [51].

6.5.2.2 Solvent Accessibilities

The accessibility of the spin label side chain toward paramagnetic probes (exchange reagents), which selectively partition in different environments of the system under investigation, can be employed to establish the location of spin label with respect to the protein/water/membrane boundaries. The accessibility of the nitroxide spin label side chain is defined by its Heisenberg exchange frequency, *Wex*, with an exchange reagent diffusing in its environment. Water-soluble metal ion complexes such as Ni(II)ethylenediaminediacetic acid (NiEDDA) or chromium oxalate are often used to quantify the accessibility from the bulk water phase, whereas molecular oxygen or hydrophobic organic radicals that mainly partition in the hydrophobic part of the lipid bilayer define the accessibility from the lipid phase. The concentration gradients of NiEDDA and molecular oxygen along the membrane normal are used to obtain the immersion depth of the spin label side chain into the lipid bilayer. CW power saturation and saturation recovery (SR) (see Section 5.3) techniques are widely used to determine the nitroxide's accessibility toward the paramagnetic probes.

Most commonly, Heisenberg exchange rates for nitroxide spin label side chains in proteins are measured using CW power saturation. In this method, the EPR signal amplitude is monitored as a function of the incident microwave power in the absence and presence of the paramagnetic quencher. From the saturation behavior of the nitroxide, an accessibility parameter, Π, can be extracted that is proportional to W_{ex}.

SP-EPR, which allows measuring the spin-lattice relaxation time T_{1e}, is also connected to the Heisenberg exchange frequency W_{ex} directly. Here, the saturating microwave pulses are applied to the sample in the absence and in the presence of exchange reagents, and the recovery of the z magnetization is monitored as a function of time. Analyses of the recovery curves provide T_{1e} and thus the accessibility for the respective exchange reagent.

$$W_{ex} = \Delta(1/T_{1e})_R \quad (6.35)$$

6.5.2.3 Polarity and Proticity of the Spin Label Microenvironment

Polarity and proticity of the spin label microenvironment are reflected in the hyperfine component A_{zz} and the g_{xx} values. A polar environment gives higher values for A_{zz}, whereas the tensor component g_{xx} ($g_{xx} = h\nu/\mu_B B$) is decreased. The A_{zz} component can be obtained from CW X-band EPR spectra of spin-labeled proteins in frozen samples. However, the principal g-tensor components and their anisotropy can be well resolved and readily determined with high accuracy using high-field EPR techniques due to the enhanced Zeeman resolution. In regular secondary structure elements with anisotropic solvation (e.g., surface exposed α-helices), the water density and hence the tensor component values A_{zz} and g_{xx} are a periodic function of amino acid residue number. Hence, similar to accessibility measurements with water-soluble exchange reagents, these EPR data can be used to obtain structural as well as topological information. Therefore, the polarity of the spin label environment can reveal details on the folding of protein.

6.5.2.4 Interspin Label Interactions and Their Distance Measurements

If two spin label side chains are introduced into a biomolecule or two singly labeled molecules are in a stable macromolecular complex, using SDSL method, the EPR can provide a vast amount of including protein/DNA/RNA/membrane interactions. Spin labeling method can also provide detailed information about the binding interface not only on the structural level but also give insights into kinetic and thermodynamic aspects of the interaction. EPR also allows determination of distances (see Section 5.10.1) between two nitroxides in double SDSL in the range from ~10 to 80 Å with accuracies down to less than 1 Å, thereby covering a range of sizes including also large multidomain proteins and protein complexes.

Spin–spin interaction is composed of static dipolar interaction, modulation of the dipolar interaction by the residual motion of the spin label side chains, and exchange interaction. The static dipolar interaction in an unordered immobilized sample leads to considerable broadening of the CW EPR spectrum if the interspin distance is less than 2 nm. Hence, the interspin distances can be observed by a detailed lineshape analysis of EPR spectra of frozen protein samples or proteins in solutions of high viscosity. Pulse EPR techniques are recently used to expand the accessible distance range up to 8 nm. Two major protocols viz., the 4-pulse double electron–electron resonance or 4-pulse pulsed electron–electron double resonance (see Chapter 5) and the double quantum coherence approaches have been successfully applied for the distance measurements.

6.6 ELECTRON PARAMAGNETIC RESONANCE DOSIMETRY

Dosimetry based on the detection by EPR spectroscopy of ionizing radiation-induced radicals is an established method for the retrospective dosimetry of past exposures and the dosimetry of potentially exposed persons in radiological emergencies. The dose is generally estimated by measuring the physical damage induced in materials contained in objects placed on or next to the potentially exposed person.

EPR dosimetry is based on the detection and the measurement of the concentration of stable radiation-induced radicals obtained from reactions triggered in matter by ionizing radiation using EPR spectrometry. In solids, the radicals become trapped (i.e., they are not "free") and in some cases they can be stable over long periods. EPR is then the choice of method for direct detection of radiation-induced radicals in solids. Ionizing radiations induce radicals in most materials, but the yield of production varies significantly. Therefore, the suitability of materials as dosimeters is mainly related to their radiation sensitivity and signal stability in time. Materials that have already been investigated are, for example, sugar, plastics, glass, wool, and cotton. Stable radicals are also generated in solid biological tissues (nails, hair, and calcified tissues). The calibration of EPR signal intensity as a function of absorbed dose can be done for each sample by delivering additional radiation doses with a laboratory source and then EPR measurement.

Measurements are typically done using X-band microwaves (around 9 GHz). It is applied, e.g., for dosimetry in radiation processing by measuring radicals of the amino acid alanine, determination of age in archaeological or paleontological dating by measuring radicals in quartz and carbonates, and retrospective dosimetry by measuring radicals in calcified tissues. Radical species can also be produced by agents other than ionizing radiation, typically UV light and mechanical operations during sample collection. These radicals are usually different from those generated by ionizing radiation [52].

REFERENCES

[1] A. Lund, M. Shiotani, S. Shimada, Principles and Applications of ESR Spectroscopy, Springer, New York, 2011.

[2] M. Yamaga, T. Yosida, B. Henderson, K.P.O. Donnell, M. Date, Electron paramagnetic resonance and optical spectra of Ti^{3+}-doped $YAlO_3$, J. Phys. Condens. Matter 4 (1992) 72–85.

[3] (a) B.N. Figgs, M.A. Hitchman, Ligand Field Theory and Its Applications, Wiley-VCN, New York, 2000;
(b) C.J. Ballahausen, Introduction to Ligand Field Theory, Mc Graw-Hill Book Co., New York, 1962.

[4] G.N. George, C.A. Kipke, R.C. Prince, R.A. Sunde, J.H. Enemark, S.P. Vramer, Structure of the active site of sulfite oxidase. X-ray absorption spectroscopy of the Mo(IV), Mo(V), Mo(VI) oxidation states, Biochemistry 28 (1989) 5075–5080.

[5] L.E. Bevers, P.-L. Hagedoorn, W.R. Hagen, The bioinorganic chemistry of tungsten, Coord. Chem. Rev. 253 (2009) 269–290.

[6] R. Kripal, M.G. Misra, EPR and optical absorption study of VO^{2+} ions doped potassium Hexaaquazinc(II) sulfate single crystals, Appl. Magn. Reson. 44 (2013) 759–779.

[7] W.P.M. Johannes, Van Schijndel, E.G.M. Vollenbroek, R. Wever, The chloroperoxidase from the fungus *Curvularia inaequalis*: a novel vanadium enzyme, Biochim. Biophys. Acta 1161 (1993) 249–256.

[8] N.O. Gopal, K.V. Narasimhulu, C.S. Sunandana, J. Lakshmana Rao, EPR and optical absorption spectra studies of Cr^{3+} ions doped in nickel maleate tetrahydrate single crystals, Phys. B 348 (2004) 335–340.

[9] R. Murugesan, S. Subramanian, EPR study of Mn(II) in a cubic ligand field, Mn(II) in hexaimidazole-cadmium(II) nitrate, Mol. Phys. 52 (1984) 281–288.

[10] K.A. Campbell, E. Yikilmaz, C.V. Grant, W. Gregor, A.-F. Miller, R. David Britt, Parallel polarization EPR characterization of the Mn(III) center of oxidized manganese superoxide dismutase, J. Am. Chem. Soc. 121 (1999) 4714–4715.
[11] F.A. Taiwo, Electron paramagnetic resonance spectroscopic studies of iron and copper proteins, Spectroscopy 17 (2003) 53–63.
[12] G. Palmer, The electron paramagnetic resonance of metalloproteins, Biochem. Soc. Trans. 13 (1985) 548–560.
[13] W.R. Hagen, EPR spectroscopy as a probe of metal centers in biological systems, Dalton Trans. 37 (2006) 4415–4434.
[14] R.A. Pinkowitz, P. Aisen, Zero-field splitting of iron complexes of transferrins, J. Biol. Chem. 247 (1972) 7830–7834.
[15] T. Kubiak, R. Krzyminiewski, B. Dobosz, EPR study of paramagnetic centers in human blood, Curr. Top. Biophys. 36 (2013) 7–13.
[16] W.R. Hagen, Metallomic EPR spectroscopy, Metallomics 1 (2009) 384–439.
[17] S. Murielle Roux, J. Cove, The iron-containing superoxide dismutase of *Ralstonia metallidurans* CH34, FEMS Microbiol. Lett. 210 (2002) 129–133.
[18] H. Fujiii, D. Yamaki, T. Ogura, M. Hada, The functional role of the dioxoisobacteriochlorin in the catalytic site of cytochrome cd1 for the reduction of nitrite, Chem. Sci. 7 (2016) 2896–2906.
[19] R. Cammack, D.S. Patil, V.M. Fernandez, Electron-spin-resonance/electron-paramagnetic-resonance spectroscopy of iron-sulphur enzymes, Biochem. Soc. Trans. 13 (1985) 572–578.
[20] P. Pietrzyk, M. Srebro, M. Radon, Z. Sojka, A. Michalak, Spin ground state and magnetic properties of cobalt(II): relativistic DFT calculations guided by EPR measurements of bis(2,4-acetylacetonate) cobalt(II)-based complexes, J. Phys. Chem. A 115 (2011) 2316–2324.
[21] M. Baumgarten, C.J. Winscom, W. Lubitz, Probing the surrounding of a cobalt(II) porphyrin and its superoxox complex by EPR techniques, Appl. Magn. Reson. 20 (2001) 35–70.
[22] K. Tahara, H. Shimakoshi, A. Tanaka, Y. Hisaeda, Synthesis, characterization and catalytic function of a B_{12} hyper branched polymer, Dalton Trans. 39 (2010) 3035–3042.
[23] A. Verberckmoes, B.M. Weckhuysen, R.A. Schoonheydt, Spectroscopy and coordination chemistry of cobalt in molecular sieves, Microporous Mesoporous Mater. 22 (1998) 165–178.
[24] K.A. Kozhanov, M.P. Bubnov, V.K. Cherkasov, G.A. Abakumov, EPR study of spin labeled nickel NCN-pincer complexes, Dokl. Chem. 407 (2006) 35–38.
[25] J. Telser, A perspective on applications of ligand-field analysis: inspiration from electron paramagnetic resonance spectroscopy of coordination complexes of transition metal ions, J. Braz. Chem. Soc. 17 (2006) 1501–1515.
[26] D.P. Barondeau, C.J. Kassmann, C.K. Bruns, J.A. Tainer, E.D. Getzoff, Nickel superoxide dismutase structure and mechanism, Biochemistry 43 (2004) 8038–8047.
[27] C. Karunakaran, K.R. Justin Thomas, A. Shanmugasundaram, R. Murugesan, EPR of an exchange-coupled, hydrogen bridged one-dimensional Cu(II) complex containing both octahedral and square pyramidal geometries in the same unit cell, Mol. Phys. 100 (2002) 287–295.
[28] C. Karunakaran, K.R. Justin Thomas, A. Shunmugasundaram, R. Murugesan, EPR of Cu(II) doped seven-coordinate inclusion compounds, $(M(stpy)_3(NO_3)_2$ ½ stpy =trans-4-styrylpyridine): low symmetry effects in admixture of ground states, Spectrochim. Acta A Mol. Biomol. Spectrosc. 57 (2001) 441–449.
[29] R. Kripal, M.G. Misra, EPR and optical absorption study of Cu^{2+}-doped diammonium hexaaqua magnesium sulphate single crystals, Appl. Magn. Reson 348 (2004) 335–340.

[30] A.C. Rosenzweigh, Bioinorganic chemistry:zeroing in on a new copper site, Nat. Chem. 1 (2009) 684–685.
[31] W.E. Antholine, P.M. Hanna, D.R. McMillin, Low frequency EPR of *Pseudomonas aerginosa* azurin, analysis of ligand super hyperfine structure from a type 1 copper site, Biophys. J. 64 (1993) 267–272.
[32] E.I. Solomon, R.G. Hadt, Recent advances in understanding blue copper proteins, coordination chemistry reviews, 255 (2011) 774–789.
[33] J.D. Lee, Concise Inorganic Chemistry, fourth ed., Chapman & Hall, 1991.
[34] R.S. Drago, Physical Methods in Inorganic Chemistry, Affiliated East-West Press Pvt. Ltd., New Delhi, 2012.
[35] L. González-Sebastián, V.M. Ugalde-Saldívar, E. Mijangos, M.R. Mendoza-Quijano, L. Ortiz-Frade, L. Gasque, Solvent and pH effects on the redox behavior and catecholase activity of a dicopper complex with distant metal centers, J. Inorg. Biochem. 104 (2010) 1112–1118.
[36] I.L. McConnell, V.M. Grigoryants, C.P. Scholes, W.K. Myers, P.-Y. Chen, J.W. Whittaker, G.W. Brudvig, EPR-endor characterization of (^{17}O, ^{1}H, ^{2}H) water in manganese catalase and its relevance to the oxygen-evolving complex of photosystem II, J. Am. Chem. Soc. 134 (2012) 1504–1512.
[37] R.E. Duderstadt, P.S. Brereton, M.W.W. Adams, M.K. Johnson, A pure S = 3/2 [Fe_4S_4]. Cluster in the A33Y variant of *Pyrococcus furiosus* ferredoxin, FEBS Lett. 454 (1999) 21–26.
[38] B. Biswas, S. Salunke-Gawali, T. Weyhermüller, V. Bachler, E. Bill, P. Chaudhuri, Metal-complexes as ligands to generate asymmetric homo-and heterodinuclear $M_A^{III}M_B^{II}$ species: a magneto-structural and spectroscopic comparison of imidazole-*N* versus pyridine-*N*, Inorg. Chem. 49 (2009) 626–641.
[39] H.D. Connor, B.E. Sturgeon, C. Mottley, H.J. Sipe Jr., R.P. Mason, L-tryptophan radical cation electron spin resonance studies: connecting solution-derived hyperfine coupling constants with protein spectral interpretations, J. Am. Chem. Soc. 130 (2008) 6381–6387.
[40] K. Ranguelova, S. Girotto, G.J. Gerfer, S. Yu, J. Suarez, L. Metlitsky, R.S. Magilozzo, Radical sites in mycobacterium tuberculosis KatG identified using electron paramagnetic resonance spectroscopy, the three-dimensional crystal structure and electron transfer couplings, J. Biol. Chem. 282 (2007) 6255.
[41] A. Gräslund, M. Sahlin, Annu. Rev. Biophys. Biomol. Struct. 25 (1996) 259–286.
[42] C. Karunakaran, H. Zhang, J.P. Crow, W.E. Antholine, B. Kalyanaraman, Direct probing of copper active site and free radical formed during bicarbonate-dependent peroxidase activity of bovine and human copper, zinc-superoxide dismutases, low-temperature electron paramagnetic resonance and electron nuclear double resonance studies, J. Biol. Chem. 279 (2004) 32534–32540.
[43] F.A. Villamena, J.L. Zweier, Detection of reactive oxygen and nitrogen species by EPR spin trapping, Antioxid. Redox Signal 6 (2004) 619–629.
[44] H. Zhao, J. Joseph, H. Zhang, H. Karoui, B. Kalyanaraman, Synthesis and biochemical applications of a solid cyclic nitrone spin trap: a relatively superior trap for detecting superoxide anions and glutathiyl radicals, Free Radic. Biol. Med. 31 (2001) 599–606.
[45] C. Karunakaran, Thiol oxidase activity of copper, zinc superoxide dismutase stimulates bicarbonate-dependent peroxidase activity via formation of a carbonate radical, Chem. Res. Toxicol. (2005) 494–500.
[46] J. Stubbe, W.A. van Der Donk, Protein radicals in enzyme catalysis, Chem. Rev. 98 (1998) 705–762.
[47] P. Majumdar, R. Nomula, J. Zhao, Activatable triplet photosensitizers: magic bullets for targeted photodynamic therapy, J. Mater. Chem. C 2 (2014) 5982–5997.

[48] C.S. Klug, J.B. Feix, Methods and applications of site-directed spin labeling EPR spectroscopy, Methods Cell Biol. 84 (2008) 617–658.
[49] W.L. Hubbell, D.S. Cafiso, C. Altenbach, Identifying conformational changes with site-directed spin labeling, Nat. Struct. Mol. Biol. 7 (2000) 735–739.
[50] A. Czogalla, A. Pieciul, A. Jezierski, A.F. Sikorski, Attaching a spin to a protein – site-directed spin labeling in structural biology, Acta Biochem. Pol. 54 (2007) 235–244.
[51] T. Stockner, A. Mullen, F. Macmillan, Investigating the dynamic nature of the ABC transporters ABCB1 and Msba as examples for the potential synergies of MD theory and EPR applications, Biochem. Soc. Trans. 43 (2015) 1023–1032.
[52] F. Trompier, C. Bassinet, A. Wieser, C. De Angelis, D. Viscomi, P. Fattibene, Radiation-induced signals analysed by EPR spectrometry applied to fortuitous dosimetry, Ann. Ist. Super. Sanità 45 (2009) 287–296.

INDEX

Note: 'Page numbers followed by "f" indicate figures and "t" indicate tables.'

F

G

H

N

T

MIX
Papier | Fördert gute Waldnutzung
FSC® C083411

Kolibri 360 GmbH
Ferdinand-Jühlke-Straße 7
99095 Erfurt, Deutschland
produktsicherheit@kolibri360.de